GENETIC PERSPECTIVES IN BIOLOGY AND MEDICINE

GENETIC PERSPECTIVES IN BIOLOGY AND MEDICINE

Edited by Edward D. Garber

The University of Chicago Press
Chicago and London

The essays in this volume originally appeared in past issues of
Perspectives in Biology and Medicine.

The University of Chicago Press, Chicago 60637
The University of Chicago Press, Ltd., London

Library of Congress Cataloging in Publication Data

Main entry under title:

Genetic perspectives in biology and medicine.

   Selected essays from Perspectives in biology and
medicine.
   Includes index.
   1. Genetics—Addresses, essays, lectures.
2. Medical genetics—Addresses, essays, lectures.
I. Garber, Edward D. (Edward David), 1918–
II. Perspectives in biology and medicine. [DNLM:
1. Genetics—biography. 2. Genetics—history—collected
works. 3. Genetics, Medical—history—collected works.
4. Genetics, Medical—trends—collected works.
QZ 11.1 G328]
QH438.G44   1985   575.1   85-8454
ISBN 0-226-28215-5
ISBN 0-226-28216-3 (pbk.)

CONTENTS

FOREWORD

Perspectives in Biology and Medicine, a University of Chicago publication, was created by Dwight Ingle 27 years ago. Its purpose was to provide a forum for the expression of biomedical concepts by publishing scholarly essays on all aspects of life ranging from fundamental science to human behavior. The fare has included biographic and autobiographic sketches of accomplished scientists and physicians and original articles calculated to broaden the knowledge and interests of all readers. At a time when the natural developments of biology and medicine were forcing specialization, *Perspectives* was conceived as a potential force to help serious scholars maintain a desirable breadth of interests.

During the quarter-century of *Perspectives'* publication, the dramatically exciting developments in genetics, including evolutionary biology, have been outstanding. As a result, we are beginning to understand what we are and how we arrived at this stage of development. It is, thus, not surprising to find important articles on these subjects scattered through every volume of *Perspectives.*

Professor Edward Garber has read all of these compositions and selected for this reprinting those essays which most effectively tell the story of latter twentieth-century advances. He has assembled these selections in a rational manner and composed succinct, pungent notes placing each contribution in the appropriate perspective in a remarkably lucid fashion. Individually, the pieces stand alone as cogent reviews of significant milestones. With Garber's comments, the collection tells a cohesive and exciting story.

RICHARD L. LANDAU
Editor, Perspectives in Biology and Medicine

CLIFFORD W. GURNEY
Associate Editor, Perspectives in Biology and Medicine

INTRODUCTION

The rediscovery of the Mendelian principles of heredity at the turn of the century stimulated biologists to get evidence supporting or denying the apparently simplistic algebra of inheritance. T. H. Morgan, the first Nobel laureate in genetics, left a distinguished career in embryology to study heredity and introduced the fruit fly (*Drosophila melanogaster*) as the species of choice. His discovery of the chromosomal basis of sex-linked inheritance and guidance of a distinguished group of graduate students established cytogenetics, the basis of the gene-chromosome relationship. An equally remarkable group of graduate students in maize cytogenetics, including future Nobel laureates G. W. Beadle and B. McClintock, under the leadership of R. A. Emerson, rivaled the "fly group."

As the principles of transmission genetics were extended to numerous species from mold to man, biochemistry was experiencing a ferment comparable to that of genetics in studying the role of enzymes in metabolism. After Beadle and Tatum presented the one gene–one enzyme relationship based on evidence from a genetic-biochemical approach in the mold *Neurospora crassa*, biochemists and geneticists merged their interests to initiate the hybrid discipline of biochemical genetics, the forerunner of molecular genetics.

Transmission geneticists defined the gene in terms of mutation, recombination, and function but could not identify the genetic material. Yet, they opened the way to molecular genetics by demonstrating recombination in bacteria and their viruses. This unexpected discovery of "sex" in different bacterial species uncovered novel mechanisms of recombination, thereby releasing geneticists from the orthodoxy of eukaryotic genetics.

After World War II, technologies were developed to extract, separate, and characterize macromolecules; industry supplied the equipment and material; and national agencies provided funding on an unprecedented scale. The war had diminished the ranks of American, European, and Japanese geneticists and biochemists, and the new generation embraced the New Genetics. Symposia and summer courses at Cold Spring Harbor created an active marketplace for recruiting and the rapid communication of ideas and techniques among an international group of eukaryotic geneticists, biochemists, and even physicists. It was common knowledge that the time was ripe to solve two fundamental problems: the machinery of protein synthesis and the identification and physicochemical

| 1

organization of the genetic material. It is difficult to document the avalanche of publications that resulted in defining the machinery of protein synthesis or to describe the impact of the double helix model for DNA. The confluence of these accomplishments is the Jacob-Monod model that provided a molecular basis for the regulation of protein synthesis.

As geneticists and biochemists prepared to reap the harvest of the New Genetics, they were presented with recombinant DNA, cloning, and nucleotide sequencing. The daydreams of geneticists had become realities—but at the expense of losing their identity. Specialization in the biological and medical sciences had created compartments of provincial interests and attitudes. The Newest Genetics restored the old order of biology and medicine: broad interests in the basic unity of life. In the future, bureaucratic approaches to solutions of biomedical problems need not resort to semantic gymnastics or conflicts. The proper blend and critical mass of organismic-cellular and molecular biomedical scientists should encourage the interdisciplinary mix needed to ensure progress rather than momentum. "There are more things in heaven and earth, Horatio, / Than are dreamt of in your philosophy" (Hamlet, I:5).

EDWARD D. GARBER

HISTORY: INTRODUCTION

Early genetics was the province of European biologists until T. H. Morgan of Columbia University introduced the fruit fly (*Drosophila melanogaster*) and R. A. Emerson of Cornell University used maize (*Zea mays*) as experimental species. As European biologists continued their broad survey, American biologists concentrated on a few species and actively recruited graduate students and colleagues. Perhaps the academic structure of American universities fostered more interdisciplinary conversation and research in the years between the great wars.

The introduction of bacteria and viruses as experimental species for genetic study coincided with the development of new technologies by biochemists and physicists. The emphasis shifted from transmission and biochemical genetics to an assault on the nature of the genetic material, culminating in 1953 with the double helix of DNA. While recombinant DNA and nucleotide sequencing are the current style, they are rooted in transmission genetics. Contrary to the general opinion of the new molecular geneticists, genetics did not begin in 1953.

The collection of essays and articles on the history of genetics was selected to provide an overview, a weaving together of all of the threads. The date of publication of each essay or article was as important as the content. Even the best-informed geneticists could not predict the pace or direction of their science.

Darwin's theory of evolution demanded a parallel theory to account for the origin and transmission of variation. Dunn presents the speculations for living units of heredity proposed during 1864–1900, the years of Mendel's dormant paper with the answer. The fusion of breeding data and chromosome structure and behavior during mitosis and meiosis marks the rise of American genetics. Ravins's historical overview of genetics in America connects the theoretical and practical aspects of genetics as they relate to the improvement of crops, livestock, and man. The application of mathematical theory to genetic variation is the beginning of population genetics and the genetics of continuous variation. Haldane defends "beanbag genetics," the collection of wild-type and mutant genes in the individuals of a population. This article is included to present a style of writing that is, unfortunately, no longer in vogue among current scientists.

Wyatt examines the discovery of DNA as genetic material in the context of a premature disclosure, one that could not be demonstrated with other microbial species at the time. It is noteworthy that Mendel could not extend his result to another plant species suggested by his academic correspondent. Crick records his reactions to the spectacular progress in the New Genetics during the 3 decades following the publication of the double helix of DNA.

For some time, there has been a widening gulf between transmission geneticists and molecular geneticists. Gilbert presents the case for a comparable situation involving biochemists and molecular biologists. A potential fusion of molecular geneticists and molecular biologists into one academic unit with a doctoral program should provide an interesting challenge, particularly in the selection of an appropriate name for it. The effect of genetics on the biological and medical sciences and the rise of molecular genetics have altered the composition of the formerly relatively homogeneous company of geneticists. Wallace contrasts the style of the "old" and the "new" geneticists in finding explanations for apparently simple observations. There is a real possibility that each discipline in the biological and medical sciences may require one or more geneticists-in-residence, and the need for a separate Department of Genetics may vanish in the academic bureaucracy.

IDEAS ABOUT LIVING UNITS, 1864–1909: A CHAPTER IN THE HISTORY OF GENETICS

L. C. DUNN*

The concept of the gene as a self-replicating unit of heredity has been one of the most influential ideas of twentieth-century biology. It was derived from Mendel's experimental proof of the segregation of alternative forms of a gene (now known as alleles) into different gametes in the formation of the germ cells of a hybrid. Although the principle was demonstrated in Mendel's paper of 1866, it did not begin to influence biological theories until after its rediscovery, confirmation, and extension in the first years of the twentieth century. That period, say from 1900–1906, can now be seen as the chief break in the continuity of ideas about the transmission system of heredity. What had happened prior to that time— except, of course, for the original publication of Mendel's pea experiments—had little genetic connection with the later course of development of these ideas. There was no such sharp break in studies of evolution or of development; in fact, many students of the latter field would say that it has not yet been affected in a major way by what happened in 1900. Views about the mechanisms of evolution began to change more gradually in the 1920's and 1930's.

It is, of course, not surprising that the effects of the break should be felt first in the territory, so to speak, in which it occurred—that of the transmission mechanism of heredity. It is interesting to reflect, from the vantage point of 1965, that that "territory" was not recognized as such until after the break occurred. Heredity, with the definiteness given to it by modern genetics, now seems to be a well-marked field, but in the nineteenth century this was not so. Then efforts to develop general theories of heredity were judged primarily by their applicability to problems of varia-

* Columbia University, Nevis Biological Station, Irvington-on-Hudson, New York. This will form a chapter in a forthcoming book, *Genetics, 1864–1939: The Development of Some of the Main Ideas*, to be published by McGraw-Hill Book Co.

tion, evolution, and development, and we see signs today, after nearly one hundred years, of a return of that attitude. Then it was due to the absence of precise questions about the transmission mechanism—what I referred to above as non-recognition of the territory. Now genetics threatens to disappear as a separate field because the solution of some of its basic problems has led to its absorption into the general fabric of biological knowledge. The gene, whatever its fate as an "ultimate" unit in the structure and functioning of living material, has become a basic part of the description of that organization.

The nature of the radical change that occurred around 1900 can be studied by examining some of the questions of chief interest to biologists in the immediately preceding period. Three main groups of questions can be distinguished: those concerning the nature of species differences and the effects of hybridization on domesticated plants; those having to do with morphology, development, and cytology, chiefly in animals; and those specifically concerned with theories of heredity. The last took the form of ideas about living units as elements of hereditary transmission and control of the development of form and function. Needless to say, all were directed toward general questions raised as the result of the intense interest in the mechanism of evolution evoked by the publication of *Origin of Species* in 1859.

I shall deal first with the last set of questions and shall consider the history of ideas concerning living units in the nineteenth century and the state of such theories today. Many of the documents describing the development of theories of heredity and of living units in the nineteenth century have been brought together, with commentary, by Alfred Barthelmess in his valuable book *Vererbungswissenschaft* [1]. The source of ideas about living units could, of course, be traced to much earlier times; Jean Rostand has made an attempt of this sort in his *Esquisse d'une histoire de l'atomisme en Biologie* [2]. But the particles imagined by Buffon, Diderot, and others in the period before experimental study began were purely speculative and did not lead to the discovery of the true *atomisme* which Mendel proved. The "elementary particles" of Maupertuis (1751), however, were invoked to account for the segregation of a dominant gene for polydactyly in a human family which he followed through several generations. Maupertuis, as Glass [3] has shown, was on the right track, although he seems to have had no immediate followers.

Beginning with the "physiological units" on which Herbert Spencer in 1864 based his views of the organization of protoplasm, many thinkers arrived by similar means at similar conceptions. Darwin's "gemmules," Nägeli's "idioplasm," the hierarchy of "biophores" and "ids" which Weismann conceived as arranged in "idants" corresponding to visible chromosomes, the "pangenes" of de Vries—all belong in this succession, from the last of which in 1909 Johannsen formed the term "gene." The question is whether the nineteenth-century units represented the same idea as that contained in our present term.

The living units of Spencer [4] and of Darwin were invented to explain facts of reproduction and, especially, of differentiation and development. We should remember that seventy-five years later theories of individual development had still not been satisfactorily accommodated by the theory of the gene. It was especially the repetition of all the details of structure in parts such as a limb or a tail which had been lost and then regenerated that called forth Spencer's idea of physiological units in the remainder of the body which "possess the property of arranging themselves into the special structures of the organisms to which they belong." It is suggestive that the idea of physiological units appears first in the "Principles of Biology" in the section on "Waste and Repair." According to Spencer, "The germ cells are essentially nothing more than vehicles in which are contained small groups of the physiological units in a fit state for obeying their proclivity towards the structural arrangement of the species they belong to." He imagined them to be units above the level of inorganic molecules and below the lowest level of morphological organization, the cell. The units within any one species were at first thought to be all alike and each to possess the "polarity" for arrangement into the structures of that species. They could thus be vehicles of transmission as well as directors of development. And since they were assumed to be self-reproducing and to circulate throughout the organism, modifications of bodily parts would result in modifications of the units and thus account for inheritance of acquired characters. This in the mid-nineteenth century was an essential requirement of a theory of heredity.

Darwin's "Provisional Hypothesis of Pangenesis" was set forth in chapter 27 of *The Variation of Animals and Plants under Domestication*. Although published four years after Spencer's theory—i.e., in 1868—it does not base itself on the earlier theory but was Darwin's response to the great variety of facts on variation which he had collected and studied:

In the previous chapters, large classes of facts, such as those bearing on bud-variation, have been discussed; and it is obvious that these subjects, as well as the several modes of reproduction stand in some sort of relation to one another. I have been led, or rather forced, to form a view which to a certain extent connects these facts by a tangible method [5].

These were facts of asexual and sexual reproduction which he considered to involve essentially the same principles—facts of regeneration, development, graft hybrids, inheritance, effects of use and disuse, reversion, and the transmission of characters in latent form. The hypothesis which was to connect these facts assumed that the units known to increase by self-division—i.e., the cells—"throw out minute granules which are dispersed throughout the whole system; that these supplied with proper nutriment, multiply by self-division and are ultimately developed into units like those from which they were originally derived. These granules may be called gemmules." These were thought to be extremely small and numerous, circulating in the body and collected into sex cells in which they could be transmitted in a dormant state.

Galton by a simple transfusion experiment with rabbits disproved that part of Darwin's hypothesis by which gemmules were supposed to circulate in the body and then be collected in the germ cells. However, this did not seem to Darwin to be a fatal objection. "I certainly should have expected that gemmules would have been present in the blood, but this is no necessary part of the hypothesis, which manifestly applies to plants and the lowest mammals" [5]. In its most general form it would, in fact, have been impossible to disprove Darwin's hypothesis, the essence of which was that new individuals are generated by invisible units rather than by the reproductive organs or buds. "Thus an organism does not generate its kind as a whole but each separate unit generates its kind" [5]. These units remained purely hypothetical, and this accounts for the withering away of pangenesis as Darwin proposed it. The word itself, however, lived on in the new and different meaning given to it some twenty years later by de Vries in his *Intracelluläre Pangenesis*.

Galton [6] produced the first experimental evidence against the transport feature of Darwin's hypothesis of pangenesis. Transfusions of blood were made between rabbits of different colors. Subsequent offspring of these revealed no evidence of having received foreign gemmules. But Galton retained the gemmule part of the hypothesis by assuming that the whole collection of units representing cells of different kinds constituted an hereditary material known as the "stirp," which was distinct from the body cells.

This sounds very much like Weismann's concept of the distinction between an immortal germ plasm and a mortal somatoplasm, and Galton may well have been the first to express such a distinction. Weismann did not, however, concede Galton's priority, for eventually in order to explain some cases of the inheritance of acquired characters, Galton had later to assume that some cells give off gemmules which get into other cells and even into germ cells. Moreover, said Weismann, the true germ plasm theory is independent of any assumption about sexual reproduction, while Galton's stirp is limited to bisexual forms.

It was clear to Johannsen, however, that the essential idea had come from Galton [7], as shown by the following passage from Johannsen's stimulating and prophetic book *Om Arvelighed og Variabilitet* [8]:

> In the two preceeding chapters we have had to dissociate ourselves from Weismann's speculations; now we must for the third time warn against "Weismannism," which here shows its dogmatic nature in its most beguiling [*besnaerende*] form, namely as the much discussed theory of the Continuity of the Germplasm. The sound conceptions on which this remarkable theory is based are quite old, and so far as we know were first put in a clear form by Galton some 20 years ago, in connection with some now abandoned ideas of Darwin. [Translation by L. C. D.]

Johannsen then described Galton's stirp theory as the direct ancestor of the germ plasm theory, attributing to Galton the essential idea that the sex cells, egg and sperm, are independent of the body cells and have their chief function in inheritance as producers of the germ cells of subsequent generations. As a collective term for the elements responsible for transmission, Galton had coined the term "stirp" (Latin *stirpes*, root). According to Galton some of the elements (germs) produce the body, others produce more germs to form the stirp of the offspring. Hereditary continuity is thus from stirp to stirp and not from body to body, and this was a new idea, as Johannsen saw. Johannsen's views, which were expressed later in his fundamental conceptions of pure lines, phenotype and genotype, and gene, owed more to Galton that to Weismann. Even when we discount Johannsen's aversion to Weismannism as "abstraction and speculation," we must admit that his points were well taken.

There is no doubt that Galton caused Weismann to think harder and to refine his speculations about material units of transmission. The versatile Galton had the same effect in many fields. In genetics none of his ideas got incorporated into durable theories of heredity, but the statistical methods developed from his first attempts to deal with problems of blending inheritance proved to be extremely valuable in leading to the rejection of many hypotheses, including his own theory of ancestral inheritance.

In the meantime, Nägeli [9], one of the foremost botanists of his time, developed his *Mechanisch-physiologische Theorie der Abstammungslehre*. Its central feature was that heredity was transmitted by a substance—idioplasm—carried by the germ cells but diffused as a network throughout all the cells. Its elements were complexes of molecules—micellae—grouped into units of higher order which during development determined the differentiation of cells, tissues, and organs. These units too were born of theoretical speculation, but in the same year, 1884, Oskar Hertwig seized upon Nägeli's idea and for the first time attempted to identify the hereditary substance with the chromatin of the nucleus. Kölliker, Strasburger, and Weismann quickly came to the same view, and what could be called the "nuclear theory of heredity" came into being.

It was August Weismann (1834–1914) whose theoretical insight and ingenuity brought this view to prominence even before the behavior of the nuclear material was elucidated. In Weismann's theory, the idioplasm became the germ plasm; and as an embodiment of the physical basis of hereditary continuity, this idea exerted a strong influence upon biological thinking which has persisted to this day. Although as early as 1883 he had expressed views about the physical basis of heredity as consisting of nuclear elements, Wiesmann's mature statement of his theory was given in his famous book *Das Keimplasma*, which appeared in German and in an English translation (*The Germ Plasm, a Theory of Heredity*) almost simultaneously in 1892–93 [10].

By that time Weismann had available to him another brilliant and prophetic formulation of a theory of heredity, published in 1889 by the Dutch physiologist Hugo de Vries under the title *Intracelluläre Pangenesis* [11].

Hugo de Vries, a student of Julius Sachs at Würzburg, became professor of botany at Amsterdam in 1878 at the age of thirty. In the early 1880's he imagined a mechanism for inheritance which more closely resembled the system as we conceive it today than any other pre-Mendelian system did. It was based on living, self-replicating units, pangenes, which provided the model for the gene. Later Johannsen took the name "gene" as well as the concept from de Vries, who paradoxically had borrowed the name from Darwin's pangenesis but had given it an entirely different meaning, which he developed fully in his book of 1889. It was this imaginative model which led de Vries to the experimental breeding work on which his later mutation theory was based and to the analysis of the inheritance of discontinuous characters in fifteen different plant genera. This work was carried out between 1892 and 1899. There is no doubt that in this period de Vries independently discovered Mendel's law of segregation and verified it in many plant species.

The genesis of de Vries's pangene idea is to be found in his interest in evolution and what he called "the species problem." In the introduction to *Intracellular Pangenesis* he said:

> If one considers the species characters in the light of the doctrine of descent, it then quickly appears that they are composed of separate more or less independent factors. Almost every one of these is found in numerous species, and their changing combinations and association with rarer factors determine the extraordinary variety of the world of organisms. . . . These factors are the units which the science of heredity has to investigate. Just as physics and chemistry are based on molecules and atoms, even so the biological sciences must penetrate to these units in order to explain by their combinations the phenomena of the living world.

Many of the features of de Vries's hypothesis sound remarkably modern. Numerous different kinds of small living particles representing separate hereditary predispositions (*anlagen*) concentrated in the nucleus of all cells; representative particles entering the cytoplasm to influence reactions there; the properties of self-replication, accompanied by occasional errors like mutation, which gave rise to the variety of pangenes—this all sounds very much like genes. Especially prophetic was de Vries's view of the species as a specific assemblage of individual characters subject to recombination.

The hypothesis was a great feat of imagination and erudition, but because of its fundamentally deductive character, it suffered the same disability as its predecessors: it did not include either the prescription or the clear possibility of proof or disproof by experimental means, although de Vries realized that further study called for crossing experiments. The difference between de Vries's pangene of 1889 and the gene as named by Johannsen twenty years later was that the former did not include the concept of alternative states or alleles. This, as we know, was derived directly by Mendel from the behavior of the two forms of a gene in segregation; the only genes whose existence could be inferred were those of which more than one alternative form had arisen by mutation.

But although de Vries's theoretical model could not predict the existence of segregating elements, it served to guide him to study mutation by means of experimental plant breeding and thus to the independent discovery of the principle of segregation. In this sense de Vries's hypothesis of "intracellular pangenesis" might be called the starting point of modern genetics [cf. 12, 13], and he more than any other should share with Mendel the credit for the foundations of the science. He himself did not lay claim to this position and appeared to set more store by his mutation theory than by his discovery of a basic principle of heredity.

The three chief contributions of de Vries—the ideas of pangenes, of mutation, and of segregation—were so interconnected in his mind as to form parts of one concept. A few citations from his writings will illustrate this.

First, in *Intracellular Pangenesis* as translated by Gager [11] we read:

In the first division we arrived at the conclusion that hereditary qualities are independent units from the numerous and various groupings of which specific characters originate. Each of these units can vary independently from the others; each one can of itself become the object of experimental treatment in our culture experiments. . . . The pangens are not chemical molecules but morphological structures, each built up of numerous molecules. They are like life units, the characters of which can be explained in an historical way only.

We must simply look for the chief life attributes in them, without being able to explain them. We must therefore assume that they assimilate and take nourishment and thereby grow, and then multiply by division, two new pangens, like the original one, usually originating at each cleavage. Deviations from this rule form a starting point for the origin of varieties and species.

At each cell-division every kind of pangen present is, as a rule, transmitted to the two daughter cells.

At the end of this summary section occurs the following sentence: "In a word: An altered numerical relation of the pangens already present, and the formation of new kinds of pangens must form the two main factors of variability." In a footnote to this sentence we find: "In a note to the translator, the author says: 'That sentence has since become the basis of the experiments described in my *Mutationstheorie.*' "

The principle of segregation (which he called *Spaltungsgesetz* or *loi de disjonction*) and his pangenesis theory were so closely connected in de Vries's mind that the first sentences in his first paper of 1900 (*Sur la loi de disjonction des hybrides*), in Aloha Hannah's translation [14], were as follows: "According to the principles which I have expressed elsewhere (*Intracellu-*

läre Pangenesis, 1889) the specific characters of organisms are composed of separate units. One is able to study, experimentally, these units either by the phenomena of variability and mutability or by the production of hybrids." After citing the F_2 results of single-character segregation in eleven species of plants, he concluded: "The totality of these experiments establishes the law of the segregation of hybrids and confirms the principles that I have expressed concerning the specific characters considered as being distinct units."

It is clear that de Vries was not a "re-discoverer" but a creator of broad general principles and that his work extended Mendel's in at least three ways. It proved the operation of segregation in a wide variety of plant species; it made material units an essential feature of the theory; and it introduced the concept of mutability as the source of variety in the units. If the stature of de Vries has steadily increased as genetics has developed, it is perhaps in part because genetics grew to resemble the ideal he entertained in the 1880's—that the organic world was expressed through the properties and activities of material units as the prime operators in the transmission of heredity, in evolution and in development.

I return now to the other great theorist of the pre-1900 period, August Weismann. His influence on the development of genetics lies mainly along another track, which led to the chromosome theory as it matured in the 1920's and 1930's. To him, the importance of living units was that they constituted a hierarchy of ascending size culminating in material bodies which could be seen under the microscope. His concept of the continuity of the germ plasm guided and inspired his thinking as the idea of pangenes had done for de Vries, but it led him toward cytological observation, not toward experiment as in de Vries's case. He was, however, frustrated by an eye disease, and his energies turned to speculating and constructing models. His essay of 1883 [15] began with Nussbaum's view of the continuity of the germ cells and developed it into the germ plasm theory, which embodied a new and fruitful view of the process of inheritance. According to this view, the offspring does not inherit its characters from the body of the parent but from the germ cell, which has derived its properties and elements from the ancestral germ cells which preceded it. The body, or soma, was considered the temporary and moral custodian of the continuous line of germ cells. Samuel Butler described this idea: "It has, I believe, been often remarked that a hen is only an egg's way of making another egg." To place chief emphasis on the germ cells and the hereditary material itself was a fundamental advance in the direction taken by modern genetics. Today we say, "We do not inherit characters, but genes, which in the course of development and in interaction with the environment produce characters." Johannsen's distinction between phenotype and genotype and Woltereck's conception of the norm of reaction, now seen as views essential to an understanding of heredity, both trace ultimately to Weismann's and Nussbaum's distinction; and behind them, according to Johannsen's view, is Galton.

It was the germ plasm theory which led Weismann to reject the Lamarckian theory of the inheritance of acquired characters. This had the effect of directing evolutionary thinking into paths which led toward genetics. It led too to the construction of the elaborate hierarchy of material bodies beginning with biophores, of molecular order of size, equivalent more or less to de Vries's pangenes. These in turn grouped into determinants which as "pieces of inheritance" (*Vererbungsstücke*) determine the forms of cells and cell groups and thus the inherited characters. Determinants were seen as grouped into a third level of organization in "ids," the name intended to sound like Nägeli's "idioplasm." Each id was viewed as containing the whole architecture of the germ plasm. They were arranged in linear sequence to constitute the "idant," which was Weismann's term for the chromosome. This view of the newly discovered chromosome expressed its function in transmission and led directly to Weismann's assumption of a reducing division. If idants replicated as mitosis showed they did, then something must intervene to hold in check "the excessive accumulation of different kinds of hereditary tendencies or germ plasms." How Weismann's prediction of a reduction division during mitosis was fulfilled belongs to the history of cytology. Similarly, the defeat of his attempt to apply the determinant theory to development—by postulating that mitoses were qualitative in the sense that determinants were progressively sorted out to different cells and organs—that belongs to a history of embryology.

It is interesting that many of the speculations involving living units— e.g., those from Spencer in 1864 to Weismann in 1883—died because they failed to account for the observed facts of development and regeneration. Yet the theory of the gene, which also fails to account for development, was accepted because of the compelling nature of the evidence from experimental breeding, backed up by cytological observation. The theory of the gene was concerned with units only as vehicles of transmission and, in Morgan's words [16], "states nothing with respect to the way in which the genes are connected with the end product or character." Nevertheless, the arguments for the restricted theory were so persuasive as to be accepted by most biologists in spite of the paradox that the mechanism proposed gives rise to the same variety of units in all cells although the cells themselves assume different forms and functions.

The break at 1900 between the preceding construction of speculative schemes of heredity by purely deductive reasoning and the succeeding experimental investigation was indeed a sharp one. There was one man, however, who bridged the gap since he worked by both methods. Hugo de Vries constructed before 1889 a system of intracellular units which he considered to derive from Darwin's intercellular pangenesis although it was fundamentally different. He also reached, inductively, by planned experiments, the same rule which had been derived by Mendel. None of his contemporaries (he was born in 1848) or fellow thinkers was able to make this transition. Galton, born in the same year as Mendel (1822), Spencer (born 1820), Weismann (1834), and Haeckel (1834) were perhaps too old.

Living units, defined in operational terms by their behavior in transmission as revealed by breeding experiments, were to become elements in a new biology, which a new generation was to establish. Carl Correns (born 1864) and William Bateson (born 1861) were the first to recognize the wider implications of the unit which later was named gene, and to see that the kind of experimental design which had been introduced by Mendel was competent to elucidate the nature of such elements. Bateson, who must in a formal sense be considered the founder of genetics since he first formulated its principles and gave it its name, expressed the revolt of the new period against the old. Weismann, who had no enthusiasm for Mendel's work and expressed grave doubts about it, published in 1904 a two-volume work, *The Evolution Theory* [17]. Bateson, reviewing it in 1905 under the title "Evolution for Amateurs," said: "It should have appeared 30 years ago, when variation and heredity were unexplored territories" [18]. By 1904 the newly discovered principles of heredity, based on genes, "at a stroke destroyed most of the speculations which in the pre-experimental era ran an illustrious course" [18]. Bateson reserved his special scorn for Weismann's theory of germinal selection, the struggle for survival among the determinants in the germ cells: "Natural selection, being here in her economical mood, is to eliminate the parts which if allowed to develop, would not have pulled their weight. We thus meet a proleptic Natural Selection, dealing in futures—as transcendental a conception as any Nägelian could desire."

"Evolution," he said, "has passed out of the speculative stage." And so, he might have added, had heredity. The experimental proof of the useful existence of stable living units as the prime operators of the transmission system of heredity had opened a new era in biological thought.

REFERENCES

1. A. BARTHELMESS. Vererbungswissenschaft. Freiburg/München: Orbis Academicus, Verlag Karl Albers, 1952.
2. J. ROSTAND. Esquisse d'une histoire de l'atomisme en biologie. Paris: Gallimard, 1956.
3. H. B. GLASS. Quart. Rev. Biol., 22:196, 1947.
4. H. SPENCER. Principles of biology, London: Williams, 1864.
5. C. DARWIN. The variation of animals and plants under domestication (1868), 2d Amer. ed. rev., vol. 2, pp. 349, 398. New York: D. Appleton and Co., 1876.
6. F. GALTON. Experiments in pangenesis. Proc. Roy. Soc. [Biol.], 19:393, 1871.
7. ———. J. Anthro. Inst., 5:329, 1876.
8. W. JOHANNSEN. Om Arvelighed og Variabilitet. Köbenhavn: Det Schuboteske Forlag, 1896.
9. K. NÄGELI. Mechanisch-physiologische Theorie der Abstammungslehre. München/Leipzig: R. Oldenburg, 1884.
10. A. WEISMANN. W. N. PARKER and HARRIETT RÖNNFELDT (trans.). The germ-plasm, a theory of heredity. London: Walter Scott, Ltd., 1893.
11. H. DE VRIES. C. S. GAGER (trans.). Intracellular pangenesis (1889). Chicago: University of Chicago Press, 1910.
12. T. J. STOMPS. J. Hered., 45:293, 1954.
13. J. HEIMANS. Amer. Natur., 96:93, 1962.
14. A. HANNAH. (Trans. of H. DE VRIES, "Sur la loi de disjonction des hybrides" [1900].) Genetics, 35:suppl. no. 5, part 2, pp. 30–32, 1950.
15. A. WEISMANN. Über die Vererbung. Jena: G. Fischer, 1883.
16. T. H. MORGAN. The theory of the gene, p. 26. New Haven: Yale University Press, 1926.
17. A. WEISMANN. J. A. THOMSON and M. R. THOMSON (trans.). The evolution theory, 2 vols. London: Arnold, 1904.
18. W. BATESON. The Speaker, June 24, 1905; reprinted in: B. BATESON. William Bateson, naturalist, pp. 449, 451, 454. London: Cambridge University Press, 1928.

GENETICS IN AMERICA: A HISTORICAL OVERVIEW*

ARNOLD W. RAVIN†

The headlines of today's newspapers speak of sharp disagreement among geneticists regarding the bounty to be derived from our knowledge of genetic material [1]. The scientific strife is mirrored in public apprehension about the good of genetic research and public concern about its social control. Current bewilderment and disillusionment are a far cry from the eager and optimistic anticipation with which the science of genetics was heralded at its inception at the turn of the century. In many ways the history of genetics parallels the rise and fall of optimistic progressivism in our nation. Never before has the connection between scientific change and sociopolitical evolution seemed so clear. Of course, genetics is not alone among the scientific disciplines to have contributed to our current malaise; the advent of nuclear power, the growth of chemical industry, and the advances in biomedical technology have added their own strains to the situation. Yet it is worthwhile to look closely at the history of one field in order to ask, in its case, what went wrong. We shall see, I believe, that confidence was justified in the capacity of the embryonic field of genetics to develop into a powerful theoretical discipline that would at once unify and propel all of biology. This confidence was coupled, however, with philosophical and political naiveté about the limits of scientific knowledge on the one hand and the values that guide and constrain the scientific enterprise on the other.

Fortunately, for ease of historical investigation, we can limit our survey of genetics for the most part to the American scene. With all due credit to the European founders of Mendelism—De Vries, Correns, Tschermak, Bateson, and Johannsen—and to the recent developers of molecular biology in France, England, Germany, and Japan, genetics is to a great extent an American science. For it was in America especially that genetic theory developed rapidly. More than that, it was in America

*Presented at a symposium on "The New Genetics and Society" held at Albion College, Albion Michigan, March 23–25, 1977.

†Department of Biology, University of Chicago, 1103 East 57th Street, Chicago, Illinois 60637.

that the applicability of genetics to human welfare was seized upon, first in respect to animal and plant breeding for human purposes and not much later in respect to the guidance of human evolution itself. Indeed, two related themes weave their way through this paper. The first is the growth of genetics primarily as a theoretical structure, while the second is the use of genetics. Theory and application never got very far apart in this pragmatic nation of ours.

News of the rediscovery of Mendel's principles had a welcome reception among certain American biologists. Among the greeters was Jacques Loeb, himself a recent émigré to this country. Much respected for his studies of animal tropisms and artificial parthenogenesis, and for his ardent advocacy of a reductive mechanistic approach to complex biological phenomena, Loeb published in *Science* an address given in 1904 in which he predicted the future course of biological research [2]. Referring to "The Recent Development of Biology," Loeb saw much hope in two significant discoveries—Buchner's separation of enzymes from the living cell and Mendelism. The capacity of agents to perform their cellular function outside the organized state of the cell fit Loeb's mechanistic predilections. Moreover, the capacity of enzymes in minute amounts to speed up selective metabolic reactions under mild physiological conditions without being significantly consumed themselves augured well for a complete understanding of the organism in physicochemical terms. What Loeb liked about Mendelism was its exactness, the quantitative precision of its laws, and its experimental testability. He called Mendel's treatise ". . . one of the most prominent papers ever published in biology." Loeb was later to write extensively about enzymes in *The Dynamic State of Living Matter,* an influential book published in 1906 [3]. He was also instrumental in stimulating the earliest attempts to associate genes and enzymes [4]. This approach took a circuitous path which cannot be detailed here, but it had two happy outcomes. The first was that gene changes were observed by others to be correlated with altered rates of metabolic reactions, indications of enzymatic deficiencies that were to be reformulated as the one gene:one enzyme relationship in the late 1930s and early 1940s. The second outcome was that still other scientists came to think of genes themselves as autocatalysts, an early step in the development of a theory of gene replication according to which the information for the structure of the gene descendants lies in the parental genes themselves. Loeb was to become a member of the Rockefeller Institute for Medical Research in 1910. Domineering and articulate as he was, he had an enormous influence on the formation of scientists at that institute. We can only guess the effect he had on a medical microbiologist such as Oswald Avery. According to his biographer, René Dubos, Avery learned much from the scientific "giants" of the institute, especially those who were chemically inclined [5]. Avery's search for the agent responsible for the specificity of pneumococcal type and later for the agent responsible for genetic transformation of type surely owed something to the emphasis placed by Loeb upon the young fields of biochemistry and genetics.

Not every biologist of the 1900s was immediately caught up in the excitement of genetics, however. While Jacques Loeb's enthusiasm for an experimental biology characterized by critical design and precise, quantitative results was shared by the zoologist Thomas Hunt Morgan, Morgan's strong penchant for the empirical made him distrustful of theory that got too far away from experimental findings. For this reason he was skeptical for a long time about Weismann's theories and even Darwin's theory of natural selection. Little surprise then that in 1905 Morgan was questioning "the so-called purity of the germ cells" [6], and as late as 1909 he was questioning the reality of Mendelian factors or genes. Trained as an embryologist, Morgan saw ". . . the condition of two alternative characters . . . as the outcome of alternative states of stability (or of conditions) that stand for the characters that make up the individual" [7, p. 366]. American embryologists were typically averse to Mendelism. Morgan saw in Mendelian factors a recrudescence of discredited preformationism, and at the University of Chicago C. M. Child inveighed against the tactic of trying to explain such organismal properties as growth, development, and reproduction by assigning those properties to invisible, hypothetical units.

Nevertheless, as a member of Columbia University's zoology department, Morgan could hardly ignore the work of his distinguished colleague, E. B. Wilson. Wilson and his student Walter Sutton had really laid the ground for Morgan's conversion.

In the first edition (1896) of that remarkable monograph, *The Cell in Development and Inheritance,* Wilson, following a thorough examination of the cytological literature, came to accept the basic tenets of Weismann. "In its physiological aspect," Wilson wrote, ". . . inheritance is the recurrence, in successive generations, of like forms of metabolism; and this is effected through the transmission from generation to generation of a specific substance or idioplasm which we have seen reason to identify with chromatin" [8, p. 326].

Chromatin was of course contained in the chromosomes, and Wilson was therefore prepared to admit the Weismannian doctrine that the hereditary potentialities are somehow contained in the chromosomes without committing himself as to the exact material form in which those potentialities were present. When Mendelism broke upon the scene, Wilson was prepared. He recognized as early as 1902 the relation between Mendelian segregation and the mechanics of chromosomal disjunction during germ-cell formation. He referred to the published observations of Montgomery (1901) and to the unpublished work of his student Walter Sutton, both of which pointed to the conclusion that ". . . the [second meiotic] or reducing division . . . leads to the separation of paternal and maternal elements and their ultimate isolation in separate germ-cells" [9, p. 992].

Hence, the genetic purity of the gametes described by Mendel had an obvious physical model in the separation of maternal and paternal chromosomes. When Sutton reported his data a year later, he went considerably further [10]. Supposing each chromosomal type in the haploid complement to be unique in its genic composition, Sutton was able to show how independent assortment of inherited characters would come about when the genes affecting those characters were located on different (i.e., nonhomologous) chromosomes. Indeed, he postulated gene linkage in advance of its observation on the grounds that some nonallelic genes must be located on the same chromosome and hence be inherited in a nonindependent manner. The surmise of Sutton's that Mendel's hypothetical genes had a physical residence on the cytologically observable chromosomes was shared in Germany by Theodor Boveri, who reached a similar conclusion from the behavior of chromosomes and its developmental consequences in polyspermically fertilized sea urchin eggs.

I can only suppose that Morgan, given his skepticism about genes, was intending to challenge the Mendelian edifice and possibly vitiate the significance of the Sutton-Boveri hypothesis when, around 1910, he undertook breeding experiments with the fruit fly *Drosophila*. William E. Castle of Harvard University, another pioneer of American genetics with whom we shall deal later, had described in 1906 some experiments in breeding fruit flies in glass-covered tumblers containing bananas for food. Morgan improved somewhat on the techniques for *Drosophila* crosses and soon was plunged into investigations of fruit-fly heredity. Unfortunately for Morgan's probable intent, and interestingly for the future of genetics, one of the first mutant characters Morgan observed was that of white eye, or absent pigmentation in the eye [11]. The results of crossing white-eyed males to normal red-eyed females were exceptional in that, while a Mendelian ratio of ¾ red-eyed to ¼ white-eyed progeny was observed in the second filial generation, the white-eyed character was relegated to the male members. Subsequently, when white-eyed females were crossed to red-eyed males, the results were even more bizarre: The alternative characters switched sexes in the first filial generation, and each sex was represented by equal numbers of red-eyed and white-eyed individuals in the second generation. The significant features of these findings were not their deviations from previously reported Mendelian ratios but their precision and reproducibility. In terms of the proportions of flies or sexes with given characters, there was no gainsaying a regular distribution. Morgan was bound to realize that something equally precise had to be behind the precise transmission of characters as well as the unique distribution between the sexes. Wilson's compendium of sex-chromosome lore must have been handy to Morgan at this stage. If Sutton and Boveri were right, a transmission of characters linked to sex might very well be due to the unusual way sex chromosomes are transmitted. When Morgan discovered a way of accounting for sex-linked inheritance by assigning the responsible genes to the X chromosomes and leaving the male's Y chromosome genically empty, he must then have been well on his way to conversion.

The history of genetic research in Morgan's laboratory thereafter followed a characteristic course. Exceptions were treasured, to use Bateson's phrase, and the exceptions found to add further support to the Sutton-Boveri hypothesis. First, Morgan observed what Sutton had merely predicted, namely, gene linkage. The linkage was not absolute, however, and the explanation of relative linkage was sought in physical exchange, or crossing over, between homologous chromosomes at the first meiotic division. Morgan's student, Alfred Sturtevant, showed that the frequencies of exchange between linked genes were very nearly additive, thus making possible the linear ordering or mapping of gene loci on chromosomes. In short, it was possible to indicate the relative position of a gene on the chromosome bearing it. Such mapping studies, begun first with the sex chromosome, were continued with autosomal genes. In this work H. J. Muller, another student, took part until it was evident that the number of mapped linkage groups was equal to the haploid number of chromosomes, a result expected if Sutton's premises were correct. Another exceptional result, that of rare deviations from the normal result in crosses involving sex-linked genes, was found by another student of Morgan's, C. B. Bridges, to be due to the exceptional phenomenon of rare failure of homologous chromosomes to disjoin at meiosis. Thus, abnormal chromosomal transmission was associated with the abnormal transmission of the genes carried by the chromosomes. All in all, a splendid fit was being observed between the expectations of the chromosome theory of heredity and a variety of experimental findings. In 1915 Morgan, Sturtevant, Muller, and Bridges presented in *The Mechanism of Mendelian Heredity* the accumulated results of their research as impressive support of the theory [12].

Indeed, from this date forward opposition to both Mendelism and the Sutton-Boveri hypothesis gradually dwindled not only in America but in Europe as well. Nevertheless, it is worth pointing out that the chromosomal theory of heredity has never been proved in a logically closed manner, despite such language as contained in the title of Bridges's paper, "Non-Disjunction as Proof of the Chromosomal Theory of Heredity." In fact, few scientific theories are established in this way, and it is important to understand this aspect of the evolution of science.

Morgan himself, you will recall, had asserted in 1909 that segregation could ". . . equally well be imagined as the outcome of alternative states of stability" [7, p. 366]. Ironically, after Morgan converted to belief in the chromosomal hypothesis of gene transmission, William Bateson, the early English advocate of Mendelism and coiner of the word "genetics," continued to object to this hypothesis on the grounds that alternative models were compatible with the data [13]. After all, Bateson pointed out, the only thing shown by Morgan's group was that certain patterns of hereditary transmission were correlated in a precise manner with certain patterns of chromosomal transmission. Both patterns could, in a logical sense, be the result of some common underlying state or process. Bateson preferred to believe in vibrational states in the cell rather than particulate genes within the chromosomes as the basis of heredity. All that one can say today is that the particulate model is certainly economical of assumptions. Bateson's model would have to be burdened by a host of complex assumptions to explain not only what Morgan's team observed but also such later observations as the absence of specific chromosomal bands associated with specific gene deficiencies, the alteration of linkage relationships associated with chromosomal translocations, and the correlation of chromosomal exchange with recombination of linked genes. Still, it is useful for students of science to understand the nature of scientific change: The "truth" of theories is rarely established as a result of logical proof but rather as a result of consensus on the part of the community of scientists. When no simpler testable explanations can be put forward, the community appears, by its behavior, to agree to the simplest or most elegant explanation. When minority voices continue to dispute the reigning hypothesis as unproven, they are at last ignored. Such was the way with William Bateson, who never accepted the particulate theory in an unequivocal manner.

I stress the consensual tradition of science as opposed to the "truth-establishing" mode because it is typical and yet unavowed. More recently, the hypothesis that deoxyribonucleic acid is the chemical material of genes faced similar opposition. When Avery, MacLeod, and McCarty's classic work on pneumoccocal transformation was published in 1944 [14], such critics as Alfred Mirsky (who later became a vigorous champion of the DNA theory) pointed out that the transformation experiments showed only that DNA was essential but not necessarily sufficient for the specificity of genetic change [15]: Small amounts of protein in bacterial transforming preparations could be the real vectors of genetic information and be unstable in the absence of DNA. Such arguments were difficult to disprove and came up as late as 1954 in an Oak Ridge symposium [16] and again in the arguments of Barry Commoner in 1964 [17]. In fact, the case that DNA carries genetic information in and of itself was won before any logically compelling proof was in hand; the critics were ignored—"outvoted" in a sense—because an accounting of all the available facts in terms that exclude DNA as the genetic material appeared too tortured for most biologists to countenance.

Genetics advanced rapidly in America, at any rate, because of the early confidence expressed in this country in the potential of the science to unify biology, through the connections between genetics and cytology on the one hand and between genetics and biochemistry on the other. But genetic knowledge also grew swiftly here because of its apparent utility in human affairs. In actuality, the application of genetics to human purposes was much in the thoughts of the earliest geneticists. As Castle remarked in his contribution to the volume celebrating the fiftieth anniversary of genetics, as early as 1900 several scientists associated with the U.S. Department of Agriculture or with state agricultural experiment stations ". . . saw in Mendelism a new tool for the production of new and improved varieties of plants and animals" [18, p. 60]. An organization called the American Breeders' Association was formed in 1903 for the express purpose of ". . . turn[ing] for a time from the interesting problems of historical evolution to the needs of artificial evolution."

These are the words of W. M. Hays, founder of the association and its first permanent secretary. The newly formed association, he announced, ". . . invited the breeders and the students of heredity to associate themselves together for their mutual benefit and for the common good of the country and the world. It thus recognized that the wonderful potencies in what we are wont to call heredity may in greater part be under the control and direction of man, as are the great physical forces of nature" [19, p. 62].

Such strong terms were typical of the general progressivist spirit that characterized America, and indeed most of the Western world, at the turn of the century. Western society had come through the Industrial Revolution believing in man's unique ability to guide his fate by controlling the forces of nature through, by and large, the power of scientific knowledge. Genetics appeared to be a new power, and the pragmatists in this nation did not lose any time to exploit it. There is, indeed, no gainsaying the rapid advances made by the breeders. In part, they contributed to the development of fundamental theory and in part to practical innovations. The agricultural scientist R. A. Emerson joined E. M. East at Harvard where they collaborated on their classic study of quantitative inheritance in maize. Later at Cornell University, Emerson made maize a powerful tool for the study of genetics, and from his school such outstanding scientists as Barbara McClintock, Marcus Rhoades, and George Beadle emerged. While these students of Emerson are best known for their fundamental discoveries, the impact of genetic discovery on practical breeding was nevertheless considerable. In 1909

George H. Shull reported to the American Breeders' Association a method of hybridization to produce corn of high productiveness. Other improvements in the breeding of wheat, cotton, and other plants were soon forthcoming, as well as the development of lines genetically resistant to bacterial and fungal infection. Domesticated animals were also improved by the use of genetic methods, stimulated by the prior Mendelian analysis conducted by C. B. Davenport in poultry and by W. E. Castle in rabbits, guinea pigs, and mice.

The very success of the methods of breeders gave encouragement to those geneticists who saw in their science not only the means for making more and better food available to mankind but also the means for improving, so to speak, the stock of mankind itself. Early in the gatherings of the American Breeders' Association, a Committee on Eugenics was organized. The committee included such diverse types as geneticists Castle and Davenport, inventor Alexander Graham Bell, and practical botanist Luther Burbank. Its chairman, David Starr Jordan, who was at the time president of Stanford University, submitted the committee's first report in 1908. In this report Jordan argued that ". . . in some cases social progress and biological progress are opposed" [20, p. 204]. He quoted Karl Pearson: "Consciously or unconsciously, we have suspended the racial purgation maintained in less developed communities by natural selection. We return our criminals after penance, our insane and tuberculous after 'recovery,' to their old lives and we leave the mentally defective flotsam on the flood tide of primordial passions" [20, p. 208].

A mild debate went on in the ranks of the committee with such persons as Alexander Graham Bell arguing against coercive restraint of human procreation but Charles R. Henderson of the University of Chicago advocating segregation and sterilization of the "obviously unfit." By 1911 the report of the committee, made by C. B. Davenport, had leaned to Henderson's side. The committee saw its duties as that of investigation, education, and legislation. "Since the weak and the criminal," Davenport reported, "will not be guided in their matings by patriotism or family pride, more powerful influence or restraints must be exerted as the case requires. And as for the idiots, low imbeciles, incurable and dangerous criminals, they may under appropriate restrictions be prevented from procreation—either by segregation during the reproductive period or even by sterilization. Society must protect itself. . . . Here is where appropriate legislation will aid in eugenics and create a healthier, saner society for the future" [21, p. 94].

In their zeal for the genetic improvement of man, the early American eugenicists were excessive in attributing control of human characters to rigidly determined genetic mechanisms, thereby supporting the political philosophy of America's "social Darwinists." Criminality, poverty, and low intellectual ability tended to be viewed as the outcome of one's genetic nature and not one's nurture, to say nothing of an interaction between nature and nurture. Moreover, the fervent advocacy by these eugenicists began to have some effect. Sterilization laws were promulgated in many states under the prodding of the eugenically minded.

In a different tradition, but still concerned with the human control of human heredity, was the genetically more sophisticated but politically still naive scientist **Hermann J. Muller.** The descendant of German socialists, Muller became dedicated early in his life to the socialist ideal of perfecting human society. He described outings with his father to the American Museum of Natural History where, on learning of the dramatic changes wrought in species by evolution, he dreamed of the possibility of man guiding his own evolution. This possibility took more concrete shape for him first as an undergraduate and then as a graduate student at Columbia University. There he learned of the new science of genetics and vowed to put genetics in the service of human society. For the socialist-humanist that Muller was, this meant using genetic knowledge to create a harmonious, cooperative society of intelligent, healthy, and compassionate individuals. Muller's thesis, and I believe his guiding philosophy, is set forth in his youthful, optimistic, and idealistic book, *Out of the Night,* now unfortunately out of print [22]. Although first published in 1936, Muller admitted that the book was essentially completed by 1925 but was kept for the most part under wraps until he left the United States for what would prove to be shattering experiences in the Germany and the USSR of the early 1930s. Muller was opposed to coercive measures to control human breeding. Moreover, he was aware of the environmental contribution to the development of human characters. Nevertheless, Muller argued that genes also contributed to all of these characters, and believed that once human beings were fully aware of their genetic heritage they would voluntarily take pains to breed positively for the most desirable characters: mental and physical fitness and brotherly love. Moreover, they would act best in this idealistic and progressive spirit within a society that had already evolved from selfish capitalism to altruistic socialism. Hence, he believed that prior social change would be helpful, if not essential, for desirable social guidance of human reproduction. He advocated ". . . the spread of birth-control techniques, . . . the easing of the burdens imposed by the bearing and rearing of children, with the downfall of the traditional ideas regarding woman's status. . . . and dicta regarding the family relationship and sex relationships in general" [22, p. 134]. All this long before women's lib! In this way, Muller argued, ". . . the way will be opened for the direction of such matters by considerations of reason and love, in the true interests, on the one hand, of the future generation, and, on the other hand, of the participating parties, rather than in obedience to the commandments of a supposed supernatural dictator or to a mystical set of absolute ethical axioms." At the same time, Muller continued, "the discoveries and inventions of advancing biology in the fields of reproduction and development must sooner or later give us radical powers of control over what has hitherto been the female's role in child-production, which will greatly extend the possibilities of eugenics and our ability to order these processes in the interests of mother and child" [22, p. 134].

The techniques Muller foresaw in 1936 were in vitro fertilization and embryo transfer as well as artificial insemination by donor. The donor should preferably be "some transcendently estimable man" [22, p. 138], so that the achievement of genetically sound progeny could be separated from the personal love life of the mother, a separation that only "social inertia and popular ignorance now hold us . . . from putting into effect" [22, p. 139]. Given this emancipation of the spirit, individuals would choose to select as germinal material for their offspring that obtained from the "greatest living men of mind, body, or 'spirit.' " Muller calculated that

. . . in the course of a paltry century or two it would be possible for the majority of the population to become of the innate quality of such men as Lenin, Newton, Leonardo, Pasteur, Beethoven, Omar Khayyam, Pushkin, Sun Yat Sen, Marx . . . or even to possess their varied facilities combined. . . . We do not wish to imply that these men owed their greatness only to genetic causes, but certainly they must have stood exceptionally high genetically; and if, as now seems certain, we can in the future make the social and material environment favorable for the development of the latent powers of men in general, then, by securing for them the favorable genes at the same time, we should be able to raise virtually all mankind to or beyond levels heretofore attained only by the most remarkably gifted. [22, p. 141]

With these words a youthful Muller sounded the clarion call in behalf of guidance of human evolution by conscious genetic intervention, a policy he never abandoned despite the disenchantments he later experienced in Nazi Germany and the Soviet Union. In subsequent publications, however, he appeared to have altered his views of estimable men in that Lenin was no longer included in his list of those from whom liberated and altruistic mothers would seek the sperm to fertilize their ova.

Genetic theory and genetic engineering were never far apart in Muller's mind, and even if Muller is atypical in his advocacy of intervention, my guess is that most geneticists would take the position that genetics should serve human ends. The problems concern the definition of those ends and the moral acceptability of the methods to attain them. These are problems that are very much with us today, and I will return to them briefly in closing.

While his socialist outlook made him an advocate of socially applied genetics, it also provided a materialistic bias in his way of considering genetic transmission. Geneticists of the first two decades of this century were often loath to speculate upon the nature of the gene [4]. For breeders like E. M. East it was sufficient that the concept of genes provided a practical calculus for breeding experiments; one need go no further than a symbolic representation of Mendelian factors. Even Johannsen, who made the signal distinction between genotype and phenotype, abhorred those who insisted upon a material or physical interpretation of genotype. Finally, Morgan himself, the establisher of a connection between abstract genes and physical chromosomes, believed it premature in 1917 to speculate upon the nature of the genes from their behavior in crosses. In one sense, Morgan was undoubtedly right, but this approach was too conservative for his student Hermann Muller. As I indicated earlier, Jacques Loeb was responsible in America for calling attention to the possible relation of genes and enzymes. In England William Bateson was noting the same relationship in Garrod's studies of alkaptonuria. Not holding a material interpretation of the gene, however, Bateson was unprepared to couple genes and enzymes too closely. In the United States, on the other hand, Loeb strongly hinted at a much closer association, and Richard Goldschmidt, another émigré, went so far as to equate the two.

A psychologist and an outsider to genetics, Leonard Troland at Harvard, stirred by the impressive specificity of both genes and enzymes, urged that genes be thought of as catalysts that could catalyze their own formation as well as other reactions [23]. This view, published in *American Naturalist* in 1917, caught the attention of the youthful Muller. One of the constant themes of Muller's later theoretical writings was the autocatalytic capacity of the gene. It was difficult for him to conceive of such an agent being other than material, and he predicted in 1921 ". . . that perhaps we may be able to grind genes in a mortar and cook them in a beaker after all" [24, p. 48].

This material view of the gene was also responsible, I believe, for the research program he undertook after leaving Columbia University for Texas. What better way to demonstrate the materiality of the gene than to use physical agents to change its state! The relation between dose of ionizing radiation and genetic mutation strongly implied a particulate target. Muller found his outlook shared by the physicist Max Delbruck whom he encountered in Timofeef-Ressovsky's laboratory in Germany.

Delbruck was later to emigrate to the United States in flight from Nazism and to join with another émigré, Salvatore Luria, a refugee from Italian Fascism, in founding the much celebrated bacteriophage school of genetics. Sure that study of relatively simple organisms like bacterial viruses would hasten understanding of the nature of the gene, impressed by the "quantum" nature of gene mutation as pointed out by Erwin Schrodinger, and guided by the vague sense that knowledge of the gene would lead to new physical laws of revolutionary scope [25], Max Delbruck was influential in expanding interest in the bacteria and their viruses as experimental material for genetic analysis. At the same time George Beadle was developing interest in the ascomycete *Neurospora* after demonstrations by B. O. Dodge and C. C. Lindegren of this mold's advantages for genetic research. Beadle's interest in *Neurospora* was as deliberate as Delbruck's turn to bacteriophage and stemmed from a preoccupation abetted by experiences at the Kerckhoff Laboratories of the California Institute of Technology which the renowned Thomas Hunt Morgan came to head in 1929 [26]. The young Beadle, looking for new fields to plow, left Cornell for Cal Tech and there met another young pilgrim to the new mecca of genetic biology. This pilgrim, Boris Ephrussi, had arrived from France and engaged Beadle in long discussions about ways to attack the outstanding question of the gene's role in development. Ephrussi invited Beadle to Paris in 1935 and there, after some faulty starts with cultures of *Drosophila* tissue, learned to transplant eye imaginal disks into the abdomens of larvae. In this manner they came to the conclusion that substances from wildtype body fluids could compensate for developmental deficiencies in mutants, and that a specific genetic mutant for a particular developmental pathway lacked a specific substance. Here it was that Beadle and Ephrussi hit upon the idea of the specificity of genetic control of metabolic reactions, a specificity that could be explained if each gene were responsible somehow for production of an enzyme catalyzing a specific reaction in a long chain of enzymatically catalyzed reactions.

On his return to the United States, Beadle vowed to establish the correctness of this one gene:one enzyme hypothesis, but *Drosophila* in the late 1930s proved refractory to enzyme studies. Quite deliberately, therefore, he sought an organism that would have the combined advantages of rapid growth, ease of biochemical manipulation, and a mode of reproduction allowing simple genetic analysis. These advantages, he re called, were contained in the mold *Neurospora*, and he enticed biochemist Edward Tatum at Stanford University to join him in the now famous crusade to confirm the one gene:one enzyme conception.

The *Neurospora* research, taken together with the work on viruses sparked by Delbruck, created a new field of microbial genetics. The bacteria joined viruses and molds when Joshua Lederberg, learning of the gene-enzyme relation and of Avery's discovery of DNA as genetic transformer of bacteria, set out as a young student to demonstrate the existence of genes in bacteria. The discovery by Lederberg of bacterial conjugation and recombination of bacterial genes followed swiftly in the laboratory at Yale University to which Tatum had moved. Rapid developments in the new field of microbial genetics culminated in the eventual triumphs of molecular genetics with which most students are familiar today. Far from ushering in new physical laws, the work of microbial geneticists led to the now-prevalent view that the nature of the gene is adequately explained with the use of existing physical principles, an apparent validation of Loeb's faith in the reducibility of life to physicochemical terms and of the reductionist's approach in general.

A historical overview can do little more than telescope the remarkable history of molecular genetics. Fortunately, excellent historical and personal accounts are available to those interested [27, 28]. We shall have to be content with a brief summary in which the principal contributions of microbial and molecular studies to modern genetic theory stand out. First of all, these studies led to a change in the concept of the gene which can hardly be overemphasized. While data from higher organisms were already suggesting a breakdown of the traditional view of the gene as an indivisible atom, the results obtained with viruses and bacteria made the shift in thinking inevitable. Because of the enormous resolving power of the methods devised for selecting rare microbial mutants and re-combinants, the gene was shown to be complex: It is made up of in-dependently mutable and separable parts, the combination of which is responsible for a unitary function. The function of the complex gene was in turn shown to consist in specifying the sequence of a cellular macromolecule, often a particular polypeptide, the basic component of a protein. Moreover, the parts of the genes are arranged in a linear order, cointegral with the linear order of the chromosome of which the genes are a part. The order within the gene is colinear with the order of the amino acids in the polypeptide it specifies. Inasmuch as the work of Avery. and his colleagues on the pneumococcus, and of Hershey and Chase on bacterial virus T4, were taken to mean that DNA is the mate-rial of which genes are composed, the structure of DNA is now seen as the ultimate basis of genetic replication and specification of other ma-cromolecular structures, the twin features of the genetic material earlier stressed by Leonard Troland. The genetic significance of DNA moti-vated, of course, the quest for knowledge of DNA's structure, a quest that culminated in the model of James Watson and Francis Crick now familiar to almost all high school students. Suffice it to say, validation of the Watson-Crick model led quickly to insight into replication, mutation, and expression of the genetic material. In particular, the structural components of the gene that had been dissected by prior recombination analysis are now equated with the individual nucleotides of DNA; their heritable alteration is equated with mutation, their removal and ex-change with recombination, and the decoding of successive nucleotide triplets with phenotypic expression. Today the molecular mechanisms of gene mutation and recombination are being studied in fine detail, and plant and animal development is being illuminated by investigation of the various means by which the phenotypic expression of genes is regulated. Moreover, molecular genetics has even been of use in the study of evolution. Borrowing techniques for studying protein variation, students of evolution have been measuring changes in the primary structure of proteins correlated with the evolution of new species. Probably the most exciting and at the same time most bewildering discovery in recent years, however, has been the uncovering of unanticipated, extensive genetically determined variation in protein structure *within* species. Such intraspecific variation has been observed in a wide variety of plant and animal populations, including those of man. The search is now on for the mechanisms that can and do support such extensive genetic polymorphisms.

Confidence in modern genetic theory would ordinarily be reinforced by the very speed with which that theory is being translated into practices of great human concern: Genes are being isolated, synthesized, recombined in vitro, cloned, and transferred between species. Why then the disenchantment we are currently experiencing? The disenchantment derives, I believe, from experience with our own fallibility. The powers that molecular genetics confers upon us come at a time when we recognize our own limitations both as scientists and as social interventionists. Students of the history and philosophical foundations of science have shown us, if we geneticists were not already aware, the open-ended nature of our enterprise. As I have already indicated, modern genetic theory is not proved beyond a shadow of a doubt; it is reasonable, in good harmony with empirical data, and accepted by the vast majority of biologists. But it is subject to change, just as genetic theory itself underwent considerable change in the course of the present century. It was not expected in 1940, for example, that DNA would prove to be the material that genes are made of; catalytic specificity seemed to belong to proteins at that time. Yet the study of the pathogenic virulence of pneumococci, an investigation not directly related to the study of genetic material, led to our current view. Nor was it anticipated that the gene would have a substructure; it took a radical shift in thinking to displace the notion of an indivisible gene like a bead on a chromosomal string. Nor have earlier ideas of genotypic evolution prepared us for the teeming genic variety that we now find to be maintained in natural populations. Even as I write, the news is abroad that studies of synthetic genes are revealing the violability of the one gene:one polypeptide rule. In short, the best we can do in any science is to assert our model or theory's compatibility with the available evidence in the simplest and most directly testable manner of all models or theories yet proposed. We have no way of denying the possibility of as yet unconceived theories that will fit the available and forthcoming data in a more satisfying way.

We cannot help the fact that our knowledge is incomplete at any given time, and yet we also cannot help acting on the basis of what knowledge we do possess. This imperfect situation can lead to unpredicted consequences of human importance. Thus, for example, the discovery of antibiotics was heralded in the 1940s as a powerful means of combatting infectious diseases. Through their use, however, we have come to learn of the widespread existence in bacteria of genetic agents, called plasmids, that can reproduce independently of the bacterial chromosome and that can be transmitted quite promiscuously between bacteria of different species. Their spread in newly invaded bacterial populations makes possible the rapid transfer of genes they have acquired by recombination with the chromosomes of previous hosts. In this way, rare bacterial mutations to antibiotic resistence can be mobilized, combined in multiply potent ways, and spread through natural populations of bacteria. As an unforseen result of the use of antibiotics, therefore, we have had to deal increasingly with pathogenic bacteria that have become genetically resistant to a wide variety of antibiotics to which they had been previously sensitive. Our lack of knowledge thus contributes to the origin of new human problems with which we have to contend. Of course, that lack also leads to new knowledge, such as the isolation of plasmids themselves and their use in constructing artificially recombinant DNA in vitro. Proponents of research on gene splicing via plasmids point to the possible advantages that may be gained: By infecting bacteria with plasmids containing human genes coding for insulin or an enzyme that is lacking in an hereditary illness and by taking advantage of the universality of the genetic code, a means is secured for producing large amounts of desirable genes or gene products. The gene products may be used therapeutically and, when the techniques become available, the genes themselves transplanted in human beings lacking them. Opponents of this research cite the dangers of unpredictable effects from letting loose, deliberately or unintentionally, artificial gene combinations through the bacterial universe. They also warn of possible dangerous effects of inserting foreign and artificially constructed gene sequences into the human genome. Neither side is wholly without honor or reason. The point is we cannot count upon a completed body of unchanging, true knowledge from which to draw a single correct answer. This does not mean that science makes us incapable of acting, but it does mean that we cannot act without risk. Thus, we have moved a long way from supposing, as late nineteenth- and early twentieth-century thinkers were wont to do, that science is an unerring guide for human progress. That move must make for cautious humility as opposed to overconfident arrogance.

The evolution of science, like biological evolution, is open ended and manifests indeterminate or fortuitous elements. I have tried to give some examples in this essay of the role of chance in scientific change. Recall the significance of Morgan's association with Wilson in the conversion of Morgan to Mendelism. Consider the synergistic effects of attracting to the same institution, at the right time of their scientific development, such persons as George Beadle and Boris Ephrussi, or Jim Watson and Francis Crick. Consider the particular training American students received and the special advantages for genetics in America resulting from the emigration to this country of such influential scientists as Jacques Loeb, Max Delbruck, and Salvatore Luria. We cannot be certain that what we believe today about heredity would have been far different had it not been for these historical accidents, but we can be sure that the rate and the specific form taken by the evolution of genetics would have been different had not they, but other accidents, occurred.

In this light science is not the inevitable unfolding of a preformed flower but a highly creative process in which the human participants do much of the creating. Affecting scientific creativity is the peculiar bias, metaphysical underpinning, or ideological commitment with which the individual scientist approaches his problems. These biases or prior commitments may be more or less conscious, but without them scientific problems lack significance and fail to motivate the scientist. The reductionist commitment of Jacques Loeb made him look upon biological problems in a certain way and directed his teaching. Similarly, the intuitive expectations of new physical laws directed Delbruck to biology, and genetics in particular. Morgan's antipreformationism excited his entry into genetics, while Muller's bias in favor of a material gene can be claimed to have motivated his mutation research.

The subjective is not only a necessary element in scientific research, it influences the activities of the scientist as social agent. For scientific knowledge is rarely abstract, entirely separated from this world of human needs and concerns, and scientists are well aware of the social potential of new knowledge. Nowhere has this been more evident than in the history of genetics in America. Some of the first geneticists in this country saw their science as a means of improving the human species. Unfortunately, their views about what constituted a superior human being were deeply affected by their own "social Darwinist" bias. Their social policy was based upon the deeply ingrained notions that the inferiority of inferior beings was chiefly owing to inferior genes, and that human societies were unduly mitigating the effects of natural selection in weeding out the genetically inferior. Hermann J. Muller, despite his socialism and humanism, retained much of the evangelical eugenic fervor. He too feared a degeneration of the human race as a result of our technical ingenuity in allowing the genetically inferior to reproduce and proposed a policy of choosing germ plasm for future generations from those individuals with desired characters. Overlooked or understressed in eugenic advocacy is the understanding that the very definition of the inferior is relative to culture, environment, and prevailing resources. Clearly the diabetic is not unfit, and whatever genes responsible for his condition not inferior, so long as the diabetic condition is easily and cheaply overcome. Whatever resources a human community wishes to commit for the continued survival and reproduction of some of its members will depend, in the final analysis, on what it cherishes as human and values as good. These values are cultural and are not exclusively determined by our biological makeup.

Perhaps this is the most important lesson of all for biologists in general, and geneticists in particular, to learn. Biologists, by the nature of their profession, come to see success in terms of genetic continuity and to regard nature, and not man, as the arbiter of the fit. This prejudice leads to a new type of reductionism, indulged by contemporary sociobiologists, according to which human evolution is entirely explicable in terms of natural selection. Sociobiologists would have us believe that human behavior has been determined by a selective process in which continuity of the genotype is the ultimate criterion, that human altruism fades with the genetic distance between interacting individuals. This viewpoint would ignore that culture in man makes possible a new form of transmission, the transmission of ideas, among which is the idea of humanity [29]. We humans are guided by ideas of how men *ought* to relate to each other, thereby creating patterns of relationships that are not necessarily coincident with genetic relationships [30]. Indeed, genetic continuity irrespective of means may not be the highest human value. Such ethical behavior may be risky so far as continuity is concerned, but it is questionable whether the prospect of a species of beings descended from us but devoid of all properties of human compassion and morality would be preferable to risk.

REFERENCES

1. H. M. SCHMECK, JR. New York Times, pp. 19, A26, March 10, 1977.
2. J. LOEB. Science, **20**:777, 1904.
3. J. LOEB. The dynamic state of living matter. New York: Macmillan, 1906.
4. A. W. RAVIN. Stud. Hist. Biol., **1**:1, 1977.
5. R. DUBOS. The professor, the institute, and DNA. New York: Rockefeller Univ. Press, 1976.
6. T. H. MORGAN. Science, **22**:877, 1905.
7. T. H. MORGAN. Rep. Am. Breeders' Assoc., **6**:365, 1909.
8. E. B. WILSON. The cell in development and inheritance. New York: Macmillan, 1896.
9. E. B. WILSON. Science, **16**:991, 1902.
10. W. S. SUTTON. Biol. Bull., **4**:231, 1903.
11. T. H. MORGAN. Science, **32**:120, 1910.
12. T. H. MORGAN et al. The mechanism of heredity. New York: Holt, 1915.
13. W. COLEMAN. Centaurus, **15**:228, 1970.

14. O. T. Avery, C. M. MacLeod, and M. McCarty. J. Exp. Med., **79**:137, 1944.
15. A. Mirsky. Cold Spring Harbor Sympos. Quant. Biol., **12**:15, 1947.
16. K. W. Cooper, R. D. Hotchkiss, and S. Spiegelman. J. Cell. Comp. Physiol., **45**(suppl. 2): 18, 1955.
17. B. Commoner. Am. Sci., **52**:365, 1964.
18. W. E. Castle. *In:* L. C. Dunn (ed.). Genetics in the 20th century. New York: Macmillan, 1951.
19. W. M. Hays. *In:* [18].
20. D. S. Jordan. Rep. Am. Breeders' Assoc., **4**:201, 1908.
21. C. B. Davenport. Rep. Am. Breeders' Assoc., **6**:91, 1911.
22. H. J. Muller. Out of the night. London: Gollancz, 1936.
23. L. T. Troland. Am. Nat., **51**:321, 1917.
24. H. J. Muller. Am. Nat., **56**:32, 1922.
25. M. Delbruck. Trans. Conn. Acad. Arts Sci., **38**:173, 1949.
26. G. W. Beadle. Am. Rev. Biochem., **43**:1, 1974.
27. R. Olby. The path of the double helix. London: Macmillan, 1974.
28. J. D. Watson. The double helix. New York: Athenaeum, 1968.
29. A. W. Ravin. Zygon, **12**:27, 1977.
30. M. Sahlins. The use and abuse of biology. Ann Arbor: Univ. Michigan Press, 1976.

A DEFENSE OF BEANBAG GENETICS

J. B. S. HALDANE*

My friend Professor Ernst Mayr, of Harvard University, in his recent book *Animal Species and Evolution* [1], which I find admirable, though I disagree with quite a lot of it, has the following sentences on page 263.

The Mendelian was apt to compare the genetic contents of a population to a bag full of colored beans. Mutation was the exchange of one kind of bean for another. This conceptualization has been referred to as "beanbag genetics." Work in population and developmental genetics has shown, however, that the thinking of beanbag genetics is in many ways quite misleading. To consider genes as independent units is meaningless from the physiological as well as the evolutionary viewpoint.

Any kind of thinking whatever is misleading out of its context. Thus ethical thinking involves the concept of duty, or some equivalent, such as righteousness or *dharma*. Without such a concept one is lost in the present world, and, according to the religions, in the next also. Joule, in his classical papers on the mechanical equivalent of heat, wrote of the duty of a steam engine. We now write of its horsepower. It is of course possible that ethical conceptions will in future be applied to electronic calculators, which may be given built-in consciences!

In another place [2] Mayr made a more specific challenge. He stated that Fisher, Wright, and I "have worked out an impressive mathematical theory of genetical variation and evolutionary change. But what, precisely, has been the contribution of this mathematical school to evolutionary theory, if I may be permitted to ask such a provocative question?" "However," he continued in the next paragraph, "I should perhaps leave it to Fisher, Wright, and Haldane to point out what they consider their major contributions." While Mayr may certainly ask this question, I may not answer it at Cold Spring Harbor, as I have been officially informed that

* Address: Genetics and Biometry Laboratory, Government of Orissa, Bhubaneswar-3, Orissa, India.

I am ineligible for a visa for entering the United States.[1] Fisher is dead, but when alive preferred attack to defense. Wright is one of the gentlest men I have ever met, and if he defends himself, will not counterattack. This leaves me to hold the fort, and that by writing rather than speech.

Now, in the first place I deny that the mathematical theory of population genetics is at all impressive, at least to a mathematician. On the contrary, Wright, Fisher, and I all made simplifying assumptions which allowed us to pose problems soluble by the elementary mathematics at our disposal, and even then did not always fully solve the simple problems we set ourselves. Our mathematics may impress zoologists but do not greatly impress mathematicians. Let me give a simple example. We want to know how the frequency of a gene in a population changes under natural selection. I made the following simplifying assumptions [3]:

1) The population is infinite, so the frequency in each generation is exactly that calculated, not just somewhere near it.

2) Generations are separate. This is true for a minority only of animal and plant species. Thus even in so-called annual plants a few seeds can survive for several years.

3) Mating is at random. In fact, it was not hard to allow for inbreeding once Wright had given a quantitative measure of it.

4) The gene is completely recessive as regards fitness. Again it is not hard to allow for incomplete dominance. Only two alleles at one locus are considered.

5) Mendelian segregation is perfect. There is no mutation, non-disjunction, gametic selection, or similar complications.

6) Selection acts so that the fraction of recessives breeding per dominant is constant from one generation to another. This fraction is the same in the two sexes.

With all these assumptions, we get a fairly simple equation. If q_n is the frequency of the recessive gene, and a fraction k of recessives is killed off when the corresponding dominants survive, then

$$q_{n+1} = \frac{q_n - k q_n^2}{1 - k q_n^2}.$$

[1] In spite of this ineligibility I have, since writing this article, been granted an American visa, for which I must thank the federal government. However, I am not permitted to lecture in North Carolina, and perhaps in other states, without answering a question which I refuse to answer. Legislation to this effect does not, in my opinion, help American science.

Norton gave an equation equivalent to this in 1910, and in 1924 I gave a rough solution when selection is slow, that is to say k small. But one might hope that such a simple-looking equation would yield a simple relation between q_n and n; if not as simple as $s = \frac{1}{2} gt^2$ for fall in a uniform gravitational field, then as simple as Kepler's laws of planetary motion. Haldane and Jayakar [4] have solved this equation in terms of what are called automorphic functions of a kind which were fashionable in Paris around 1920, but have never been studied in detail, like sines, logarithms, Gamma and Polygamma functions, and so on. Until the requisite functions have been tabulated, geneticists will be faced with as much work as if a surveyor, after measuring an angle, had to calculate its cosine or whatever trigonometrical function he needed. The mathematics are not much worse when we allow for inbreeding and incomplete dominance. But they are very much stiffer when selection is of variable intensity from year to year and from place to place (as it always is) or when its intensity changes gradually with time. If we had solved such problems, our work would be impressive.

Let me add that the few professional mathematicians who have interested themselves in such matters have been singularly unhelpful. They are apt to devote themselves to what are called existence theorems, showing that problems have solutions. If they hadn't, we shouldn't be here, for evolution would not have occurred.

Now let me try to show that what little we have done is of some use, even if we have done a good deal less serious mathematics than Mayr believes. It may be well to cite the first formulation of beanbag genetics. This was by the great Roman poet Titus Lucretius Carus just over two thousand years ago (*De rerum natura*, IV, l. 1220):

> Propterea quia multa modis primordia multis
> Mixta suo celant in corpore saepe parentes
> Quae patribus patres tradunt ab stirpe profecta,
> Inde Venus varia producit sorte figuras
> Maiorumque refert vultus vocesque comasque.

A free rendering is: "Since parents often hide in their bodies many genes mixed in many ways, which fathers hand down to fathers from their ancestry; from them Venus produces patterns by varying chance, and brings back the faces, voices, and hair of ancestors." Very probably the great materialistic (but not atheistic) philosopher Epicurus had expressed the theory more exactly, if less poetically, in one of his lost books. Lucretius elsewhere described genes as "genitalia corpora" and claimed that they were immutable. What is important is that whether he called them primordia or even seeds, he always thought of them as a set of separable material bodies. When Mendel discovered most of the laws according to which Venus picks out the hidden genes from the mixture, and Bateson and Punnett further discovered linkage, we could get going; and it was Punnett [5] who first calculated the long-term effect of a very simple program of selection.

Now let me begin boasting. So competent a biologist as Professor L. T. Hogben [6] has recently written, "The mutation of chromosomes or of single genes is admittedly the pace-maker of evolution." A strong verbal argument could be made out for this statement. In racing, a "pacemaker" runs particularly fast, but I suppose Hogben means that mutation determines the rate of evolution, which would be faster if mutation were more frequent. The verbal argument might run as follows: "Evolution is the resultant of a number of processes, including adaptation of individuals during their development, migration, segregation, natural selection, and mutation. Now in this list the slowest process is mutation. The probability that a gene will mutate in one generation rarely exceeds one hundred thousandth, and may be much less than a millionth. Whereas selective advantages of one in ten are quite common, a species may spread over a continent in a few centuries, and so on. Since mutation is the slowest process, it must set the pace, or be the 'rate-determining process,' for the remainder." This is quite as good an argument as those on which most human ethical and political decisions are based. When Muller had determined a few mutation rates, Wright and I, around 1930, began to calculate the evolutionary effects of mutation. We showed that in a species with several hundred thousand members mutations could not be a pacemaker. Almost all mutations occurred several times in a generation in one member or another of a species. But this again is a verbal argument. Only algebraical argument can be decisive in such a case. No doubt Wright's original "model" or hypothesis was too simple, but it was, I believe, near enough to the truth. I put in some rather ugly algebra to show that it made no appreciable difference whether selection occurred before or after mutation in a life cycle. I do not regret this effort. It is necessary to test all sorts of possibilities in such a case. I was trying to build a mathematical theory of natural selection. In doing so I calculated the equilibria between mutation of various types of genes and selection against them. As soon as this was done it became possible to estimate human mutation rates, and I did so [7]. Later on I improved this estimate, and since then many others have done it better. The estimation of human mutation rates, which is a by-product of my mathematical work, has since assumed some political importance. Had I devoted my life to research and propaganda in this field, rather than to expanding the bounds of human knowledge, I should doubtless be a world-famous "expert." I believe that the estimation of the rate at which X-rays, gamma rays, neutrons, and so on, produce mutations in animals could be vastly improved. But what I believe to be the most accurate method [8, 9] has not been given a serious trial, probably because it involves a good deal of mathematics. However, the work of Carter [10] and of Muramutsu, Sugahara, and Okazawa [11] shows that it is practicable, but expensive.

Now, Professor Mayr might say, "We must thank Haldane for the first estimate of a human mutation rate, but his argument is very simple indeed; in his own words, 'the rates of production by mutation and elimination by natural selection [of a harmful gene] must about balance.' So if we can find out how many people die of hemophilia or sex-linked muscular dystrophy per year, we can find out how many genes for these conditions arise by mutation." Anyone can understand this argument, and it has been used to estimate many human mutation rates, even though one estimate, based on years of careful work, is out by a factor of 2 through an elementary mathematical error. But as it stands it is no better than most political arguments. Selection and mutation must balance in the long run, but how long is that? In two rather complicated mathematical papers [12, 13] I showed that while harmful dominants and sex-linked recessives reach equilibrium fairly quickly, the time needed for the frequency of an autosomal recessive to get halfway to equilibrium after a change in the mutation rate, the selective disadvantage, or the mating system, may be several thousand generations. In fact, the verbal argument is liable to be fallacious. As few people have read my papers on the spread or diminution of autosomal recessives, and still fewer understood them, the "balance" method, which I invented, is applied to situations where I claim that it leads to false conclusions.

I am in substantial agreement with David Hume when he wrote (*A trea-*

tise of human nature, Book 1, Part 3, Section 1): "There remain therefore algebra and arithmetic as the only sciences, in which we can carry on a chain of reasoning to any degree of intricacy, and yet preserve a perfect exactness and certainty." Not only is algebraic reasoning exact; it imposes an exactness on the verbal postulates made before algebra can start which is usually lacking in the first verbal formulations of scientific principles.

Let me take another example from my own work. From the records of the spread of the autosomal gene for melanism in the moth *Biston betularia* in English industrial districts, I calculated [3] that it conferred a selective advantage of about 50 per cent on its carriers. Few or no biologists accepted this conclusion. They were accustomed to think, if they thought quantitatively at all, of advantages of the order of 1 per cent or less. Kettlewell [14] has now made it probable that, in one particular wood, the melanics have at least double the fitness of the original type. As Kettlewell very properly chose a highly smoke-blackened wood where selection was likely to be intense, I do not think his result contradicts mine. The mathematics on which my conclusion was based are not difficult, but they are clearly beyond the grasp of some biologists. In a recent book [15] it was stated that this melanism must originally have been recessive, in which case even the large advantage found by Kettlewell would have taken some thousands of years to produce the changes observed in fifty years. I suspect this curious mistake is due to the fact that in an elementary exposition one may produce an argument which ignores dominance and gives a result of the right order of magnitude. But such an exposition may not stress that the argument breaks down when applied to rare recessives. I think that in this particular instance Professor Mayr may have unwittingly been a little less than fair to us beanbaggers. On his page 191 [1] he says that my "classical" calculations in a book published in 1932 were deliberately based on very small selective intensities and implies that I only reached the same conclusion for industrial melanism in 1957. In fact, it was not till 1957 that biologists took my calculation of 1924 seriously. I did not stress it in 1932 because I thought such intense selection was so unusual as to be unimportant for evolution. If biologists had had a little more respect for algebra and arithmetic, they would have accepted the existence of such intense selection thirty years before they actually did so.

When Landsteiner and Wiener discovered the genetical basis of human fetal erythroblastosis, I pointed out [16] that the death of Rh-positive babies born to Rh-negative mothers could not yield a stable equilibrium and suggested that the modern populations of Europe were the result of crossing between peoples who, like all peoples then known, possessed a majority of Rh-positive genes and peoples who had a majority of Rh-negative genes. A distinguished colleague had calculated an equilibrium but had not dipped far enough into the bag to notice that it was unstable. Since then two relict populations have been discovered, in northern Spain and in one canton of Switzerland, with a majority of Rh-negative genes. If the mortality of the babies were higher, such differences would constitute a barrier to crossing, and I do not doubt that differences of this sort play a part in preventing hybridization between mammalian species. They can, for example, kill baby mules. I therefore regard the above paper as a contribution both to anthropology and to general evolution theory.

Once one has developed a set of mathematical tools, one looks for quantitative data on which to try them out. There are perhaps three main lines of such machine tool design, which may be called the Tectonic (from Greek τέκτων, a Wright), the Halieutic (from Greek ἁλιεύτης, a Fisher), and my own. Morton and C. A. B. Smith are developing a fourth, for use in human genetics. P. A. P. Moran [17] may be starting a fifth, or he may merely have made a hard road into an impassable swamp. A worker looks for numerical data on which his own favorite tools will bite. Thus Wright has collected data on small more or less isolated populations to which his theory of genetic drift is applicable. Fisher was probably at his best with samples from somewhat larger populations, for example his brilliant demonstration [18] of natural selection in Nabours' samples of wild *Paratettix texanus*, which is still perhaps the best evidence for heterosis in wild populations. Perhaps I am at my best with still larger populations. Thus I was, I think, the first to estimate quantitatively the rate of morphological change in evolving species [19]. My estimates are of the right order of magnitude, but based on estimates of geological time less reliable than those of Simpson [20]. I therefore fully accept Simpson's emendations (his pp. 10–17) of my figures. The question was the rate at which the mean of a morphological character changes. For one tooth measurement on fossil Equidae, paracone height, the rate of increase of the mean per million years ranged from 2.4 per cent to 7.9 per cent; for another, ectoloph length, from 0.6 per cent to 3.4 per cent. The rate of increase of the ratio of these lengths, which is of greater evolutionary importance, ranged from 0.9

per cent to 5.5 per cent. The total time covered was about 50 million years. On the other hand, I suggested that human skull height had increased by over 50 per cent per million years during the Pleistocene. The fossil data could have been so analyzed earlier. If I was the first to do so it was because, as the result of my mathematical work, such numbers had come to have more meaning for me than for others.

We can now come back to the justification of mathematical genetics. I leave out the body of mathematics which has grown up around human genetics. Here we cannot experiment and must squeeze all the information out of available figures, whereas where experiment is possible, not only is experiment often easier than calculation, but its results are more certain. In the consideration of evolution, a mathematical theory may be regarded as a kind of scaffolding within which a reasonably secure theory expressible in words may be built up. I have given examples to show that without such a scaffolding verbal arguments are insecure. Let me take an example from astronomy. I do not doubt that when Newton enunciated his gravitational theory of planetary movement many people said that if the sun attracted the planets they would fall into it. This is not so naïve as might be supposed. Cotes, of whose early death Newton wrote, "If Mr. Cotes had lived, we might have known something," showed that if the system "of planets, struggling fierce towards heaven's free wilderness," as Shelley put it, were attracted by the sun with a force varying as the inverse cube of the distance, they would move in spirals, and either fall into the sun or freeze in the free wilderness. Newton felt that he had to show not only that the inverse-square law led to stable elliptic motion, but that spheres, whose density at any point was a function of distance from their center, attracted one another as if they were particles. If he had not done so, he was aware that someone might readily disprove his highly ambitious theory. This does not mean that in explaining Newtonian gravitational theory to students one need go into these or many other details.

It is, in my opinion, worth while devoting some energy to proving the obvious. Thus, suppose a population consists of two genotypes A and B, of which B is fitter than A so long as it is rare. For example, B could be a mimic only advantageous when rare compared with its model, or a self-sterile but interfertile genotype of a plant species. It is intuitively obvious that B will spread through a population till its mean fitness falls to equal that of A, and a stable equilibrium will result. But is it sure that this equilibrium will be stable? Every physicist and cybernetician knows that if regulation is too intense a system may overshoot its equilibrium and go into oscillations of increasing amplitude. Haldane and Jayakar [21] found that in several cases investigated by them there was no danger of such instability. In a microfilm on population genetics circulated in A.D. 2000 we may either find the statement, "Haldane and Jayakar showed that such equilibria are almost always stable," or, "Haldane and Jayakar believed that they had demonstrated the stability of such equilibria. They overlooked the investigations of X on termites, where, as Y later showed, the equilibrium is unstable." But even if we have given the wrong answer, we deserve a modicum of credit for asking the right question.

I could give many more examples. Thus, posterity may or may not think that my concept of the cost of natural selection—that is to say, the number of genetic deaths required to bring about an evolutionary change [22]—is important. I think it defines one of the factors, perhaps the main one, determining the speed of evolution. It has been accepted by some and criticized by others. If it is shown to be false, the demonstration of its falsity will probably reveal the truth, or at least a closer approximation to the truth. And so I could continue on a large scale. If I were on trial for wasting my life, my defense would at least be prolonged, even if unsuccessful; for I have published over 90 papers on beanbag genetics, of which over 50 contained some original statements, whether or not they were important or true, besides 200 papers on other scientific topics.

The existing theories of population genetics will no doubt be simplified and systematized. Many of them will have no more final importance than a good deal of nineteenth-century dynamical theory. This does not mean that they have been a useless exercise of algebraical ingenuity. One must try many possibilities before one reaches even partial truth. There is, however, a danger that when a mathematical investigation shows a possible cause of a phenomenon, it is assumed to be the only possible cause. Thus Fisher [23] showed that if heterozygotes for a pair of autosomal alleles are fitter than either homozygote, there will be stable polymorphism, and later work has extended this theorem to multiple alleles. Numerous cases have been discovered where such heterosis, both at single loci and for chromosomal segments, has been observed in nature. It has therefore been assumed that, except where rarity confers an obvious advantage, the Halieutic mechanism is at work. Now Haldane and Jayakar [24] have shown that, without any superiority of heterozygotes, selection of fluctuating direction will sometimes preserve polymorphism. There is no reason to think that this often happens, but it may sometimes do so. However, if I had made this calculation in 1920, as I might have done, while Fisher had published his work somewhat later, my explanation, which I do not doubt is more rarely true than Fisher's, might have been accepted as the usual explanation of stable polymorphism. It seems likely that this has happened in other cases, though naturally I do not know what these are. The best way to avoid such contingencies is to investigate mathematically the consequences following from a number of hypotheses which may seem rather farfetched and, if they would lead to observed results, looking in nature or the laboratory for evidence of their truth or falsehood.

One such possibility is the origin of "new" genes in higher animals or plants by viral transduction from species with which hybridization is impossible, conceivably even from members of a different phylum. While no doubt exaggerated claims have been made by the Michurinist school, some of its claims, such as the facilitation of hybridization by grafting, have been verified outside the Soviet Union. Transduction could account for some grafting effects which could not be regularly repeated. In terms of orthodox American genetics, such transduction would be described as a mutation leading to a neomorph. It is obvious that transduction could help to explain some cases of parallel evolution.

Let me be clear that I think the above hypothesis is improbable. But it serves to underline a fundamental point. Let us suppose that it had been proved that all evolutionary events observed in the fossil record and deduced from comparative morphology, embryology, and biochemistry could be explained on the basis of the generally accepted "synthetic theory"; this would not demonstrate that other causes were not operating. I think we have come near to showing that the synthetic theory will account for observed evolution and that a number of other superficially plausible theories, such as those of Lamarck, Osborn, and de Vries, will not do so. This does not exclude the possibility that other agencies are at work too. To take an example from astronomy, it was believed until recently that celestial mechanics were almost wholly dominated by gravitational forces. It is now believed that cosmic magnetic fields are also important.

Of course, Mayr is correct in stating that beanbag genetics do not explain the physiological interaction of genes and the interaction of genotype and environment. If they did so they would not be a branch of biology. They would be biology. The beanbag geneticist need not know how a particular gene determines resistance of wheat to a particular type of rust, or hydrocephalus in mice, or how it blocks the growth of certain pollen tubes in tobacco, still less why various genotypes are fitter, in a particular environment, than others. If he is a good geneticist he may try to find out, but in so doing he will become a physiological geneticist. If the beanbag geneticist knows that, in a given environment, genotype P produces 10 per cent more seeds than Q, though their capacity for germination is only 95 per cent of those of Q, he can deduce the evolutionary consequence of these facts, given further numbers as to the mating system, seed dispersal, and so on. Similarly, the paleontologist can describe evolution even if he does not know why the skulls of labyrinthodonts got progressively flatter. He is perhaps likely to describe the flattening more objectively if he has no theory as to why it happened.

The next probable development of beanbag genetics is of interest. Sakai [25] described competition between rice plants. A plant of genotype P planted in the neighborhood of plants of genotype Q may produce more seeds than when planted in pure stand, while its neighbors of genotype Q produce less. Roy [26] has described cases of this kind but also cases where, when P and Q are grown in mixture, both P and Q produce more seed. In such a case, if the mixed seed is harvested and sown, P may supplant Q, or a balanced polymorphism may result. Of course if P and Q interbreed, the results will be very complicated. But I have no doubt that such cases occur in nature, and are of evolutionary importance. Given quantitative data on yields of mixed crops, a beanbag geneticist can work out the consequences of such interaction, even if he does not know its causes.

Another probable development is this. It is likely [27] that as the result of duplications one locus in an ancestor can be represented by several in descendants. If so, this is one of the important evolutionary processes, and its precise "beanbag" genetics will require investigation, even though the relative fitnesses of the various types, and the reasons for them, besides the causes of duplication, are matters of physiological genetics.

I would like to make one more claim for beanbag genetics. It has been of some value to philosophy. I consider that the theory of path coefficients invented by Sewall Wright may replace our old notions of causation. A path coefficient answers the question, "To what extent is a set of events

B determined by another set A?" Path coefficients were invented to deal with problems such as the determination of piebaldness and otocephaly in guinea pigs, which are beyond the present scope of beanbag genetics. But Wright showed how to calculate them exactly in the case of inbreeding, which he treated on "beanbag" principles. Again Haldane [28] discussed how to argue back from a handful of beans to the composition of the bag. Jeffreys [29] made this paper the basis of his theory of inverse probability. Jeffreys is generally regarded as a heresiarch and takes my theory more seriously than I do myself. Nevertheless, there is presumably some measure of truth in it, and even if Birnbaum [30] has shown how to do without it, I may take some credit for stimulating him to lay a stronger foundation than my own for the theory of inverse probability.

The dichotomy between physiological and beanbag genetics is one of the clearest examples of the contrast between what my wife, Spurway [31], calls Vaiṣṇava and Śaiva[2] biology. Modern Hindus can, on the whole, be divided into Vaiṣṇavas—that is to say, worshippers of Viṣṇu, usually in one or other of his most important incarnations, Rama and Kriṣṇa—and Śaivas, or worshippers of Śiva. Viṣṇu has, on the whole, been concerned with preservation and Śiva with change by destruction and generation. This is a very superficial account. Spurway may be consulted for further details. Devotees of Viṣṇu do not deny the existence of Śiva, nor conversely are they necessarily exclusive in their worship, and many state that both deities are aspects of the same Being. Neither sect has actively persecuted the other. Roughly speaking, Darwin was a Śaiva when he wrote on natural selection and a Vaiṣṇava when he wrote on the adaptations of plants for cross-pollination, climbing, and so on. A biologist who is always a Śaiva, and does not worry about how living organisms achieve internal harmony and adaptation to their environment, is as narrow as a Vaiṣṇava who takes an organism as given and does not interest himself in its evolutionary past or its success in competition with other members of its species. It is very difficult to combine the two approaches in one's thought at the same moment. It may be easier a century hence. Thus, we know that human sugar metabolism depends on the antagonistic action of pancreatic insulin and one or more diabetogenic hormones from the anterior pituitary. Insulin production and anterior pituitary function are both under genetic control, but we do not know enough about this even to speculate

[2] Ś and ṣ are both near to the English *sh*.

fruitfully on the level of beanbag genetics, except to say that several different genotypes may achieve good homeostasis, while other combinations of the genes concerned are less well adapted for homeostasis, though they may have other advantages. Even this is a mere speculation. There may be only one adaptive peak in Wright's sense.

As I happen to be responsible for some of the mathematical groundwork of enzyme chemistry [32], I can say that the mathematical basis of physiological genetics is about fifty years behind that of beanbag genetics. If a metabolic process depends on four enzymes acting on the same substrate in succession, one can calculate what will happen if the amount of one of them is halved, provided that one is working with enzymes in solution in a bottle. We know far too little of the structural organization of living cells at the molecular level to predict what will happen if the amount is halved in a cell, as it is in some heterozygotes. If the enzyme molecules are arranged in organelles containing just one of each kind, the rate of the metabolic process will probably be halved. But if they are in a random or a more complicated arrangement, it may be diminished to a slight extent, or even increased; for the activities of some enzymes are inhibited by an excess of their substrate. This is a conceivable cause of heterosis, though I do not think it is likely to be common.

Now let me pass over to a counterattack. One of the central theses of Mayr's book is that speciation is rarely if ever sympatric. One species can only split into two as the result of isolation by a geographic barrier, save perhaps in very rare cases. Let me say at once that Mayr's arguments have convinced me that sympatric speciation is much rarer than some authors have believed, and a few still believe. But when, in his chapter 15, he discusses other authors' hypotheses as to how sympatric speciation might occur, his arguments are always verbal rather than algebraic. And sometimes I find his verbal arguments very hard to follow. Thus, on page 473 he makes seven assumptions, of which (1) is "Let A live only on plant species 1," and (4) is "Let A be ill adapted to plant species 2." These two assumptions seem to me to be almost contradictory. If A lives only on species 1, the fact that it is ill adapted to species 2 is irrelevant. If emus only live in Australia, the fact that they are ill adapted to the Antarctic has no influence on their evolution. If the assumptions had been "(1) Let A females only lay eggs on species 1," and "(4) Let A larvae (not all produced by A mothers) be ill adapted to species 2," I could have applied mathematical analysis to the resulting model. I propose to do so in the next few years. But I hope I have given enough examples to justify my complete mistrust of verbal arguments where algebraic arguments are possible, and my skepticism when not enough facts are known to permit of algebraic arguments.

In earlier chapters Mayr seems to show a considerable ignorance of the earlier literature of beanbag genetics. Thus, on page 215 he writes that "the classical theory of genetics took it for granted that superior mutations would be incorporated into the genotype of the species while the inferior ones would be eliminated." The earliest post-Mendelian geneticists, such as Bateson and Correns, wrote very little about this matter. Fisher [23] pointed out that if a heterozygote for two alleles was fitter than either homozygote, neither allele would be eliminated. He may well have been anticipated by Wright or some other geneticists, but at least since 1922 this has been a well established conclusion of beanbag genetics. In my first paper on the mathematical theory of natural selection [3], I ignored Fisher's result as I was dealing with complete dominance; in my second [33] I referred to it and, as I think, extended it slightly. As Mayr cites neither of these papers of mine, he can hardly mean that the first was classical and the second post-classical! I agree with him that when I first read Fisher's 1922 paper I probably did not think this conclusion as important as I now do, and that many writers on beanbag genetics ignored it for some years. But were they classical?

Mayr devotes a good deal of space to such notions as "genetic cohesion," "the coadapted harmony of the gene pool," and so on. These apparently became explicable "once the genetics of integrated gene complexes had replaced the old beanbag genetics." So far as I can see, Mayr attempts to describe this replacement in his chapter 10, on the unity of the genotype. This does not mention Fisher's fundamental paper [34] on "The correlation between relatives on the supposition of Mendelian inheritance," in which, for example, epistatic interaction between different loci concerned in determining a continuously variable character was discussed. This chapter contains a large number of enthusiastic statements about the biological advantages of large populations which, in my opinion, are unproved and not very probable. The plain fact is that small human isolates, whether derived from one "race," like the Hutterites, or two, like the Pitcairn Islanders, can be quite successful. I have no doubt that some of the statements in Mayr's chapter 10 are true. If so, they can be proved by the methods of beanbag genetics, though the needed mathematics will be exceedingly stiff. Fisher and Wright have both gone further than Mayr believes toward proving some of them. The genetic structure of a species depends largely on local selective intensities, on the one hand, and migration between different areas, on the other. If there is much dispersal, local races cannot develop; if there is less, there may be clines; if still less, local races. The "success" of a species can be judged both from its present geographical distribution and numerical frequency and from its assumed capacity for surviving environmental changes and for further evolution. I do not think that in any species we have enough knowledge to say whether it would be benefited by more or less "cohesion" or gene flow from one area to another. We certainly have not such knowledge for our own species. If inter-caste marriages in India become common, various undesirable recessive characters will become rarer; but so may some desirable ones, and the frequency of the undesirable recessive genes, though not of the homozygous genotypes, will increase. Since there is little doubt that extinction is the usual fate of every species, even if it has evolved into one or more new species, the optimism of chapter 10 does not seem justified. Sewall Wright has been the main mathematical worker in this field, and I do not think Mayr has followed his arguments. Here Wright is perhaps to blame. So far as I know, he has never given an exposition of his views which did not require some mathematical knowledge to follow. His defense could be that any such exposition would be misleading. I have given examples above to illustrate this possibility.

I am reviewing Mayr's book in the *Journal of Genetics*, and my review will, on the whole, be favorable. But if challenged, I am liable to defend myself, and have done so in this article. If I have not defended Sewall Wright, this is largely because I should like to read his defense. In my opinion, beanbag genetics, so far from being obsolete, has hardly begun its triumphant career. It has at least proved certain far from obvious facts. But it needs an arsenal of mathematical tools like the numerous functions discovered or invented to supply the needs of mathematical physics. Of course, it also needs accurate numerical data, and these do not yet exist, except in a very few cases. The reason is simple enough. Suppose we expect equal numbers of two genotypes, say, normal males and colorblind males, from a set of matings and find 51 per cent and 49 per cent; then if we are sure that this difference is meaningful, it will have evolutionary effects which are very rapid on a geological time scale. But to make sure that the difference exceeded twice its standard error (which it would do by chance once in twenty-two trials), we should have to examine 10,000 males. To achieve reasonable certainty, we should have to examine 25,000. We often base our notions of the selective advantage of a gene on mortality from some special cause. Thus babies differing from their mothers in respect of certain antigens are liable to die around the time of their birth. But this may well be balanced wholly or in part by greater fitness in some other part of their life cycle. If it were found that color-blind males had a 10 per cent higher mortality than normals from traffic accidents, this could be balanced by a very slightly greater fertility or frequency of implantation as blastocysts. One of the important functions of beanbag genetics is to show what kind of numerical data are needed. Their collection will be expensive. Insofar as Professor Mayr succeeds in convincing the politicians and business executives who control research grants that beanbag genetics are misleading, we shall not get the data. Perhaps a future historian may write, "If Fisher, Wright, Kimura, and Haldane had devoted more energy to exposition and less to algebraical acrobatics, American, British, and Japanese genetics would not have been eclipsed by those of Cambodia and Nigeria about A.D. 2000." I have tried in this essay to ward off such a verdict.

Meanwhile, I have retired to a one-storied "ivory tower" provided for me by the Government of Orissa in this earthly paradise of Bhubaneswar and hope to devote my remaining years largely to beanbag genetics.

REFERENCES

1. E. MAYR. Animal species and evolution. Cambridge, Mass.: Harvard University Press, 1963.
2. ———. Sympos. Quant. Biol., 24:1, 1959.
3. J. B. S. HALDANE. Trans. Cambridge Phil. Soc., 23:19, 1924.
4. J. B. S. HALDANE and S. D. JAYAKAR. J. Genet., 58:291, 1963.
5. R. C. PUNNETT. J. Hered., 8:464, 1917.
6. L. T. HOGBEN. In: M. P. BANTON (ed.). Darwinism and the study of society. London: Travistock Publications, 1961; Chicago: Quadrangle Books, 1961.
7. J. B. S. HALDANE. The causes of evolution. London: Longmans Green, 1932.
8. ———. J. Genet., 54:327, 1956.
9. ———. J. Genet., 57:131, 1960.
10. G. S. CARTER. J. Genet., 56:353, 1959.
11. S. MURAMUTSU, T. SUGAHARA, and Y. OKAZAWA. Int. J. Radiat. Biol., 6:49, 1963.

12. J. B. S. Haldane. Ann. Eugen., 9:400, 1939.

13. ———. Ann. Eugen., 10:417, 1940.

14. H. B. D. Kettlewell. Proc. Roy. Soc. [B.], 297:303, 1956.

15. R. G. King. Genetics. New York: Oxford University Press, 1962.

16. J. B. S. Haldane. Ann. Eugen., 11:333, 1942.

17. P. A. P. Moran. The statistical processes of evolutionary theory. Oxford: Clarendon Press, 1962.

18. R. A. Fisher. Ann. Eugen., 9:109, 1939.

19. J. B. S. Haldane. Evolution, 3:51, 1949.

20. G. G. Simpson. The major features of evolution. New York: Columbia University Press, 1953.

21. J. B. S. Haldane and S. D. Jayakar. J. Genet., 58:318, 1963.

22. J. B. S. Haldane. J. Genet., 55:511, 1957.

23. R. A. Fisher. Proc. Roy. Soc. Edin., 42:321, 1922.

24. J. B. S. Haldane and S. D. Jayakar. J. Genet., 58:237, 1963.

25. K. Sakai. J. Genet., 55:227, 1957.

26. S. K. Roy. J. Genet., 57:137, 1960.

27. V. M. Ingram. Fed. Proc., 21:1053, 1962.

28. J. B. S. Haldane. Proc. Cambridge Phil. Soc., 38:55, 1932.

29. H. Jeffreys. The theory of probability. London: Oxford University Press, 1948.

30. A. Birnbaum. J. Amer. Stat. Ass., 57:269, 1962.

31. H. Spurway. Zoölogische Jährbücher., 88:107, 1960.

32. J. B. S. Haldane. Enzymes. London: Longmans Green, 1930.

33. ———. Proc. Cambridge Phil. Soc., 23:363, 1926.

34. R. A. Fisher. Trans. Roy. Soc. Edin., 52:399, 1918.

KNOWLEDGE AND PREMATURITY: THE JOURNEY FROM TRANSFORMATION TO DNA

H. V. WYATT*

The scientific work which led to the recognition that DNA is the carrier of genetic information continues to fascinate both those who were alive at the time and those who are now students. Griffith is commemorated by the memorial lectures of the Society for General Microbiology, while memories of Avery continue to warm those who knew him. Yet there are still many aspects which are puzzling. Why was it that Griffith's discovery of transformation led eventually to DNA, whereas Avery's demonstration of DNA as the genetic material did not lead immediately to a new paradigm?

Stent has claimed that in Avery's case the demonstration was premature [1]. I should like to suggest that in the case of both Griffith and Avery, there were two elements, one technical, the other concerned with the acceptance of a new idea. A parallel case is penicillin, where the idea of an antibiotic was not acceptable until after the discovery of sulphonamides. Fleming did, however, persist but was defeated because it was not technically possible, at least in his circle of associates, to purify the crude samples. It was only after the acceptance of the idea that it was thought worthwhile to solve the technical problems. Thus we may extend Stent's use of "premature" and my use of "knowledge and information" [2] if we include a new concept: discovery can be premature if it is not capable of being extended experimentally because of technical reasons. Griffith's discovery of transformation was extended because this was technically feasible, and in the next 10 years steady progress was

*Department of biology, University of Bradford, Bradford BD7 1DP, United Kingdom.
I am deeply grateful to the many persons who wrote to me following my paper in *Nature*, and I am particularly grateful to those who wrote giving personal reminiscences and details, especially Dr. R. D. Hotchkiss and Dr. R. M. Herriott. It is a very striking tribute to Avery that so many of his former associates wrote so warmly about him.

made by Avery's associates at the Rockefeller Institute. Each step brought Avery nearer to the identity of transforming principle. It was only when a new paradigm was desirable in 1944 that the story seemed less clear.

Transformation and DNA

If genetic recombination in bacteria is rare, it is not easy to see with ordinary culture techniques. Demonstration requires conditions which kill or prevent growth of the parents while allowing growth of the re-combinant. Such selective systems can be deliberately produced, as by Tatum and Lederberg [3], but do not often occur except by design.

Tatum and Lederberg used a selective system to test whether sexual conjugation occurred in bacteria: a reasonable if remote possibility. To test for the possibility that small pieces of genetic material could enter a bacterium and change its genotype would not have been an experiment which would have occurred to a bacteriologist. Yet transformation was discovered accidently before 1928 by Griffith. This was due to a remark-able combination of circumstances. He intended that the mouse should act as a selective-agent medium and should kill the avirulent bacteria. If, however, normally avirulent bacteria could be made to produce a cap-sule, the bacteria would multiply and the mice would die. He performed the experiment to test whether avirulent (capsuleless) bacteria would produce capsules if provided with sufficient concentration of polysac-charide to restore some kind of autocatalytic process [4].

In some experiments, however, the capsule type produced by the previously avirulent type could be identified as belonging not to that of the avirulent pneumococcus but to the same capsule type as the dead virulent pneumococcus which had been the source of the heat-treated polysaccharide. This Griffith called transformation.

The gene or genes responsible for capsule synthesis could be recog-nised not by one character but by three interrelated characters: (1) in the mouse by virulence; (2) on agar medium by the gross morphologic ap-pearance of the colony: smooth as opposed to rough; (3) by the presence of a capsule under the microscope.

Moreover, the capsular material could be distinguished by the use of specific antibody both in vivo and in vitro and classified into definite and different types. Thus, the experimental system was almost perfect! Selection and characterisation could be made by several different easy methods in very different circumstances. At a time when pneumococcal pneumonia was still a great killer [5], others were bound to repeat Griffith's work and extend it. In vitro transformation, accomplished by Dawson and Sia [6], was detected by colony morphology, identified by type-specific serum agglutination, and confirmed by virulence tests.

A further property of the system was used initially but later proved unnecessary: growth of the untransformed cells could be selected in liquid medium by the presence of anti-R species specific serum containing antibodies. Selection by this method was used later still by Austrian and MacLeod [7]. Dawson and Sia were unable to effect transformation with cell-free extracts of pneumococci [6], but Alloway succeeded in inducing transformation in vitro with cell-free purified and filtered extracts [8] and was able to show the heat stability of the transforming substance. The range of its heat stability almost precluded the possibility that it was either protein or type-specific polysaccharide as such. It is no wonder that Avery, who always thought in terms of specificity [5], should have remarked in conversation as early as 1936 that he thought it would be nucleic acid [9]. Even if none (except Avery!) knew what to do with this phenomenon, it was widely known. Landsteiner devoted a page to transformation in his book of serological reactions [10] and compared the transforming substance(s) with plant viruses, phages, etc., which were endowed like genes.

Identification of DNA

The publication of Avery, MacLeod, and McCarty [11] was immediately recognized, and, in spite of the limitations of the information content of the paper [2], its contents were widely noted. Avery also presented the results at a meeting of the American Association for the Advancement of Science held in Cleveland in September 1944. Among those who were there were Sewell Wright [2] and D. F. Poulson, a friend of G. E. Hutchinson, to whom Poulson talked about Avery's paper [2; Hutchinson, personal communication]. It is possible that at this meeting and at the staff meeting at the Rockefeller Institute Avery was rather more explicit than he was in the published paper.

In his address on the awarding of the Copley Medal in 1945 to Avery, Sir Henry Dale said: "And now only last year . . . which if we were dealing with higher organisms we should afford the status of a genetic variation; and the substance inducing it—the gene in solution, one is tempted to call it—appears to be a nucleic acid of the desoxyribose type. Whatever it be, it is something which should be capable of complete description in terms of structural chemistry" [12]. Peyton Rous, the editor of the *Journal of Experimental Medicine*, who had printed Avery's paper with no hint of its genetic significance, nevertheless instructed his secretary to file the reprint under "Genetics" [13]. The difficulty lay in translating the information into a form which could lead to further experiments and could be assimilated into a meaningful scientific paradigm. After Avery's official retirement from the Rockefeller Institute in 1943, the work on transformation was continued and extended by Hotchkiss and by Austrian and Harriet Taylor, all connected for a time with the three authors of the 1944 paper.

Even after Alloway, pneumococcal transformation in vitro required difficult and stringent experimental procedures. It was perhaps unfortunate for transformation but lucky for other aspects of biology that some of the early attempts to extend transformation to other systems did not succeed for technical reasons, although we now know that they are feasible. As Lederberg has fortunately now recounted, his first experiments with F. J. Ryan were an unsuccessful attempt at transformation with *Neurospora,* following the biochemical work on that fungus by Beadle and Tatum [14]. Later, when Lederberg and Tatum showed recombination in *Escherichia coli,* they were able to show by several experiments that this was not another example of transformation (see [15] for references). Curiously, the techniques which led to the discovery of bacterial conjugation originated in observations from an unsuccessful attempt at transformation. Seymour Cohen also tried unsuccessfully to transform bacteriophages with phage DNA by using bacterial cells already infected with another phage strain [16]. No changes could be detected among the three recipient strains. He remarked that possibly the added DNA did not penetrate the bacterial host: infection with naked phage DNA was only accomplished much later.

Thus Avery's discovery was premature because the technical means were not yet available to extend the work into other systems and confirm the universal nature of the phenomena. Other techniques were available to the plant virologists, but they did not do the crucial tests. Microbiologists were meanwhile busy with conjugation, and it was the use of new techniques which brought confirmation from the phage workers.

Plant Viruses

During the 1930s Bawden and Pirie carried out an extensive investigation of plant viruses. They confirmed that these viruses contained protein and nucleic acid, and they showed that nitrous acid and X-rays destroyed the infectivity of their preparations but had no effect on the serology, whereas other treatments also affected the serology. They isolated the nucleic acid and investigated it chemically but found that their nucleic acid was not infective [17]. Pirie has recently written that in the circumstances of their working at different places, the residual infectivity of material as fragile as TMV nucleic acid could not have been observed [18].

In 1948, Markham and co-workers isolated protein, nucleic acid, and nucleoprotein of turnip yellow mosaic virus. The protein fraction was not infective and the nucleoprotein was, but they did not report on the infectivity of the nucleic acid [19]. Infective nucleic acid of TMV was finally discovered by Gierer and Schramm in 1956, subsequent to the Blender Experiment [20]. It is ironic that the workers with plant viruses did not test their preparations of nucleic acid or, if they did, found no evidence of infectivity. Markham has since said that they accepted Avery's work [R. Markham, personal communication]; yet they nevertheless failed to report on the infectivity of their nucleic acid.

Phage

Herriott, who was working at the Rockefeller laboratories at Princeton, had no doubt of the importance of Avery's work and persuaded Hotchkiss and Avery to prepare and assay pneumococcus transforming principle for him. He found that the preparation, which was almost pure DNA with no demonstrable protein, was the most sensitive material which he had yet tested for inactivation by mustard gas (table 1). This supported Herriott's idea that mustard gas was far more damaging to biologically active DNA—in cells and DNA viruses—than to enzymes or RNA viruses. This was followed by his observation that osmotic shock released the phage DNA into solution and that the ghosts could still absorb to bacteria and lyse them [21]. He pointed out that this was a beginning toward the identification of viral functions with viral substances (quoted in [22]). Northrop, from the Rockefeller laboratories at Berkeley, in a discussion the same year 1951 said that the assumption that the phage particle lost its identity when in the cell appeared to be the simplest explanation of the observed results, for example, ultraviolet inactivation [23]. He added: "The nucleic acid may be the essential autocatalytic part of the molecule as in the case of the transforming principle of the pneumococcus (Avery etc., 1944) and the protein portion may be necessary only to allow access to the cell." Herriott put the idea clearly

TABLE 1

SENSITIVITY OF CERTAIN BIOLOGICAL MATERIALS
TO MUSTARD GAS

Material	Relative Sensitivity
Pneumococcus transforming principle	30
DNA phage	20
Newcastle disease virus	12
Yeast	2
Tobacco mosaic virus (RNA-protein)	1.4
Enzymes:	
1. Pepsin	0.5
2. Hexokinase	0.2

SOURCE.—[31].

when he suggested that the phage acted as a microsyringe to inject the DNA [24].

There are therefore many clear links from Avery to the experiments of Hershey and Chase. Yet the identification of the role of DNA in phage infection would probably have occurred within a year or two of 1952 even if the chemical nature of transformation had not already been known. Whether that discovery would have been immediately accepted as proof of the genetic importance of DNA, without the previous 8 years of cumulative evidence from the transforming principle, is an interesting speculation. The coincidence of the phage group working on an organism containing only protein and DNA and the availability of specific physical markers, radioisotopes, would have led sooner rather than later to experiments on the isotopic transfer from parental to progeny phage [25].

Watson and Maaløe (published in 1953 but written in April 1952 and cited in [22]) showed that there was equal transfer (nearly 50%) of phosphorus and adenine from parent to progeny phage [26]. Similar findings had been made by others, and it was surely only a matter of time before the significance of this was tested in some experiment.

Even if the experimenter had been seeking proof that protein was the carrier of genetic information, the results would have had to lead to DNA.

Knowledge and Prematurity

I have previously suggested that, although the work of Avery et al. [11] was soon known, it was known as an isolated and interesting piece of information which could not immediately be assimilated as knowledge [2]. Gunther Stent has more recently described Avery's work as being premature and suggests that a discovery is premature if its implications cannot be connected by a series of simple logical steps to canonical, or generally accepted, knowledge [1]. The claim that Avery's work was not fully appreciated (Wyatt) or that it was premature (Stent) has been challenged by a number of persons: Carlson [27], Lederberg [15], and R. Markham [personal communication].

The difficulty can, I think, be resolved: Avery's work contained two elements. The first, which was generally recognised as an experimental tour de force, was the isolation, purification, and characterisation of the transforming principle. Whatever the nature of the substance, it behaved as "a gene in solution" [12]. It was, however, controversial whether the active substance was DNA, as clearly stated by Dale, or a protein impurity, as believed by some biochemists. The acceptance that Avery had a gene in solution was knowledge; that it was of DNA was unfortunately not recognised by some and was only information to others. What is characteristic of so many statements from 1944 on is their ambiguity about the two elements; in many cases, no doubt, this was unintentional, but it may have hindered understanding by others.

I believe that a reading of Muller's Pilgrim address of 1946 will show that he recognised the importance of transformation as gene transfer (see [27]) but was ambiguous in his views that the active principle might be DNA [2]. Even in 1950 at the meeting of Genetics in the Twentieth Century, he was still uncommitted [1, 2]. Similarly; the statement that "we have obtained direct evidence for the widely held belief that nucleic acid is essential for the multiplication of plant viruses" [19] could be interpreted as showing that the nucleic acid was structurally necessary for the functioning of the (genetically active) protein (see foregoing).

Thus I quoted an ambiguous comment by Wright [2] and was rebuked by Lederberg [15]. However, Lederberg's quote from Wright—"The results suggest chemical isolation and transfer of a gene rather than induction of mutation"—is in a section on mutation, not in the section on chromosome chemistry, and could have been taken as implicating protein rather than DNA.

Transformation was surely discovered so early because the combination of characters—virulence, morphology, capsule, and antigenicity together with the different types—provided the necessary experimental system for its detection. No other combination of easily identifiable characters is associated with other examples of transformation: specific experimental techniques have had to be devised. The very factors which led to the detection of transformation provided no clue for its application outside that limited system. Thus it was not until almost 20 years later that bacterial transformation was extended to other systems, for example, Boivin [28] in 1967 and Hotchkiss, Austrian, and Harriet Taylor in the following years [29]. Transformation in the pneumococci was the ideal system for discovery but not for analysis or interpretation. Griffith's discovery of transformation [30] was therefore "premature" after it had been exploited by Avery, and required extension. The discovery of conjugation by Tatum and Lederberg [3] was made with an experimental system which was immediately capable of wide application and refinement. However, Griffith's discovery was not premature in another respect: by rare coincidence, Avery was not only interested in the pneumococci and thus transformation but also had the necessary intellectual curiosity and background for specificity [5] for identifying the chemical nature of the transforming principle.

REFERENCES

1. G. STENT. Sci. Am., **228**(12): 84, 1972.
2. H. V. WYATT. Nature, **235**:86, 1972.

3. E. L. Tatum and J. Lederberg. J. Bacteriol., **53**:673, 1947.
4. W. Hayes. J. Gen. Microbiol., **45**:385, 1966.
5. H. V. Wyatt. New Sci., **56**(814): 29, 1972.
6. M. H. Dawson and R. H. P. Sia. J. Exp. Med., **54**:681, 1931.
7. R. Austrian and C. M. MacLeod. J. Exp. Med., **89**:439, 1949.
8. J. L. Alloway. J. Exp. Med., **55**:265, 1933.
9. R. D. Hotchkiss. Genetics, **51**:1, 1965.
10. K. Landsteiner. The specificity of serological reactions. Springfield, Ill.: Thomas, 1936.
11. O. T. Avery, C. M. MacLeod, and M. McCarty. J. Exp. Med., **79**:137, 1944.
12. H. Dale. Proc. R. Soc. Lond. [Biol.], **133**:123, 1946.
13. P. F. Cranefield. Nature, **236**:413, 1972.
14. J. Lederberg: *In:* W. First (ed.). University on the heights. New York: Doubleday, 1969.
15. ———. Nature, **239**:234, 1972.
16. S. S. Cohen. Cold Spring Harbor Sympos. Quant. Biol., **12**:35, 1947.
17. F. C. Bawden and N. W. Pirie. Proc. Roy. Soc. Lond. [Biol.], **123**:274, 1937.
18. N. W. Pirie. J. Gen. Microbiol., **72**:1, 1972.
19. R. Markham, R. E. F. Matthews, and K. M. Smith. Nature, **162**:88, 1948.
20. A. Gierer and G. Schramm. Nature, **177**:702, 1956.
21. R. M. Herriott. J. Bacteriol., **61**:752, 1951.
22. A. D. Hershey and M. Chase. J. Gen. Physiol., **36**:39, 1952.
23. J. H. Northrop. J. Gen. Physiol., **34**:715, 1951.
24. A. D. Hershey. *In:* J. Cairns, G. S. Stent, and J. D. Watson (eds.). Phage and the origins of molecular biology. New York: Cold Spring Harbor Laboratory, 1966.
25. H. V. Wyatt. Nature. **249**:803, 1974.
26. J. D. Watson and O. Maaløe. Biochim. Biophys. Acta, **10**:432, 1953.
27. E. A. Carlson. Sci. Am., **229**(3): 8, 1973.
28. A. Boivin. Cold Spring Harbor Sympos. Quant. Biol., **12**:5, 1947.
29. R. D. Hotchkiss. *In:* J. Cairns, G. S. Stent, and J. D. Watson (eds.). Phage and the origins of molecular biology. New York: Cold Spring Harbor Laboratory, 1966.
30. F. Griffith. J. Hyg., **27**:113, 1928.
31. R. M. Herriott. J. Gen. Physiol., **32**:221, 1948.

DNA TODAY

FRANCIS CRICK*

In 1953 Jim Watson and I proposed a structure for DNA, based mainly on the experimental X-ray data of Wilkins, Franklin, and their colleagues and on the base ratios measured by Erwin Chargaff. The structure was a double helix, in which the bases of the two chains were paired together in a specific manner: Adenine pairing only with Thymine and Guanine only with Cytosine.

At the time we were very excited because although the experimental evidence in favor of the structure was not decisive, the pairing of the bases suggested a neat way for the cell to duplicate the structure exactly and thus implied that the main genetic material was nucleic acid and not protein, the original favorite for that role before the discovery of the nature of the transforming factor by Avery and his colleagues. It also suggested that the genetic information was carried by the exact sequence of the four types of base.

Few had imagined that underlying the complexities of genetics there was a simple scheme of this kind, and for this reason, if no other, many people at the time felt that our ideas were too simplistic to be true. Where do things stand after an interval of nearly 30 years?

A brief answer would be that although the mechanisms of gene replication and gene expression were vastly more complicated than we then imagined, nevertheless they are indeed based on the simple ideas we put forward. Because of this basic theoretical framework, and helped by cheap, quick, and powerful experimental methods, molecular biology has made spectacular progress, far beyond anything imagined in 1953.

This progress took place in two broad phases. The initial phase, from 1953 to 1966, started slowly but advanced very rapidly from 1959 onward, till by the end of that period the universal genetic code—the little dictionary which connects the four-letter language of the nucleic acids with the 20-letter language of the proteins—was deciphered using a variety of biochemical and genetic methods. The mechanism of protein

*Salk Institute, Post Office Box 85800, San Diego, California 92138.

synthesis was understood in broad outline, and, thanks mainly to the work of Jacob and Monod, the general way genes were turned on and off was known for several genes in bacteria.

After 1966 there was a period of consolidation. More and more details were discovered for each of these mechanisms, but two problems made very slow progress: what was the actual base-sequence for any gene of interest (in viruses, bacteria, or higher organisms), and how were genes controlled in eukaryotes and especially in the higher plants and animals?

These problems appeared very difficult, but during the early seventies a number of breakthroughs took place, so that now, in 1982, we can achieve results which were almost unthinkable as little as 10 years ago. DNA can be delicately divided by restriction enzymes; lengths of DNA can be moved from one position to another, so that, grown in microorganisms, relatively large amounts of any desired stretch of DNA can be obtained with a high degree of purity. Given a suitable probe, which in some cases can be obtained without too much difficulty, it is possible to select the desired stretch of DNA out of an entire genome having some billions of base-pairs. Sanger, Gilbert, and their colleagues have developed very rapid sequencing methods so that stretches of DNA a few hundred base-pairs long can, in a properly equipped lab, be sequenced in a few days. The total rate, worldwide, of base-sequence determination is approaching a million base-pairs a year, an input of precise information so fast that special arrangements are having to be made to store the results on one or more central computers. In addition, chemical methods of synthesizing long stretches of DNA have become both accurate and rapid. Techniques in related fields, such as the amino acid sequencing of proteins on a micro scale and the development of monoclonal antibodies, have enabled most problems to be attacked from more than one angle.

The second revolution in molecular biology is already producing novel and unexpected results, not only in academic problems but also in applications to medicine and to the production of industrial chemicals. Large amounts of venture capital have been attracted to these developments, which show every sign of being both rapid and very competitive.

Amid this turmoil it is rather difficult to get a clear picture of what is going on. In the first instance we should perhaps go back right to the beginning and ask: Was the DNA structure right? Is DNA really a double helix? The answer, broadly, to both questions is yes, although other competing models have been proposed in which the two chains do not twine round one another but lie more nearly side by side. These latter suggestions we now know to be wrong, both because of powerful topological arguments based on the behavior of *circular* double-stranded DNA and because Dickerson and his colleagues have determined by

X-ray diffraction the structure of a crystal of 12 base-pairs of DNA (with a defined base-sequence) using the isomorphous replacement method and to a sufficiently high resolution. The result is very similar in its broad features to the refined double-helical model deduced in the late fifties and early sixties by Wilkins and his colleagues from much less precise X-ray diffraction data from DNA fibers. A vast body of chemical and biochemical work had previously established that the two chains are antiparallel and that the bases do pair in the expected manner.

Nevertheless there has been one surprise. Crystals of a short length of double-stranded DNA having the sequence (CG)$_n$ have been found by Rich, Dickerson, and their colleagues to have a quite different structure, forming a *left*-handed helix (normal DNA is *right*-handed) which has been called the Z form because the backbones follow a zigzag path. The base-pairing itself remains the same. It is not yet known whether this has biological significance, but since in higher organisms this is the sequence which often contains 5-methyl cytosine, it is more than likely that Nature has found some way to exploit Z DNA.

Perhaps the most surprising discovery made by the newer techniques has been the existence of introns in many of the genes of higher organisms. These are lengths of DNA, in the middle of the stretch which codes for a particular protein, which have no obvious coding function. The enzyme which makes single-stranded RNA copies of the DNA of a gene also copies the introns along with the rest, but a further process splices them out before the RNA goes to the cytoplasm. Almost all globin genes, for example, appear to contain two distinct introns, and because they are in exactly the same position in all species studied so far, these introns are surmised to have originated a long time ago. Moreover, the total length of the introns is often longer than the length of the coding part of the gene. Loosely speaking, there is more nonsense there than sense.

Preliminary results on longer stretches of DNA suggest that there is much DNA of low specificity not only within genes but between genes. There is suggestive evidence that at least some of this DNA is actively moving from one place to another in the chromosomes as evolution proceeds.

Exactly what function introns have is still a matter of controversy. It has been suggested that they make it easier to move small domains of protein around and, on an evolutionary time scale, put them together to form new proteins. This would probably be more efficient than evolving a radically new protein directly from a single existing base-sequence. Another possibility is that much of this extra DNA, either within genes or between genes, originated as "selfish" DNA, that is, DNA which replicated and spread around the chromosomes for its own sake. Much of this DNA would probably be slowly eliminated by normal natural selection, but over long periods of evolution some of these sequences, or ones closely related to them, might be adopted by the organism for its own purposes and therefore preserved.

However, not all the extra DNA is likely to have started as selfish DNA since there are some sequences which appear to be dead genes. That is, they resemble a known functioning gene, but they have accumulated so many mutations that they can no longer code for a protein. These are more likely to be genes which were once useful and then became superfluous. They may perhaps have some special function of which we are ignorant, but it seems more likely that, in time, they will be eliminated by natural selection.

There is, of course, always the possibility that some of these extra sequences are not completely useless but have some rather nonspecific function. For example, they may serve to keep a DNA domain above a certain critical size; but such hypotheses are usually rather difficult to prove experimentally.

Probably the major biological problem that this new information should help solve is how genes are turned on and off in higher organisms and in particular how groups of genes are controlled in a coordinated manner, a process which seems essential in any multicellular organism in which different groups of cells specialize in different tasks. These processes may depend in part on how the DNA is folded and packaged by combining with special proteins. The combination with the commoner proteins, the histones, is now understood in outline. Once again the underlying structure, called a nucleosome, is simpler and more regular than most people would have expected; but many of the details are obscure, especially the way the nucleosomes are assembled into even higher structures and the role played by the other, nonhistone proteins. The chromosome may be divided into "domains" (perhaps on average some 50,000 base-pairs long) whose manner of functioning may depend on exactly how they are folded. Even so, the control of gene action is likely to depend to a considerable extent on the exact base-sequence in special control regions, and the new methods should help us discover just how this process works. It would help if it could be reproduced in the test tube in an efficient and accurate manner.

Another striking achievement has been the determination by Sanger and his colleagues of the complete sequence of human and mitochondrial DNA, a total of over 16,000 base-pairs. This has shown not only that the genes are in mitochondria pressed closely together and that there are several as yet unknown genes there, but that the genetic code in mitochondria is slightly different from the "universal" one. Work by others has shown that the mitochondria in certain fungi also have slightly variant codes. Mitochondrial DNA is inherited only through the egg and not through the sperm. Allan Wilson and his colleagues have started comparative studies on it, using material from humans and other primates. It will be interesting to examine DNA sequences on the Y chromosome, which is inherited through the male line, to see how much variation that has.

Another striking, recent discovery has been the confirmation of the idea of "oncogenes," that is, cancer genes. It has been known for some time that certain small viruses can cause cancer in animals. Only parts of their nucleic acid are critical for producing cancer. In short, such viruses contain a gene (or genes) which can cause cancer. The new methods have shown that very similar genes are present in the chromosomes of a normal animal, having no cancer. The difference appears to be that when the gene is carried to the chromosomes by the virus, more product is produced than when it is in its normal position in the chromosome. In other words, a perfectly normal gene, if it works too strongly, can cause at least some types of cancer. Almost a dozen such genes are now known.

Exactly how such genes cause cancer is still unclear. The first step appears to be that the protein coded by such a gene adds phosphate to one or more tyrosine groups in certain other proteins, thus presumably modifying their action. Exactly what this set of proteins then does to cause the cancer is the subject of very active research. It seems more likely than ever that the explanation of cancer will involve the properties of DNA. Whether a successful cure will involve DNA directly is another matter.

Once it was realized that a specific piece of DNA could be moved around almost at will—so that a human gene could be put into a bacterial cell or a yeast cell and with a little skill be made to work there and produce its protein product there—it became theoretically possible to get a particular protein in rather large amounts by putting the gene for it into a microorganism and growing large cultures of it in commercial fermenters of the sort used in brewing or for the production of anti-biotics. Thus we may expect to see many proteins, especially human versions of useful proteins like insulin or interferon, become available in amounts suitable for medical treatment. Once bulk production becomes technically and economically feasible—and there is now such a large effort being put into this that the efficiency should increase and the cost come down—many other proteins, from many different sources, are likely to become commercial products, some for medical diagnosis, some for medical treatment, and others for the production of industrial chemicals. The major problem for these latter applications is to make such proteins (which are usually used as organic catalysts or enzymes) sufficiently stable, otherwise their effective cost will be too great. If this can be overcome, then many new industrial processes for the production of basic organic chemicals, such as ethanol or fructose, are likely to come into use. The new techniques, collectively, are already so powerful that before the end of the century we may expect many applications which are unheard of today.

There is, inevitably, a certain degree of danger in introducing any new process, especially if the new entity is not just a new chemical but a new organism which, unlike a chemical, can multiply biologically. A few years ago there was considerable concern that the hazards might be rather high, but research since then has shown that it is very difficult to produce a dangerous organism, especially if the strains are derived from those enfeebled by growth in the laboratory for many years. At present the risks seem extremely small, probably much less than catching a dangerous disease from an animal by keeping a pet or by visiting the zoo.

With so many techniques already available and new ones being produced every few months it is perhaps a useful exercise to consider what we cannot do in molecular biology. There is one tremendous gap in our knowledge. Given a stretch of DNA coding for a protein we can easily derive from it the amino acid sequence of that protein by using the genetic code. What we are as yet unable to do is to put this sequence into a computer and obtain the 3D structure of the protein. If we could, we might perhaps be able to deduce its function from the 3D structure, though this itself is still an unsolved problem. This is a maddening restriction, especially since it is now relatively easy to sequence a stretch of DNA.

This problem aside, it is fair to say that the advances in our understanding of the structure and function of DNA over the past 25 years have been spectacular. It reflects great credit on organizations such as the British Medical Research Council (a government-financed organization) and the Rockefeller Foundation (based on private money) that at a very early stage they were farsighted enough to recognize the importance of fundamental research into biochemistry and genetics. The story has been told how, when the original application went to the MRC to support work on X-ray crystallography at the Cavendish Laboratory at Cambridge, the then secretary of the MRC, Sir Edward Mellanby, invited the then Cavendish Professor, Sir Lawrence Bragg, to lunch in London at the Atheneum. Over lunch Bragg explained to Mellanby that though the chance of solving a protein structure was very small, the scientific returns, if the project was successful, would be very great, a degree of candor and farsightedness not always found in senior scientists asking for money. On that basis Mellanby persuaded the MRC to support it, and it was this support, and the parallel support of their unit at Kings College, London, which made possible the discovery of the DNA double helix.

INTELLECTUAL TRADITIONS IN THE LIFE SCIENCES: MOLECULAR BIOLOGY AND BIOCHEMISTRY

SCOTT F. GILBERT*

I

Biological science is a vibrant, collective human endeavor consisting largely of controlled experiments and their critical interpretations. This does not, however, constitute an exhaustive catalog of the component parts of the life sciences. Another element is the context in which these experiments are integrated [1]. I am not speaking here of Kuhnian paradigms or of heuristic constructs, but of intellectual traditions which are not proved or disproved by experiment and which, of themselves, rarely suggest new investigations.[1] Nevertheless, these intellectual traditions subtly guide the direction of the entire enterprise of biology.

Most analyses of the intellectual traditions in biology have focused on the integration of biological sciences into the larger intellectual ferments of the eighteenth and nineteenth centuries. This essay, however, will attempt to look at the intellectual currents which led to the separation of molecular biology and biochemistry during the mid-twentieth century. Both of these disciplines attempt to understand the physical basis of life; yet molecular biology was founded by scientists who not only were untrained in biochemistry but were antagonistic to it. In particular, one of the most influential founders of molecular biology, Max Delbrück, "deprecated biochemistry," claiming that the analysis of the cell by biochemists had "stalled around in a semidescriptive manner without noticeably progressing towards a radical physical explanation" [2, p. 22]. This bias was transmitted to several of his students and colleagues. Replying in unkind, the famed nucleic acid *biochemist* Erwin Chargaff has

This paper was submitted in the first Dwight J. Ingle Memorial Young Writers' competition for authors under 35.

*Department of Biology, Swarthmore College, Swarthmore, Pennsylvania 19081.

[1]Foucault has suggested the term "episteme" for such a concept; but as he claims that only one episteme defines the condition of scientific knowledge at any time, I find that "intellectual tradition" denotes what I mean more readily.

accused the molecular biologists of "practicing biochemistry without a license" [3, p. 140] and believes that molecular biology has severely impeded the flow of scientific creativity. This animosity between biochemists and molecular biologists was especially vigorous in the 1940s and 1950s. This essay will attempt to show that biochemistry and molecular biology are the current incarnations of two complementary traditions in biological science, and that they embody radically different ideas concerning the nature of life.

II

"The ultimate goal of biochemistry" according to Fruton and Simmonds, "is to describe the phenomena that distinguish the 'living' from the 'nonliving' in the language of chemistry and physics" [4, p. 1]. But neither they nor anyone else has been able to create a set of criteria for the definition of life.

To Claude Bernard, the founder of the cell physiology, which eventually gave rise to the disciplines we will be discussing, " . . . there is no need to define life in physiology." Such attempts, he claimed, were "stamped with sterility." Bernard concluded, "It is enough to agree on the word *life* to employ it; but above all it is necessary for us to know that it is illusory and chimerical and contrary to the very spirit of science to seek an absolute definition of it. We ought to concern ourselves only with establishing its characteristics and arranging them in their natural order of rank" [5, p. 19].

In the absence of any consensual accord, biology has evolved two traditional approaches to characterize the physical basis of life. In each, the "natural order of rank" is the reverse of the other. The first tradition emphasizes the phenomena of growth and replication as the major vital characteristics. Organisms are seen to increase in size and numbers and are thus akin to crystals. The second perspective focuses on metabolism as life's prime requisite, whereby an organism retains its form and individuality despite the constant changing of its component parts. In this respect, living beings resemble waves or whirlpools. These alternative crystalline and fluid models of organisms have interacted with each other for the past hundred years. It will be shown that "classical" biochemistry largely retained the metabolic model of life whereas molecular biology, a discipline with ostensibly the same goals, formulated its research program around the crystalline model.

If the basis of living organisms is to be found, not in a "life force," but in organizations of nonliving materials, what better program is there than to look for inorganic substances which contain rudiments of living processes? To the scientists of the seventeenth and eighteenth centuries, such lifelike substances were to be found as crystals. The forces responsible for such arborescent growths of silver amalgam (*arbor Dianae*), and *arbre de Mars* (iron silicate and potassium carbonate) were often compared with those responsible for the formation of living structures [6],[2] and such analogies can be found in the writings of Bacon, Hooke, and Coxe. Nehemiah Grew (1674) wrote that the processes of growth were the same as those responsible for the crystallization of salt and that plants were composed of crystalline formations [8, p. 14]. French scientists Beaumont (1676), Maupertuis (1744), and Buffon likewise employed crystals as useful analogues to living organisms; Guéneau de Montbéliard presented spontaneous generation as a crystalline process, and de la Métherie (1811) saw in crystallization the mechanism of organic growth and development: "The seeds of the male and of the female, being mixed, act as two salts would and the result is the crystallization of the fetus."[3] The culmination of this crystalline model of organic development was the cell theory of Theodor Schwann and Matthias Schleiden. In 1839, Schwann published his researches on plant and animal structure including his "theory of organisms." Herein he established a new "mode of explanation" which sought to explain organic phenomena solely through physicochemical causes and to drive vitalism out of biology [10]. In so doing, Schwann took any discussion of life from the level of the total organism to the level of the cell and hypothesized that cell formation and growth could be explained by the same processes which govern the formation of inorganic crystals. "Organisms," said Schwann, "are nothing but the form under which substances capable of imbibition crystallize" ([11, p. 257] quoted in Mendelsohn [10]). In this view, the nucleus (cytoblast) would condense out of solution around the nucleolus and then the cell substances are deposited on the nucleus in the manner of a growing crystal [6].

Nevertheless, Schwann accepted that crystals represent only an instructive model to study living cells.

The attractive power of the cells manifests a certain degree of election in its operation; it does not attract every substance present in the cytoblastema, but only particular ones; and here a muscle cell, there a fat cell is generated from the same fluid, the blood. Yet crystals afford us an example of a precisely similar phenomenon, and one which has already been frequently adduced as analogous to assimilation. If a crystal of nitre be placed in a solution of nitre and sulphate of soda, only the nitre crystallizes; when a crystal of sulphate of soda is put in, only the sulphate of soda crystallizes. Here, therefore, there occurs just the same selection of the substances to be attracted. [6]

[2]The analogy that the earth generates metals as a mother produces an embryo has been detailed in M. Eliade [7].
[3]The history of the view that a fetus "precipitates" from a fluid mixture is given in Needham [9]. Needham, as we shall see, was extremely influential in the reintroduction of the crystal analogy into embryology.

And although Schwann has reservations concerning the identity of cellular and crystalline processes, he concludes that "the process of crystallization in inorganic nature ...is, therefore, the nearest analogue to the formation of cells." In her study of organic metaphors in twentieth-century developmental biology, Donna Haraway has seen the critical ways in which the analogy of crystallization has influenced embryology. "If one sees the world in atomistic terms (metaphysically and methodologically), the crystal is a smaller, simpler version of the organism in a nearly literal sense. If one sees the world in terms of hierarchically organized levels (the organism becomes the primary metaphor), the crystal becomes an intermediate state of organization" [12, p. 11]. Although Haraway is analyzing twentieth-century embryology, crystals were being used in the same way by the physiologists and cytologists at the turn of the century. Several cytologists analyzed protoplasm in terms of a geometrical space lattice, and entire classifications of organisms were constructed around this type of crystalline symmetry.[4] Even Claude Bernard, who believed life is without direct analogy to any nonliving entity [5], referred to the ability of crystals to regenerate their form such that "the physical force that arranges the particles of a crystal according to the laws of a wise geometry has results analogous to those which arrange the organized substance in the form of an animal or a plant." However, Bernard was cautiously aware that "these comparisons between mineral forms and living forms certainly constitute only very different analogies and it would be imprudent to exaggerate them. It suffices to mention them."

III

By this time, however, an alternative view of life had reached maturity. This tradition held that living beings were in a constant flux and therefore analogous to waves of water—that is, just as a wave maintains its individuality while constantly changing its component parts, so do living organisms. This tradition concentrated on existing organisms rather than emerging ones, and its principle supporters were French biologists. Flux was the central focus of de Blainville's view of life, and Cuvier claimed that "the living being is a whirlpool constantly turning in the same direction, in which matter is less essential than form." Flourens paraphrased the vital whirlpool concept, stating that "life is a form served by matter" [5].

[4]The idea that crystals provided a framework to order life continued to have adherents throughout the nineteenth century. Foremost among them was Ernst Haeckel, who utilized crystalline symmetries for his studies of development and who declared that "the secure promorphological foundation makes possible a mathematical understanding for organic individuals just as crystals" [13, p. 543].

The contrast between the two models was demonstrated by Thomas Huxley, who clearly stated his preference. A unicellular organism, he stated, "is a perfect laboratory in itself, and it will act and react upon the water and the matters contained therein, converting them into new compounds resembling its own substance, and at the same time giving up portions of its own substance which have become effete. Furthermore, the Euglena will increase in size; but this increase is by no means unlimited, as the increase in crystal might be" [14].

Furthermore, following the French biologists, Huxley postulated that each individual organism was formed by the temporary combination of molecules which

stand to it in the relation of particles of water to a cascade or a whirlpool; or to a mould into which the water is poured. The form of the organism is thus determined by the inherent activities of the organic molecules of which it is composed; and as the stoppage of a whirlpool destroys nothing but a form, and leaves the molecules of water, with all their inherent properties intact, so what we call death and putrefaction of an animal or of a plant, is merely the breaking up of the forms, or manner of association, of its constituent organic molecules. [15, p. 355]

Not having a theory of intermediary metabolism to underwrite his views, he speculated on the special nature of these biologically active molecules and the remarkable interconvertibility of such molecules from species to species.

Nutrition was the paramount criterion for this view of life. Indeed, this was the vital characteristic that Bernard had ranked most highly. Huxley's view of the chemical nature of nutritive metabolism "flowed" directly from this tradition: "The particles of matter that enter the vital whirlpool are more complicated than those that emerge from it . . . . The energy set free in this fragmentation is the source of the active forces of the organism" [5].

This view is common among the founders of biochemistry, and in 1907 Emil Fischer defined the "ultimate aim of biochemistry" as the gaining of "complete insight into the unending series of changes which attend plant and animal metabolism" [6, p. 136]. As biochemistry improved, so did this model. Whereas Huxley could only intuit that life had a "tendency to disturb existing equilibrium," L. J. Henderson could relate the vital whirlpool of life to the laws of thermodynamics. His work, *The Fitness of the Environment,* popularized the view among scientists that "living things preserve, or try to preserve, an ideal form, while through them flows a steady stream of energy and matter which is ever changing . . ." [16]. Here was a context in which intermediary metabolism could be studied, and the triumphs of intermediary metabolism caused the eclipse of the crystal model for decades. The biologist-philosopher

Edmund Sinnott of Yale gave the following explanation: "There is good reason to believe that the sort of formativeness found in living things is really different from that in such lifeless ones or crystals. First, the crystal system is stable. Its molecules are at rest. Whatever change there is results from the addition of new molecules along the crystal surface. A living organism, on the contrary, is in a continual state of change . . . . An organism has a sort of fluid form like a waterfall, through which water ceaselessly is pouring, but which keeps in its descent a definite pattern" [17, pp. 116–117].

Other biochemists followed this tradition, reflecting on the flow inherent in biological systems [18]. Thus, by 1940, biochemistry had allied itself to the tradition of flux.

IV

While biochemists could use the waterfall to analogize the metastable state of the adult organism or cell, it never became fully acceptable to those scientists concerned with the genesis of living beings—that is, geneticists and embryologists. In 1924, geneticist J. Arthur Thomson noted Thomas Huxley's comparison of the living body to the great whirlpool beneath Niagara Falls. However, while agreeing that "the whirlpool of the body is a useful metaphor," Thomson criticized it for lack of completeness, especially since the body "has the power of giving rise to other whirlpools like itself" [19, p. 123].

At the same time, geneticists were having a difficult time trying to integrate the gene into the context of biochemistry. To do this, the gene was often viewed as an enzyme catalyst, yet one that could catalyze its own replication (autocatalysis). This linked genes not only to crystals but also to viruses.

One of the first attempts to equate autocatalytic genes with heterocatalytic enzymes was Leonard Troland's "enzyme theory of life," first presented in 1914 and expanded in the 1917 volume of *American Naturalist*. Linking the new biochemistry of Fischer to the new genetics of Morgan, Troland proposed that the genes of the nucleus were composed of catalytic enzymes. "Although the fundamental life property of the chromatin is that of autocatalysis, it is necessary and legitimate to suppose that the majority of them sustain specific heterocatalytic relationships to reactions occurring in living matter" [20].

This idea was seconded in a review by J. B. S. Haldane, who attempted to squeeze genes into the tradition of flux. "The ultimate goals of biochemistry may be stated as a complete account of intermediary metabolism, that is to say, of the transformations undergone by matter passing through organisms" [21]. Moreover, to Haldane, as to Troland, "The gene has two properties. It intervenes in metabolism . . . and it reproduces itself.... The most economical hypothesis is that these two processes are closely related, and that the primary products of genes during the 'resting' stage differ from the genes themselves in not being attached to the chromosomes, but perhaps in no other respect" [21]. To account for gene replication, Haldane uses a variety of crystallizations.

The growth and reproduction of large molecules are not, it may be remarked, hypothetical processes. They occur, it would seem, in certain polymerizations which are familiar to organic chemists. In my opinion, the genes in the nuclei of cells still double themselves in this way. The most familiar analogy to the process is crystallization. A crystal grows if placed in a supersaturated solution, but the precise arrangement of the molecules out of several possible arrangements depends on the arrangement found in the original crystal with which the solution is "seeded." The metaphor of seeding, used by chemists, points to an analogy with reproduction. [22, p. 156]

Haldane claimed that "the problem of gene reproduction is very similar to that of virus reproduction" [21] and eventually linked the process to crystallization [23]. "[The] gene is within the range of size of protein molecules, and may be like a virus. If so, the chemist will say, we must conceive of reproduction as follows. The gene is spread out in a flat layer, and acts as a model, another gene forming on top of it from pre-existing materials such as amino acids. This is a process similar to crystallization in the growth of a cellulose wall" [23]. In this we can see one tradition trying to break away from the other, and in 1937 Haldane predicted that "classical" biochemistry, focusing on degradative phenomenon, would be superseded by a "new branch of biochemistry which will, I believe, arise from genetics ... and its final goal will be the explanation and control of the synthesis of genes" [21]. As we will see, this is the field in which the tradition of crystalline models is strongest. Thus, while the classical biochemistry of intermediary metabolism is retained the model of flux, a "new" biochemistry was emerging which would reject that tradition.

V

The term "molecular biology" was coined in the same year that Haldane's article showed the strains within biochemistry. There was obviously a need to describe the new discipline that was ready to formulate its own research program for the physicochemical basis of life and which was receiving its input from some nontraditional sources—genetics, virology, and two new branches of physics, X-ray crystallography and quantum mechanics. The term was first used (and in large type) by Warren Weaver, the director of the Rockefeller Foundation's funding for natural science and the man responsible for the recruitment of several physicists and chemists into this new science [24]. It was used in the decision to fund a new research institute which Joseph Needham proposed, in 1935, to investigate the biological importance of crystal physics. Needham, as well as other embryologists, were aware of the recent advances in crystallography and were impressed by Bernal's beliefs that crystallography could help elucidate biological problems.[5] In his proposal, Needham stated that crystals were not only models or analogues for the living process, but are the physical components of cells. "Next, there is the profound importance of the paracrystalline state of biology. The 'liquid crystal' is not merely a model for what goes on in the living cell; it is in point of fact a state of organization actually found in the living cell" [24].

This raises the stakes considerably. We are no longer talking about speculative hypotheses that the processes of life and crystallization are governed by the same laws, or facile similies that life is somehow like a crystal. What is now being proposed is that the crystalline state is, in fact, found in the living cell. A "proof" for this was soon forthcoming, since the same year that Needham proposed this institute, Wendell M. Stanley crystallized tobacco mosaic virus.

This was a philosophical as well as biological breakthrough, similar in its significance and interpretation to Wohler's synthesis of urea, and it was recognized as such by biochemists. Stating that this feat demonstrated "that there is no definite boundary between the living and the nonliving," the authors of one textbook relate that "certain viruses have been isolated in highly purified crystalline form and studied extensively in the laboratory. Yet these substances, isolated in the laboratory and having no apparent or obvious features characteristic of living matter . . . an inanimate crystalline compound, when introduced into the proper environment, appears to behave as though it were a living viral agent" [25, p. 1].

Within a year, geneticist Hermann Muller was willing to claim that this crystallized virus was apparently a pure protein capable of "auto-synthesis" and that "it represents a certain type of gene" [24]. At the Genetical Congress in Edinburgh in 1939, no less than three crystallographers, Astbury, McKinney, and Gowen, put forth analogies between viruses and genes. Astbury said that viruses and chromosomes were the two simplest reproductive systems known. Thus, autocatalysis, which was thought to occur by crystalline duplication, was now seen to occur in viruses which could themselves be crystallized. Furthermore, since this process is also common to chromosomes, the chromosomes, too, must be crystalline in nature.

[5] Dr. Hans Oberdiek has pointed out to me that each of the three "prophets" of molecular biology (Bernal, Needham, and Haldane) was a committed Marxist. There may be, then, a further ideological reason for their demanding a chemistry of biological "production."

Thus, by 1940 the lines were being drawn. Biochemistry, concerned with intermediary metabolism and the energy that drives it, worked well within the tradition of flux and thermodynamics. However, the portion of the life sciences concerned with the transmission and expression of inherited characteristics rejected this view for the tradition of crystalline morphogenesis. Not only did the gene just not fit into the whirlpool model, but it looked as if functional genes (i.e., viruses) could even be crystallized. Whereas the principal characteristic of life for the biochemist was metabolism, life's principal characteristic for the molecular biologist was replication. Furthermore, the primary unit of life for the biochemist was the resting cell (metabolically active but not replicating), whereas the unit of life for the molecular biologist was the virus—crystalline, nonmetabolizing, and capable of enormous feats of replication (see, e.g. [26, pp. 12–13, 18]).

One approach to the study of life was proposed in 1944 when Schroedinger published his essay, "What Is Life?" The importance of this essay has been attested to by such researchers as Crick, Watson, Wilkins, Luria, Benzer, and even Chargaff. The only major dissenter appears to be Seymour Cohen, who called the book marginal. Yoxen has studied the influence of this volume and claims that it "exerted on some scientists a powerful and transient influence in suggesting and validating a particular line of research" [27]. This line of research was predicated on the idea that crystallinity is a state unifying all of matter and that "the most essential part of a living cell—the chromosome fiber—may suitably be called *an aperiodic crystal.*"

Schroedinger's notions were derived largely from the work of another physicist who had emigrated into biology, Max Delbrück. Delbrück had worked on *Drosophila, Neurospora,* and bacteriophage viruses in order to study how genes replicate exactly and transmit themselves flawlessly in each generation. One of Delbrück's most important contributions in this regard was a paper co-authored with N. W. Timoféef-Ressovsky and K. G. Zimmer [28]. This was an analysis of gene size and the mechanism of mutation by ionizing radiation, and it gave rise to what has been termed the quantum mechanical model of the gene. The stability of the gene was seen to be due to forces linking the atoms of the gene and could be overcome by causing a quantum jump from one stable configuration to another. This paper was "rediscovered" for Schroedinger by the crystallographer P. P. Ewald, and it became the basis of Schroedinger's analysis of the gene as crystal.

The study of Delbrück's entry into bacteriophage research and the subsequent recruitment of the "phage" group has been told several times and will not be repeated here [24, 29]. Suffice it to say that Delbrück felt that in focusing on the metabolic concerns of life, biochemists had misrepresented the cell as "a sack full of enzymes acting on substrates, converting them through various intermediate stages either into cell substance or waste products" [6]. He also believed that "the field of bacterial viruses is a fine playground for serious children who ask ambitious questions." These questions arise from the replication of viral genes. "He will say: 'How come one particle becomes 100 particles of the same kind in 20 minutes? This is very interesting. Let us find out how it happens! How does the particle get into the bacterium? How does it multiply? Does it multiply like a bacterium? . . . Does it have to go inside the bacterium to do this multiplying, or can we squash the bacterium and have the multiplication go on as before?'" [24]. Delbrück was convinced that the problem of autocatalytic synthesis would "turn out to be simple, and essentially the same for all viruses as well as genes. The bacterial viruses should serve well to find the solution, because their growth can be studied with ease quantitatively and under controlled conditions. The study of bacterial viruses may thus prove the key to basic problems in biology" [24].

Thus, by 1945 one finds two rival claimants to the study of the physical basis of life: the older biochemical school conducting its research with cells in the tradition of flux, and the aggressive molecular biologists working with viruses in the newly resurrected tradition of the crystal nature of life. The 1950s became a golden age of research for both groups. Molecular biologists Hershey and Chase used phage to demonstrate that DNA was the basis of gene structure, and Watson, Crick, Wilkins, and Franklin used X-ray crystallography to determine the structure and chemical nature of DNA. Soon to follow were the in vitro synthesis of nucleic acids (Kornberg, Ochoa), the research into the nature of genetic mutation (Ingram), the redefining of the gene concept (Benzer), and the decipherment of the genetic code (Nirenberg and colleagues). Meanwhile, in biochemistry, the concept of allosteric interactions, the use of radioisotopes to elucidate metabolic pathways, and new techniques of isolation and separation leading to the characterization of hundreds of enzymes were highlighted in the 1950s. Thus, just as biochemistry reached its long-awaited golden age, it found its pedestal shared with molecular biology.

By the end of the 1950s, however, a new synthesis was emerging between these two disciplines. Both were needed to account for the phenomena of life. In 1953, for instance, Watson and Crick [30] hypothesized that the nitrogenous base precursors of DNA would line up ("crystallize") on their template and be zippered together. No enzyme was needed for DNA replication. The biochemists, however, had all sorts of models for the enzymatic construction of nucleic acid, most of them based on the synthesis of NAD (a dinucleotide coenzyme) or on the notion that whatever enzyme degrades a substance can also synthesize it (see [31, pp. 87–90]). When, in 1958, DNA polymerase was discovered, it was not what biochemists had expected. "Directing enzyme function with a template was unique and unanticipated by any advance in enzymology, and was for some biochemists, even after a number of years, very hard to believe" [31].

Similarly, Jacob and Monod [32] integrated biochemistry and molecular biology into a scheme whereby DNA not only coded for proteins but the proteins could bind back to specific regions of DNA. In so doing, proteins were modulating DNA for the synthesis of other proteins. Thus, the two traditions blended into each other, neither one predominating.

In this essay I have attempted to show that two rival intellectual traditions concerning the physical basis of life were manifest in the two similar disciplines, molecular biology and biochemistry, during the earlier half of this century. Today, these traditions have merged into an integrated study which characterizes far better the complex chemistry of life.

REFERENCES

1. FIGLIO, L. M. The metaphor of organization: a historiographical perspective on the biomedical sciences in the nineteenth century. *Hist. Science* 14:17–53, 1976.
2. DELBRÜCK, M. A physicist looks at biology. In *Phage and the Origins of Molecular Biology,* edited by J. CAIRNS, G. STENT, and J. D. WATSON. Cold Spring Harbor, N.Y.: Cold Spring Harbor Press, 1958.
3. CHARGAFF, E. *Heraclitean Fire.* New York: Rockefeller Univ. Press, 1978.
4. FRUTON, J. S., and SIMMONDS, S. *General Biochemistry.* New York: Wiley, 1953.
5. BERNARD, C. *Lessons on the Phenomena of Life Common to Animals and Plants,* translated by R. E. HOFF, R. GUILLEMIN, and L. GUILLEMIN. Springfield, Ill.: Thomas, 1974.
6. FLORKIN, M., and STOTZ, E. M. Crystallization and biosynthesis. *Comp. Biochem.* 32:133–143.
7. ELIADE, M. *The Forge and the Crucible: The Origins and Structures of Alchemy.* Chicago: Univ. Chicago Press, 1978.
8. RITTERBUSH, P. C. *The Art of Organic Forms.* Washington, D.C.: Smithsonian Institution Press, 1968.
9. NEEDHAM, J. *Chemical Embryology.* New York: Macmillan, 1931.
10. MENDELSOHN, E. Physical models and physiological concepts: explanation in nineteenth century biology. *Brit. J. Hist. Sci.* 2:201–218, 1965.
11. SCHWANN, T. *Mikroskopick Untersuchungen über die Uebereinstimmung in der Struktur und dem Wachstehun der Thiere und Pflanzen.* Berlin, 1839.
12. HARAWAY, D. *Crystals, Fabrics and Fields: Metaphors of Organicism in 20th Century Developmental Biology.* New Haven, Conn.: Yale Univ. Press, 1976.
13. HAECKEL, E. *Générelle morphologie,* vol. 1. Berlin: Reimer, 1866.
14. HUXLEY, T. On the educational value of natural history sciences. In *Science and Education.* New York: Appleton, 1897.
15. HUXLEY, T. Spontaneous generation. In *Lay Sermons, Addresses and Reviews.* New York: Appleton, 1877.
16. HENDERSON, L. J. *The Fitness of the Environment.* New York: Macmillan, 1913.
17. SINNOT, E. *The Biology of the Spirit.* New York: Viking, 1966.

18. BRILLOUIN, L. Life, thermodynamics, and cybernetics. *Am. Sci.* 37:544–568, 1949.
19. THOMSON, J. A. *Science and Religion.* New York: Scribner's, 1925.
20. RAVIN, A. W. The gene as catalyst, the gene as organism. *Stud. Hist. Biol.* 1:1–45, 1977.
21. HALDANE, J. B. S. The biochemistry of the individual. *Perspectives in Biochemistry,* edited by J. NEEDHAM and D. E. GREEN. Cambridge: Cambridge Univ. Press, 1937.
22. HALDANE, J. B. S. The origin of life. In *The Inequality of Man and Other Essays.* London: Chatto & Windus, 1932.
23. HALDANE, J. B. S. *New Paths in Genetics.* London: Allen & Unwin, 1941.
24. OLBY, R. *The Path to the Double Helix.* London: Macmillan, 1971.
25. WHITE, A.; HANDLER, P.; and SMITH, E. L. *Biochemistry.* New York: McGraw-Hill, 1959.
26. MONOD, J. *Chance and Necessity,* translated by A. WAINHOUSE. New York: Knopf, 1971.
27. YOXEN, E. J. Schroedinger and molecular biology. *Hist. Sci.* 17:17–52, 1979.
28. TIMOFÉEF-RESSOVSKY, N. W.; ZIMMER, K. G.; and DELBRÜCK, M. Ueber die Natur der Genmutation und der Genstruktur. *Nachr. Ges. Wiss.* 6:189–245.
29. JUDSON, H. F. *The Eighth Day of Creation.* New York: Simon & Schuster, 1979.
30. WATSON, J. D., and CRICK, F. H. C. The structure of DNA. *Cold Spring Harbor Symp. Quant. Biol.* 18:123–131, 1953.
31. KORNBERG, A. *DNA Replication.* San Francisco: W. H. Freeman, 1980.
32. JACOB, F., and MONOD, J. Genetic regulatory mechanisms in the synthesis of proteins. *J. Mol. Biol.* 3:318–356, 1961.

CHANGES IN THE GENETIC MENTALITY

BRUCE WALLACE*

Pique at indignities I recently suffered at the hands of an anonymous reviewer caused me to lament "Oh, how geneticists have changed": geneticists familiar with numbers and the meaning of numbers are disappearing. Or, so it seemed. After sober thought, however, I am now prepared to admit that the change in genetic mentality has been, contrary to my initial impression, an example of micro-, not of macro- or even of meso-, evolution. A change has occurred in the frequency of a certain type of biologist who for the moment will remain unidentified except to claim that he (and I use the bisexual "he") occurred originally almost exclusively among geneticists. This type of geneticist has probably not decreased in absolute numbers; he has decreased in frequency only because the ranks of geneticists have been swelled in recent years by workers from other branches of biology, workers who are now assigned, or claim for themselves, the title "geneticist."

The success of modern genetics can be measured by the number of biological disciplines that it has invaded, not in a trivial sense but, rather, as the hypothesis-generating member of the union: if one of these invaded disciplines is to advance, it must now do so by incorporating modern genetics into its repertoire of concepts and hypotheses.

Thus, from the former pool of physiologists and biochemists—scientists who once scarcely conversed with geneticists—have come modern molecular geneticists. From a second pool, the embryologists, have come today's developmental geneticists. And, from the ecologists of yesterday have come the population biologist and sociobiologists who spend

This paper was prepared while the author was a Fulbright lecturer at Alexandria University, Alexandria, Egypt, and while his research was supported by grant no. GM29810 from the National Institutes of Health, U.S. Public Health Service. It was presented as an invitational lecture at the 1983 annual meeting of the Genetics Society of Australia, May 20, 1983, University of New England, Armidale, New South Wales, Australia.

*Department of Biology, Virginia Polytechnic Institute and State University, Blacksburg, Virginia 24061.

considerable time predicting the genetic changes that might occur in populations under one or another of a variety of postulated conditions.

Examples illustrating the point made here—namely, that the influx of other workers into genetics (and I could have included clinical physicians, many of whom are now referred to as medical geneticists) has caused one of the original, and formerly common, types of geneticists to become extremely rare—will be given as we go along. Here, let me merely point out that before Milislav Demerec became director of both the Department of Genetics, Carnegie Institution of Washington, and the Biological Laboratories during the early 1940s, the staffs of these two physically adjacent laboratories did not collaborate. The one was staffed by geneticists, the other by physiologists. I might also cite a second example, one to which I can personally attest. While being interviewed during the 1950s concerning a possible appointment to the university's staff, I spent about an hour with Harvard's insect physiologist. When I suggested that he might profit from joining forces with a geneticist, the response was utter silence—the silence of one dumbfounded and, hence, momentarily unable to regain the thread of rational thought. That hormones should have any connection with genetics appeared as alien to his vision of the world as the notion that gene frequency analyses should be of use to an aeronautical engineer troubled by a vibrating airplane wing would have been to mine.

A look at the origins of the science itself may prove useful. Between 1865, when Mendel published his paper on the study of hybridization in peas, and 1900 when this work was rediscovered, Francis Galton developed his law of ancestral inheritance. Both Mendel and Galton sought to characterize an individual with respect to his heredity. Galton chose to study what flowed in from the past—from two parents, from four grandparents, from eight great-grandparents, and so on back through the family tree.

Mayr [1] has commented on Galton's procedure by saying that it has no predictive value and was erroneously applied to individual characters. Neither comment is wholly correct. The methods of modern quantitative genetics parallel Galton's procedures quite closely and do have predictive value. In deciding which of two bulls is likely to father the best milk-producing daughters, one examines the pedigree records of each, choosing the bull with the best milk producers in his background—especially his recent background. The fundamental reasoning in this procedure is that "like begets like"; I would suggest that some 80 percent of all agricultural gains made until now have been made by persons, prehistoric and modern, whose entire knowledge of genetics could be expressed by those three words.

The flaw with Galton's law is that it reveals what ought to be, not what is. If we imagine the individual as possessing (or consisting of) some

1,000 inherited particles, and we consider the 1,024 ancestors of 10 generations back, it would be highly unlikely that each ancestor contributed exactly one particle. By chance, nearly 40 percent of the ancestors would have contributed nothing, a similar proportion would have contributed a single particle, and the remainder would have contributed two or more particles. The fault with considering individual traits under Galton's scheme is not that a new problem has arisen but that the role of chance events becomes greater earlier. A character consisting of 64 of the 1,000 particles just mentioned would encounter the problems described above not in the tenth generation back but in the sixth where there are 64 ancestors.

Chance events render Galton's predictions statements of probability, any one of which may be either right or wrong. Such statements are more likely to be wrong under some circumstances than under others. Using Galton's logic and knowing (1) that I have spent $100 on New York State's gambling game, Lotto, and (2) that New York State law requires that one-half of all money received through the sale of Lotto tickets be returned to the players in cash prizes, someone might calculate that I have probably won $50 playing Lotto. There is, in fact, no less likely outcome. The probabilities of the game are such that most players lose every time, whereas a very few (about one per week) gain considerable wealth. No occasional player can regain precisely, or even approximately, one-half of what he has gambled.

In contrast to Galton who looked to the ancestral input in characterizing the individual, Mendel examined his output. This approach allowed Mendel to make a definite (and correct) statement about every tested individual: *A, a, Aa, ABb,* and the like. These quaint looking notations are a form of mathematical shorthand in which Mendel noted that the first individual transmits only *A,* the next one only *a,* the next one-half *A* and one-half *a,* the next 100% *A* but one-half *B* and one-half *b.* The offspring of the hybrids appeared therefore under nine different forms, some of them in very unequal numbers (table 1 [2]).

The individual under study has now been characterized by what has seemingly been poured into him (as determined by a study of ancestors) and what actually comes out of him (determined by studying his progeny). Now comes William Bateson, who finally inserts symbols into the individual, himself. "Consequently if *Aa*'s breed together, the new *A* gametes may meet each other in fertilization, forming a zygote, *AA,* namely, the pure *A* variety again; similarly two *a* gametes may meet and form *aa,* or the pure *a* variety again. But if an *A* gamete meets an *a* it will once more form *Aa,* with its special character. This *Aa* is the hybrid, or 'mule' form, or, as I have elsewhere called it, the *heterozygote,* as distinguished from *AA* or *aa* the *homozygotes*" [3].

Students sometimes forget that familiar scientific terms have origins:

TABLE 1

OFFSPRING OF THE HYBRIDS

Plants	Sign
38	AB
35	Ab
28	aB
30	ab
65	ABb
68	aBb
60	AaB
67	Aab
138	AaBb

here we have Bateson saying that the individual who forms only *A* gametes is an *AA* homozygote, that the one who forms only *a* gametes is an *aa* homozygote, and that the one who forms gametes one-half of which are *A* and one-half *a* is a heterozygote, *Aa*. When I was an undergraduate, I was led to believe that Bateson had corrected a glaring defect in Mendel's work: Mendel, it seems, had neglected to provide the second allele needed by all diploid organisms.

William Bateson might be referred to as a sincere biologist. As an undergraduate, I was taught by another sincere biologist, L. C. Dunn. The term *sincere* is not used in an ironic or pejorative sense; it is used only to differentiate between Bateson's and Dunn's mentalities and that of Mendel. Bateson identified the proper number of alleles that a diploid individual ought to possess in each somatic cell; his concern was for properly representing the genetic makeup of the individual. Mendel described the individual only in terms of the probability that his gametes would contain a certain factor or certain combinations of factors. He had no pressing concern for the immediate cause of these transmission frequencies; he may even have had no basis for visualizing the hereditary composition of his pea plants. The sincere biologist, however, is not at home with such abstract thinking. On the contrary, he is gravely concerned about the physical makeup—the actual substance—of organisms. Our rare geneticist, of whom Mendel was one, is content with, or, at least, is not frightened by, theoretical concepts.

The adoption of Bateson's notation has led to the necessity of inserting gametes between the individuals of one generation and those of the next when illustrating Mendelian crosses for students. Gametes are physical entities, but their proportions reflect the transmission probabilities stressed by Mendel. Despite such educational devices, persons who see the genetic symbols as biological rather than mathematical symbols

TABLE 2

GAMETES AND OFFSPRING FROM A CROSS
BETWEEN NN × N0 INDIVIDUALS

	1N	1N
1N	2N	2N
0N	1N	1N

NOTE.—N is the nucleolar organizer "locus."

remain; these persons tend to make unnecessary calculations in demonstrating the outcomes of crosses. The example cited here (table 2 [4, p. 18]) has been taken from Lucina Barth's textbook on developmental biology. Notice that the duplication of the 1N column is redundant; all gametes of 2N individuals are of the 1N type. Dr. Barth is an embryologist.

Recently I have encountered a favorable reference to a claim that genetics has advanced from an inferential science to a visual one. I would guess that a statement lauding such a change would not have been made by an old-fashioned geneticist. If I were to admit that such a change represented an advance, for example, I would also have to admit that the hands-on, bomb-creating physicists of Los Alamos practiced an art that was a cut above Einstein and his logical inferences. I cannot bring myself to make that admission.

One may by now suspect that the rare geneticist to whom I continually allude is in some sense a mathematician. Yes, in some sense. But that is not the attribute that I would stress. Galton was an accomplished mathematician. Darwin, perhaps to the surprise of many, was also highly quantitative. He seems to have calculated continuously: the proportions of endemic species and genera on oceanic islands; the amount of soil moved in the form of earthworm droppings; and, in his massive study on the effects of self- and cross-fertilization in the vegetable kingdom, the relative growth rates and final sizes of thousands of pairs of plants— each pair containing one plant obtained by selfing and carefully matched for comparative purposes with one obtained by cross-fertilization. The example (table 3) used here seems to be the first experimental data on hybrid vigor in corn.

The elusive feature is not that of being a mathematician or even of being merely quantitative in one's outlook. Returning to Mendel's studies, we get a second hint concerning the intellectual quality we seek to identify:

Expt. 6. Position of flowers. Among 858 cases 651 had inflorescences axial and 207 terminal. Ratio, 3.14 to 1.

TABLE 3

ZEA MAYS

No. of Pot	Crossed Plants (Inches)	Self-fertilized Plants (Inches)
I	23 4/8	17 3/8
	12	20 3/8
	21	20
II	22	20
	19 1/8	18 3/8
	21 4/8	18 5/8
III	22 1/8	18 5/8
	20 3/8	15 2/8
	18 2/8	16 4/8
	21 5/8	18
	23 2/8	16 2/8
IV	21	18
	22 1/8	12 6/8
	23	15 4/8
	12	18
Total in inches	302.88	263.63

SOURCE.—[5, table 97].

Expt. 7. Length of stem. Out of 1,064 plants, in 787 cases the stem was long, and in 277 short. Hence a mutual ratio of 2.84 to 1. In this experiment the dwarfed plants were carefully lifted and transferred to a special bed. This precaution was necessary, as otherwise they would have perished through being overgrown by their tall relatives. Even in their quite young state they can be easily picked out by their compact growth and thick dark-green foliage.

If now the results of the whole of the experiments be brought together, there is found, as between the number of forms with the dominant and recessive characters, an average ratio of 2.98 to 1, or 3 to 1. [2]

In his analyses, Mendel calculated observed ratios to the second decimal: 2.84:1, 3.14:1, and 2.98:1, for example. His genius, however, was in ignoring the decimal points, stating flatly that these are 3:1 ratios, and then proceeding to develop an explanatory model on that basis. Surely, Mendel realized that if he could not understand a simple relationship, he would never understand more complex ones. Indeed, we find in his paper a comment about the height of Tt hybrid plants being greater than that of plants of the tall (TT) variety—just a passing comment, no more. Why hybrids exhibit vigor is still a largely unsolved problem; Mendel was wise in putting it aside.

A great many biologists, sensing that they may understand neither simple nor complex relations, choose the complex ones for study. This choice provides them with work for a lifetime and, if the problem is sufficiently complex (and uninteresting), virtually assures them that no other person will solve it first.

Before (and after, as well) he took up the genetic studies on *Drosophila,* T. H. Morgan was an embryologist. In a little-known article of that era (one that seemingly embarrassed him in later years), Morgan [6] took genetics and geneticists to task: "If one factor will not explain the facts, then two are invoked; if two prove insufficient, three will sometimes work out." Actually, his criticisms were well founded and should not have caused him later misgivings. The practice that he criticized was that in which an observable difference between two strains (or lines) is studied in the F_2 generation by counting the numbers of individuals showing each of the two phenotypes. If the numbers fall close to $3:1$, the difference is said to be caused by a single gene with no additional confirmation; if the ratio is closer to $15:1$, or $9:7$, the difference is said to be caused by two genes (note that $12:4$ would have been classified as $3:1$). If the observed ratio demands still another explanation, large numbers of possibilities are provided by the three-locus model.

It is to Morgan's credit that he spotted a growing weakness that had developed in the genetic research of that era and then undertook to move genetic analyses to a new plane. In doing so, he left behind persons like C. B. Davenport who fitted any Mendelian-like inheritance pattern, no matter how complex the character (nomadism or wanderlust was one; love of the sea or thalassophilia was another), to a single-gene model, if possible—otherwise, to a two-locus model (human skin pigmentation).

From Morgan's group emerged another of these rare individuals— A. H. Sturtevant. When confronted with the formation of new combinations of genes in proportions that were often far short of those expected under independent assortment, Sturtevant [7] suggested that such genes are located on the same chromosome and, further, that the proportion of recombinant pairs represents the distance between the two loci on that chromosome.

In reviewing Morgan's "The Mechanism of Mendelian Heredity," William Bateson [8], still the sincere biologist, was nearly undone by what he read. "For instance, we find the following extraordinary series given," he says, and then he lists seven sets of numbers from which map distances had been estimated by the Morgan group—numbers among which it is impossible to recognize the expected patterns of two equally numerous rare recombinant classes and two equally numerous parental ones. "The machinery [possessed by the Morgan group] for dealing with unconformable cases is extraordinarily complete," he continued. This machinery included differential viability, association of alleles with lethal factors, changes with age, a special phenomenon called "interference," and "an altogether distinct kind of crossing-over in the four-stranded stage." Sincere persons at times become righteously indignant.

Before leaving the Morgan group, I can cite a passage from E. B. Wilson [9]: "*chromatin,* is to be regarded as the physical basis of inheritance. Now, chromatin is known to be closely similar to, if not identical with, a substance known as *nuclein* . . . a tolerably definite chemical composed of a nucleic acid (a complex organic acid rich in phosphorus) and albumin. And thus we reach the remarkable conclusion that inheritance may, perhaps, be effected by the physical transmission of a particular chemical compound from parent to offspring."

This passage antedates the rediscovery of Mendel's paper by 5 years and Oswald Avery's work on pneumococcus by nearly 50. That Wilson, a cytologist, later changed his mind and said that protein contained genetic information is not relevant here. What is relevant is that Wilson, in making his original statement, made no attempt to ask about its quantitative consequences. Had he done so, either Mendel's results or Galton's or both would have been anticipated. Instead, we have the pattern of inheritance revealed first in Mendel's numerical data and then, after a lapse of some years, correlated with chromosomal behavior. In fact, the known behavior of chromosomes, coupled with Wilson's 1895 claim, should have predicted the subsequent numerical data.

Later, H. J. Muller [10] was to allude to the failure of geneticists and physiologists to view the world from similar perspectives: "Just how these genes thus determine the reaction-potentialities of the organism and so its resultant form and functioning, is another series of problems, at present practically a closed book in physiology, and one which physiologists as yet seem to have neither the means nor the desire to open." When they did achieve the means, however, they brought their earlier world view to bear on the substance of heredity; they did not adopt the world view of the early geneticists.

To illustrate that latter point (and to move into modern times), we might consider Chargaff's work in the early 1950s, especially his paper of 1950 [11] in which he reported that the proportions of adenine to thymine (A:T) in DNAs of different origins were always close to 1:1 and, similarly, that those of guanine to cytosine (G:C) were also very nearly 1:1; the proportions of A and T to C and G, he reported, took on a variety of values with DNAs of different sources differing from one another. In discussing these observations, Chargaff claimed that the apparent 1:1 ratios were probably coincidental.

An account of Watson and Crick's model-building attempts, the failures as well as the success, has been given by Watson [12] in *The Double Helix.* One day, after struggling with a wretched model, Watson sought Crick's reaction to it. Crick's reaction was negative. Why? Because Watson's proposed model did not account for Chargaff's ratios. Watson admits that he never really took the ratios seriously; in fact, he had accepted some gossip according to which, if Chargaff saw 1:1 ratios, the true ratios, then, could not possibly be 1:1 because of biases in Chargaff's chemical procedures. Here, again, I see the distinction between what I have called the sincere biologist and the rare type to whom I constantly refer. Watson in his youth had considered becoming an ornithologist; now, when confronted with 1:1 ratios, he, as Chargaff himself had done, ignored them. Crick did not.

Two biochemists, Portugal and Cohen [13], in a book entitled *A Century of DNA*, criticize Chargaff for later attempting to claim some credit in solving the mystery of DNA by having discovered the 1:1 A:T and G:C ratios. By admitting at the outset that the ratios might be coincidental, Chargaff relinquished his right to claim credit for any special insight regarding the structure of DNA; certainly he could not claim credit for discovering "base pairing." In their attempt to minimize Chargaff's belated claim, however, Portugal and Cohen state that the 1:1 ratios are (and here they quote *The Double Helix*) a *consequence* of the structure of DNA (their emphasis). This is a strange point to emphasize; in fact, it is totally without meaning. The X-ray diffraction patterns that were studied by Rosalind Franklin were also a consequence of the structure of DNA. The birefringent fibers of DNA that can be picked up on twirled glass rods owe their optical properties to the molecules' structure. One is tempted to add that the character of life on earth is a consequence of the structure of DNA. Had Chargaff taken the 1:1 ratios seriously, he could (in theory, at least) have proposed a structure for DNA for which X-ray diffraction data would have served as corroborative evidence. That, however, was not the case; the simple base-pair ratios had been brushed aside.

So, who are my heroes? Who are those persons whose counterparts in my view are becoming increasingly rare among geneticists? Mendel, Sturtevant, Muller, and Crick have been mentioned specifically. Weismann might be included merely for predicting, before the phenomenon had been seen, a reduction division during formation of germ cells.

We can return to Sturtevant once more and to his proposal that the frequency of gametes carrying new combinations of alleles corresponds to the physical distance between these alleles on the chromosome. Now, it is well known that distances computed by recombination data simply do not add up: the distance from A to C obtained in one experiment never equals the sum of the smaller distances from A to B and from B to C that are obtained in others. Castle, as one of our sincere biologists, constructed linkage maps that actually corresponded to observational data. His maps remind me of the child's string game, cat's cradle. The Morgan group was less kind; they referred to "Castle's rat-trap model." Castle's [14, 15] diagrams, however, were true to the observed data;

Sturtevant's were not. Castle approached his task as a sincere biologist; his philosophy was that of many biologists: if data do not fit the model, the data take precedence.

The time has come to characterize more accurately than heretofore the rare geneticist. He is a person who senses that simple numerical ratios reflect the existence of comprehensible underlying causes. He is able to examine such data and to visualize a plausible *explanatory* model. An explanatory model is one that explains the observations, makes predictions, and is precise (and simple) enough to be subject to rejection. (Raff and Kaufman [16, p. 345] distinguish between *explanatory* and *predictive* models. I prefer the terms *explanatory* and *descriptive* because a model that explains can also predict. If a model cannot predict, then it has not really explained.)

Explanatory models differ from descriptive ones. The latter (among them I would include the equations commonly encountered in population genetics and the models generated by systems ecologists) do not explain. Descriptive models, as the term implies, merely describe complex situations. Their form is not rigid. Lewontin tells of a conversation with Sewall Wright in which he informed Wright of the high levels of allozyme variation which, at that time, had only recently been discovered in *Drosophila*. "Then our estimates of population size have been much too small," was Wright's immediate response. The earlier calculations were not abandoned; they were merely modified to cover (i.e., to describe) a wholly unexpected development. The calculations by Kimura and Crow [17] are of the same sort: they were originally used as evidence that populations are genetically homogeneous; now that populations have proved to be heterogeneous, the calculations are used as evidence that many alleles can be found at each locus.

The situation of ecological models is similar: they describe existing situations; they do not explain these situations. If the rare *Chrysolina* beetle that feeds on the equally rare Klamath weed were to be eradicated by accident (say, by spraying forests to control spruce bud worms), the Klamath weed might once more spread over and destroy thousands of square miles of California rangeland. But that potential calamity cannot be fitted into a systems model that is designed to describe the current interrelations between and among the major features of the environment. Descriptive models describe the general and, of necessity, neglect the specifics. If the situation that has been modeled changes, however, the systems model can be modified to describe the changed situation just as accurately as it had described the previous one.

Among biologists, there are those who fear and distrust all models; in their estimation, only observations have meaning. Others, some of whom possess considerable mathematical skill, make descriptive models that reduce complex situations to a series of equations. Others—and these were the persons who were to be found with exceptional frequency among early geneticists—create explanatory models from simple numerical relationships.

Mayr [1] has criticized the choice of physics as the "role model" for biology. Biology, according to Mayr, is a probabilistic science. For the most part, he is quite correct. Male birds, to cite one of Mayr's examples, may successfully defend their nesting territories against invaders 98.7 percent of the time. This is an observational fact that has an explanation; but the one figure—0.987—is inadequate to provide an explanation for what is an exceedingly complex phenomenon. Similarly with gene frequency data. Two nearby but spatially isolated populations of flies may differ in the frequency of the Est-6-F allele: 39.6 percent in one, 52.7 percent in the other. Again, an explanation for this difference must exist, but that explanation is far too complex to be fathomed from gene-frequency data alone.

Among those who advanced explanatory models in genetics, or who exhibited an appreciation for the simple ratios on which such models rest, have been a number whose early training was in physics. Mendel, himself, was a physicist by training. Sturtevant was not, but Benzer, who has made considerable use of Sturtevant's analysis of recombination among the anatomical features of mosaic flies in constructing fate maps, was formerly a physicist. Delbruck was also a physicist; Luria and Hershey were not. Crick, who insisted that Watson's trial models must generate Chargaff's 1:1 ratios, was a physicist by training.

While it is true that the bulk of biological observations are of situations so complex that they can be described only with great difficulty, it is not true that *all* biological processes are complex. Nägeli, a sincere biologist, told Mendel (in effect), "I have studied hybrids for many years and have never encountered a simple situation; something must be wrong with your experiments. Here, try my *Heiracium,* and see for yourself." Mendel tried, and was overwhelmed by the experimental results he obtained; we now know that *Heiracium* is an apomict whose pollen does not transmit genetic material.

Nägeli's mistake was to trust complexity and to distrust simplicity—to distrust simple results for which precise explanations (right or wrong) could be advanced. Simple relationships should always be cherished; they should never be deliberately ignored.

Nägeli admitted that Mendel had encountered simple relationships but believed that the outcome of hybridization experiments of necessity must be complex. Another type of disbelief leads to an out-and-out refusal to simplify, a refusal to admit that 2.84 or 3.14 can be said to equal 3, an integer. Lysenko was an acknowledged disbeliever of Mendel's ratios. Today, armed with pocket calculators that can handle eight-digit numbers as easily as one- or two-digit ones, biologists have become loathe to round off, to simplify. I once suggested that 602 events in 1,188 represented one-half and was immediately asked if I knew how many thousands of observations would be needed to make my claim valid. My critic was right: the true value for the proportion of events I had cited could, with 95 percent probability, fall anywhere from 0.477 to 0.537. I defy anyone, however, to base an explanatory model on 0.519, one of the many possible true values; at best, one can make of such a value only a descriptive, probabilistic model of the sort cited by Mayr. Like Mendel, however, I believe that, if a phenomenon is to be understood, it must at first be assumed to be simple—consequently, I shall not abandon my original suggestion.

The outcome of this account is to claim that, although physics may indeed be a poor model for studying, interpreting, and understanding many biological phenomena, a search for explanatory (as opposed to descriptive) models should never be abandoned altogether. Simple phenomena that follow simple laws do occur in biology. A flip-flop switch in DNA, for example, has only two orientations. Those persons who see explanations for such simple phenomena are often, but not always, physicists turned biologists. They do not always have outstanding mathematical skills. They are, however, at home with abstractions that are far removed from material substances. They visualize symbols rather than mull over words. Such persons were attracted to genetics shortly after its birth, bringing genetics as near to being a theoretical biology as was possible. Physiologists did not understand early geneticists; physiologists, like many biologists, prefer to deal with physical substances. Embryologists, like early hybridizers, deal with complex developmental systems whose verbal descriptions alone require tremendous time and energy; a mere glance at the recent textbook *Embryos, Genes, and Evolution* [16] will confirm this statement. Nägeli's "syndrome," the distrust of simplicity by persons who are steeped in complexity, is, I suspect, rife among embryologists. Ecologists, many of whom even refuse to generalize from experimental or field observations, are not prone to formulating explanatory models. Recently, ecologists and population geneticists have been grouped into the field of population biology. Although gathered under a common name, the two disciplines have not become thoroughly integrated, as a perusal of population biology textbooks will reveal; such books are clearly separable into individual sections dealing with either ecology or population genetics. Much of the difference is revealed by the mathematical models of the two subdisciplines. References to natural selection and its consequences with respect to the genetic composition of populations are frequently advanced by those population biologists who are classified as sociobiologists. The complex problems that fall within the domain of sociobiology seem to preclude that models other than descriptive ones will emerge from that science.

The influx of persons of these many backgrounds into an area now generally known as genetics—albeit with subheadings such as developmental genetics, ecological genetics, medical genetics, and others—will not eliminate those special persons who seek explanations for seemingly simple observations; such persons, fortunately, will persist. They will, however, become increasingly difficult to find.

REFERENCES

1. Mayr, E. *The Growth of Biological Thought.* Cambridge, Mass.: Harvard Univ. Press, Belknap, 1982.
2. Mendel, G. *Experiments in Plant Hybridization.* Translated by William Bateson. Cambridge, Mass.: Harvard Univ. Press, 1941. (Originally published in *Verh. naturforsch. Ver. Brünn*, vol. 4, 1866.)
3. Punnett, R. C. *Scientific Papers of William Bateson*, vol. 2. Cambridge: Cambridge Univ. Press, 1928. Reprint. New York: Johnson Reprint, 1971.
4. Barth, L. *Development: Selected Topics.* Reading, Mass.: Addison-Wesley, 1964.
5. Darwin, C. *The Effects of Cross and Self Fertilization in the Vegetable Kingdom.* New York: Appleton, 1877.
6. Morgan, T. H. What are "factors" in Mendelian explanations? *Proc. Am. Breeder's Assoc.* 5:365–368.
7. Sturtevant, A. H. The linear arrangement of six sex-linked factors in *Drosophila*, as shown by their mode of association. *J. Exp. Zool.* 14:43–59, 1913.
8. Bateson, W. Review of *The Mechanism of Mendelian Heredity*, by T. H. Morgan, A. H. Sturtevant, H. J. Muller, and C. B. Bridges. *Science* 44:536–543, 1916.
9. Wilson, E. B. *An Atlas of the Fertilization and Karyokinesis of the Ovum.* New York: Columbia Univ. Press, 1895.
10. Muller, H. J. The gene as a basis of life. *Proc. Int. Congress Plant Sci.* 1:897–921, 1929.
11. Chargaff, E. Chemical specificity of nucleic acids and mechanism of their enzymatic degradation. *Experientia* 6:201–209, 1950.
12. Watson, J. D. *The Double Helix.* New York: New American Library, 1968.
13. Portugal, F. H., and Cohen, J. S. *A Century of DNA.* Cambridge, Mass.: MIT Press, 1977.
14. Castle, W. E. Is the arrangement of the genes in the chromosome linear? *PNAS* 5:25–36, 1919.
15. Castle, W. E. Model of the linkage system of eleven second chromosome genes of *Drosophila*. *PNAS* 6:73–77, 1920.
16. Raff, R. A., and Kaufman, T. C. *Embryos, Genes, and Evolution.* New York: Macmillan, 1983.
17. Kimura, M., and Crow, J. F. The number of alleles that can be maintained in a finite population. *Genetics* 49:725–738, 1964.

TWO GENETICISTS AND A FOUNDING FATHER: INTRODUCTION

The constraints of research and an academic career do not necessarily suppress the personality and style of scientists. Graduate students and junior faculty, however, usually do not penetrate the facades of famous colleagues or elder statesmen. Consider the emotional turmoil of a graduate student in the Department of Genetics at the University of California about to present a research seminar to an audience including R. B. Goldschmidt, C. Stern, R. E. Clausen, and G. L. Stebbins. Current graduate students in genetics are not likely to recognize these names.

Richard B. Goldschmidt was a product of classical German training in zoology before World War I. From the start, Goldschmidt seemed to find or to seek controversy. It was almost certain that he would enter genetics as soon as this *enfant terrible* of biology attracted both attention and debate. Here he found a second career as well as considerable controversy. C. Stern is perhaps the only one who could have written this balanced account of an eminent geneticist.

Sewall Wright is one of the quiet giants of genetics in an area that did not attract widespread attention at the peak of his career: population genetics and the mathematical basis of evolution. After a career in the U.S. Department of Agriculture, Wright had a second distinguished career at the University of Chicago and then "retired" to a third career at the University of Wisconsin. The products of Wright's fertile mind guided animal breeders and broke new ground for experimental evolutionists.

Once variant enzymes and proteins in crude extracts could be separated by electrophoresis, the levels of heterozygosity in the populations of almost any species could be determined. As the data accumulated and required analysis and interpretation, the pioneering work of Sewall Wright was available and ready. The merger of molecular biology, population genetics, and evolution presented no problem. The obvious phenotypes of progeny from wild fruit flies were not conceptually different from variant enzymes or proteins.

Crow's affectionate and perceptive view of the life and times of Sewall Wright reveals facets of this scientist that were not obvious in a modest gentle man. The story of the "guinea pig–eraser" may be apocryphal but it is cherished by all who knew him.

The identification of the transforming factor in *Pneumococcus* as DNA by Avery and his colleagues marked the beginning of the New Genetics. The work was not the result of serendipity but the successful culmination of a carefully crafted program with a practical goal: the production of an effective vaccine for bacterial pneumonia. Coburn's account of conversations with T. O. Avery and a letter from Avery to his brother, couched in the careful language of the scientist, document the fact that Avery appreciated the significance of his discovery.

RICHARD BENEDICT GOLDSCHMIDT*
April 12, 1878–April 24, 1958

CURT STERN†

Richard Goldschmidt left not only a published record of nearly sixty years of scientific activities but also a full autobiography and a detailed sketch of his life written in fulfillment of the traditional request of the National Academy of Sciences. In addition, he preserved many letters which he had received from colleagues all over the world and deposited them in the archives of the Library of the University of California at Berkeley. It would be possible to reconstruct a large part of the development of the biological sciences in the twentieth century on the basis of Goldschmidt's publications and their interaction with his contemporaries. It would also be possible to explore in depth the personality of the man as it was formed through the impact of his time. These may be worthwhile tasks for future historians. The present Memoir can present only a selection out of the multitude of his activities.

Richard Goldschmidt was born on April 12, 1878, in Frankfurt am Main, Germany. He died in Berkeley, California, on April 24, 1958. His parents came from respected, prosperous local families and the circle of his relatives included an unusually large number of well-known scientists, bankers, and philanthropists. In his own words, he grew up in "a typical German bourgeois family, comfortable but strict and even parsimonious in spite of cooks, nursemaids, and French governesses." The city itself had a proud and distinguished history and provided an atmosphere of rich

* The final form of this Memoir owes much to the critical comments of F. Baltzer, E. Caspari, E. Hadorn, H. Nachtsheim, Leonie Piternick, J. Seiler, and Evelyn Stern. Th. Bullock and R. B. Clark provided evaluations of Goldschmidt's early work on the nervous system of Ascaris, and the late F. Schrader on some of Goldschmidt's cytological investigations. Reprinted, without the Bibliography, from National Academy of Sciences, *Biographical Memoirs*, Vol. 39, with permission of the Academy and Columbia University Press.

† Department of Zoölogy, University of California, Berkeley, California 94720.

interest in all cultural pursuits. Simultaneously it was democratic in spirit. There he attended the Gymnasium with its curriculum of nine years of Latin, French, and mathematics and six years of Greek as the nucleus of education. The young boy was a voracious reader: world literature, Goethe, prehistory, archaeology, and comparative linguistics were the major areas. When he was thirteen years old he began to see himself as a future naturalist and world traveler and, three years later, the center of his interest permanently became biology. At seventeen he could read rather fluently French, English, Italian, Latin, and Greek and made abundant use of these abilities. He also tried to read the works of philosophers from Spinoza to Nietzsche, but confesses that he did not succeed in understanding them then or later.

When he entered Heidelberg University he enrolled at his parents' request as a medical student. Among his "glorious teachers" were such historical figures as Bütschli, the zoologist, Gegenbaur, the comparative anatomist, and Kossel, the biochemist. After two years of study he passed his premedical examinations and then went to Munich. At this point he abandoned further medical training and became a student of zoology under Richard Hertwig, with minors in botany, physiology, and paleontology. At the age of twenty-one he completed his first paper, a detailed account of developmental features in a tapeworm, the result of a chance finding during course work. Shortly he returned to Heidelberg where he became laboratory assistant to Bütschli, his beloved teacher and later lifelong friend. Under him he worked out his Ph.D. thesis on maturation, fertilization, and early development of the trematode Polystomum.

For a year the progress of Goldschmidt's zoological research was interrupted by the compulsory period of training in the German Army. When he returned to civilian life he followed an invitation from Richard Hertwig to join his staff at Munich, remaining there until 1914.

The first seven years of this period show Goldschmidt as the intense worker which he remained throughout his life. However, in contrast to his later periods he had not yet found a great central problem for his studies. Instead he tried his hand—and brain—at histology and neurology, cytology and protozoology, embryology and a monographic treatment of the anatomy of the "Amphioxides stages" of the lancelets. This was no mere dabbling. Although some of his publications at this time were but short reports on random findings made during his brief visits to the Mediterranean marine laboratories of Naples, Villefranche, Banyuls, and Rovigno, many were detailed accounts of elaborate studies. The most impressive of these is the sequence of papers on the nervous system of the parasitic roundworm Ascaris. It was Bütschli who suggested to Goldschmidt the study of the histology of the nematodes and particularly their nervous system. Goldschmidt discovered that in Ascaris this system is composed of a fixed, relatively small number of cells, 162 in the male and 160 in the female. He set himself the task of establishing the topography and the interrelations of every one of these cells, by making dissections and reconstructions from microtome sections. Altogether between 1903 and 1910 more than 350 printed pages were given to the results, accompanied by numerous text figures and nearly 20 plates, all drawn by the author. One cannot but admire the patience, the persistence, and the thoroughness with which these observations were made and put on paper. Some of the illustrations give fine details of cytologic and histologic nature; others, on folded charts of up to 65×37 cm, depict the paths of nerve fibres in a manner resembling the intricate system of a railroad switchyard. These studies find no equal in our knowledge of the nervous system of any other invertebrates. Unfortunately, they have borne no fruit as a basis for physiological investigations. For one reason, Goldschmidt's description of the interconnections of the nerve cells does not provide a picture of the probable routes of conduction of impulses between the sensory and motor elements; in any case, the parasitic nature of Ascaris did not seem favorable to experimental work. It is not known either how many of Goldschmidt's findings would stand up to a reinvestigation. While they were in the process of publication, they found support in the work of Martini, who confirmed and expanded Goldschmidt's discovery of cell constancy of various organs in nematodes. On the other hand, a special study of the nervous system of Ascaris by Deineka came in part to very different results. Goldschmidt responded vigorously to Deineka's publication and its defense by Dogiel, Deineka's teacher, accusing the St. Petersburg authors of having stained indiscriminately all kinds of tissues and then having regarded them all as parts of the nervous system. It is the belief of modern surveyors of the literature that Goldschmidt was probably much closer to the truth than his adversaries, but only new original studies can decide the issue. It may be added that the Ascaris work gave Goldschmidt occasion to make histological findings beyond those reported here, to participate in the discussions of the validity of Cajal's neurone theory, and to apply Koltzoff's principle of the necessity of cellular fibrils as a skeletal basis for the morphology of nonspherical cells.

The series of papers on the Amphioxides cover very different ground. Perhaps still under the influence of the comparative anatomists of his student days, Goldschmidt asked to be entrusted with the material of acrania which the German Deep Sea Expedition of 1898–1899 had collected in southern oceans. In 1905 he submitted the results of his work in a monograph of nearly 100 quarto pages and 10 plates of illustrations. One half of the work gives the details of the anatomy and histology of his forms, the other half a thorough analysis of the development of Amphioxus and the phylogenetic implications of the findings on Amphioxides. Among his various specific discoveries Goldschmidt himself ranked highly those of the solenocytes in an excretory organ of the animal, strangely enough unaware that this important cellular feature had already been described some years earlier by Goodrich who had found it in the kidney tubules of living specimens. Nevertheless, Goldschmidt's independent recognition of the annelid-type nature of the excretory organs, made on specimens which had not been too well fixed when they were collected, is testimony both to his powers of observation and to his powers of interpretation.

The 1905 monograph discusses carefully the question whether the Amphioxides really represent an independent, more primitive, or neotenic group of acrania than Amphioxus or whether they are developmental stages of Amphioxus itself. Goldschmidt originally decided in favor of taxonomic independence but, soon afterwards, on the basis of a new find recognized that his material belonged to Amphioxus. Until 1909 he added new observations on the Amphioxides forms. In 1933, after a brief stay at the Bermuda Biological Station, he returned once more to them.

While the work on Ascaris and Amphioxus was a late fruit of nineteenth-century interests, the third main area of young Goldschmidt's studies belonged directly to his time. The great discoveries of the last decades of the past century on chromosomes, mitosis, maturation, and fertilization had also made apparent the existence of many unsolved fundamental problems. Many of the outstanding biologists therefore were active in cell research and Goldschmidt joined their ranks. Two aspects attracted his attention, the chromidial apparatus of protozoan and metazoan cells and the problems of meiosis. In 1902 Richard Hertwig had published an interpretation of his observations on the presence of numerous granules in the cytoplasm of Actinospherium and other protozoa. Since these granules stained similarly to the nuclei, he assumed that they were nuclear equivalents, arising from well-formed nuclei, in some cases by breakdown, in others by being extruded from intact nuclei. In their turn, the chromidia, as the granules were called, supposedly could give rise to well-formed nuclei. Goldschmidt studied chromidial elements in the flagellated amoeba Mastigella, in the tissue cells of Ascaris, and in the so-called yolk nucleus, a cytoplasmatic element of the egg cells of spiders and other animals. He believed that he could trace the extrusion of chromidia from the nucleus, observed variations in the extent of the chromidial apparatus in relation to the functional state of cells, and devised a comprehensive theory to account for his own and a variety of other findings.

The chromatic material of all cells was assumed to consist of two kinds: that concerned with the metabolic functions and that concerned with the reproductive functions, trophochromatin and idiochromatin. We now know that the chromidial apparatus was a concept which encompassed in a single term a great many essentially different kinds of cytoplasmatic elements such as mitochondria and Golgi bodies and, in some protozoan cases, even intracellular parasites. Nevertheless, the similarity of the concept with modern knowledge of the derivation of cytoplasmatic RNA from the DNA of the nucleus is not just an analogy. The recent insights were foreshadowed by the work of Goldschmidt and a host of followers. Goldschmidt himself watched with keen interest the cytochemical and fine-structural investigations of the 1940s and 1950s. With his last Ph.D. student, T. P. Lin, he described in 1947 the Feulgen staining properties of the end parts of the Ascaris chromosomes which are, as is well known, left behind in the cytoplasm during the mitoses of the future somatic cells of the embryo.

Goldschmidt's doctoral dissertation had dealt with oögenesis and other cytological phenomena in the trematode Polystomum. In order to broaden his observations he soon turned to other trematodes: *Zoogonus mirus* (1905) and *Dicrocoelium lanceatum* (1908). In Zoogonus, he saw a most remarkable phenomenon, the reduction of the diploid chromosome number to the haploid in a single division by simple migration of one half of the chromosomes to one pole of the spindle and of the other half to the other pole. None of the elaborate chromosomal maneuvers which occur in the meiotic prophases of other organisms were seen. The term "Primärtypus" of reduction was coined for the newly discovered process, which corresponded completely to Weismann's prediction of chromosome reduction. Dicrocoelium was studied in the hope of encountering the primary type in this form also, but to his disappointment Goldschmidt found its oögenesis to be of the standard type.

The Zoogonus paper led to a dramatic sequence of events which may be reported in the words of a letter to the author (March 25, 1960) by the late Franz Schrader:

"The Zoogonus case was at one time very close to me because, as you may not remember, after Goldschmidt I was the next man to claim that meiotic reduction may sometimes occur without a preceding series of pairing maneuvers. So far as Goldschmidt was concerned, his arguments concerning Zoogonus represented one of those intellectual premonitions of which he had so many. . . . One must not forget that when Goldschmidt, in 1905, claimed to have a case of 'Primärtypus,' the cytologists and geneticists had barely succeeded in establishing the regularity of the meiotic process; any one who tried to upset the applecart again was regarded as nothing less than a fool or a criminal. Hence Goldschmidt's conclusions were opposed by all the powers that ruled the roost, among them the Schreiners and Grégoire.

"It was almost fatal to make a little mistake therefore, and G. [Goldschmidt] did make such a mistake; he miscounted the chromosome number of Zoogonus.

"But the Schreiners, who proceeded to annihilate him, made an even bigger error in their counts and, for the rest, were so prejudiced that their analysis of the meiotic prophase doesn't mean much.

"Grégoire finally established the right chromosome number, but he too was so biased that his view of the synapsis period is anything but final.

"When, in 1911, Wassermann began to re-examine Zoogonus, he concluded that Grégoire had been correct about the chromosome number but that there is something funny in the prophase that Grégoire had misinterpreted. But Wassermann never really cleared up the case which therefore still stands in abeyance."

It remains to be added that Goldschmidt had made his original slides available first by request to the Schreiners and then on his own initiative to Grégoire so that the divergent findings were based on the same material, in some instances on the same individual cells. In their criticism of Goldschmidt the Schreiners had attempted to remain within bounds but Goldschmidt's reply exceeded, in sharpness, aggressiveness, and denunciation, perhaps all other scientific controversies at a time which was rich in personal invective. It must be admitted, however, that the allegation that he was unable to distinguish 10 from 26 chromosomes not only deeply hurt Goldschmidt personally but also endangered greatly his future career as a young scientist. His relation to Grégoire remained on a different level. When the latter wrote to him of his own interpretation Goldschmidt answered that, after renewed examination of the slides, he had to stand by his original views. However, spontaneously he once more sent Grégoire his material with the authorization to publish the divergent results. Grégoire in his paper acknowledged these facts with the following sentence: "C'était mettre le comble à l'amabilité et au desintéressement scientifique."

The year 1910 terminated the first period of Goldschmidt's biological work. However, a somewhat later paper belonging to his studies in genetics will be reported here, since it resembles the Zoogonus episode in committing a serious mistake.

In 1911 de Vries published the results of his investigations on the hybrids of *Oenothera biennis* and *O. muricata*, results which, in Goldschmidt's words, belonged to the most peculiar phenomena uncovered by hybridization studies and which caused the greatest difficulties for a deeper understanding. It is not necessary here to describe de Vries's discoveries. Suffice it to say that Goldschmidt conceived an explanation of them which was as ingenious as unexpected. He assumed that in these species crosses the maternal nucleus of the fertilized egg cell degenerates and that development proceeds solely on the basis of the paternal nucleus. This assumption of "merogony" was subject to test and Goldschmidt proceeded to furnish it. At his request Renner, his younger colleague in the Department of Botany, repeated de Vries's crosses and provided the necessary stages for the cytological study made by Goldschmidt. The examination seemed to bear out his theory completely. He adduced various observations in its favor, the most decisive of which was the finding of the haploid chromosome number 7 in the cells of early hybrid embryos in contrast to the diploid number 14 in normal embryos. In spite of his pleasure in proving his theory correct Goldschmidt must have felt that the evidence which he provided was less conclusive than he had persuaded himself to believe, for he asked Renner to study the cytology anew. A year later Renner demonstrated, and Goldschmidt agreed, that grievously incorrect observations had been made, that the hybrids had 14 chromosomes like the normals and not 7, and that there was no indication of merogony.

In a draft of his autobiography, Goldschmidt refers to his Oenothera work as the single instance of which he had reason to be ashamed. "There is an excuse: at that time I was so overworked that I should not have written anything, had I been wise. But as I was not wise I had to suffer for years, when I realized what I had done." In 1916 he attempted to show the possibility that the hypothesis of merogony in slightly changed form had still a chance of being correct. Soon, however, the solution of the Oenothera riddle came from a very different approach, primarily thanks to Renner who had entered the field on the stimulus provided by Goldschmidt. In the printed list of the latter's "Publications 1900–1954" the two merogony papers cannot be found.

When one looks back at Goldschmidt's accomplishments between 1900 and 1910 one finds that the young scholar had achieved a highly prominent place in his field. He had enriched zoological knowledge by many discoveries based on an astonishing amount of hard and detailed work and had provided stimulating hypotheses and theories in various biological areas. Although he had made some serious mistakes and his position was not unequivocal, his influence was felt widely. Moreover, at the age of twenty-nine he had founded the *Archiv für Zellforschung*, which under his editorship instantly became a most important, if not the most important, international center for publications in cytology. He had also written a pioneering textbook of genetics, first to appear in 1911, which constituted a brilliant, highly original achievement. In Hertwig's institute he had been promoted stepwise to the Associate Professorship, had given lecture courses and supervised laboratory instruction, and had become the leader of a school of advanced students from all over the world.

It was during this decade that Goldschmidt was married to Else Kühnlein. Their two children, Ruth and Hans, were born in 1907 and 1908 respectively. These happy personal events found their echo in the publication of two small popular volumes, one on the protozoa and one on reproduction in the animal kingdom. The royalties for these books paid for the expenses connected with the birth of the children. Ruth later became a physician, Hans an engineer.

It was natural for an ambitious young biologist of the new century not to be satisfied permanently with purely descriptive work. In 1909, therefore, he turned to genetics, "influenced considerably by the logical and dissecting presentation of this then novel field" as given in Johannsen's *Elemente der exakten Erblichkeitslehre*, published in that year. The choice of his own experimental material was suggested by a study of the Swiss entomologist Standfuss's experiments on moths carried out mostly before the rise of genetics. Goldschmidt selected two problems, one in evolutionary genetics, the other in the area of sex determination. The first made use of the nun-moth *Lymantria monacha*, which furnishes one of the classic examples of so-called industrial melanism, the spread in historic times of melanic phenotypes particularly in areas of large-scale industrial activity. In the course of six breeding seasons Goldschmidt studied the genetic situation which distinguishes melanic from nonmelanic forms. After extrinsic delays the results were published in 1921. Goldschmidt not only gave the Mendelian analysis but discussed in astonishing depth its evolutionary implications. By applying the mathematics of the Hardy-Weinberg law (then still unnamed and hardly recognized in its importance) he showed that mutation pressure alone was unable to account for the observed increase of melanism unless demonstrably impossibly high mutation rates were assumed. He concluded that in industrial regions unknown physiological advantages were correlated with the possession of melanism. That the selective advantage of melanism consists in the protective coloration of the phenotype, as has been shown in recent years, was too naïve an explanation for the thinking of the somewhat disillusioned evolutionists of the times. The paper on the nun-moth aroused littled attention, undoubtedly because it was greatly ahead of its time and because its author did not pursue the topic much further. It is a testimony to the agility of Goldschmidt's mind that this paper shows him as a pioneer in population genetics, by using a method which was original and yet not close to his own way of thinking.

The problem of sex determination was an object of much study in Hertwig's laboratory. There it proceeded mostly on the basis of non-Mendelian approaches. Goldschmidt decided to analyze it by means of crosses of different species or races, led by a statement of Standfuss's that abnormal sex types frequently appear in such hybrids. He tried a variety of forms, among them the gypsy moth *Lymantria dispar*. In crosses of a

European with a Japanese race he obtained in F_1 normal males but instead of typical females, sex intergrades. Goldschmidt later called it "an unbelievable piece of luck" that he happened to have had available two races which would give this result although a great many other racial hybrids of Lymantria consist of the normal sexes only. One may disagree with this opinion. Goldschmidt's restless large-scale activities were bound to provide him with new discoveries and his mind was eager and ready to make the most of them. Indeed, he immediately realized the importance of the result, repeated the crosses on a much larger scale with more geographic varieties, and recognized that the variable results obtained required that many different sexually determined races must exist in Japan. "But the decisive point was that, walking home a dark night the two miles from the lab, I suddenly understood [in 1911] the amazing genetical situation, which was outside the accepted Mendelian tenets. I had found what is today called the balance theory of sex determination."

He had even found more than that. After ten years of describing one mendelizing gene after the other, in many plants and animals, biologists not yet stimulated by the just initiated work of the Drosophila group were in the process of becoming bored with these findings. Goldschmidt himself from the start of his genetic work was anxious to progress beyond the exploration of the Mendelian mechanisms and hoped to accomplish this by combining Mendelian with developmental physiological investigations. When he began raising Lymantria he saw that the caterpillars of different races had different markings and that certain hybrid larvae at first were light like the light parent and then stepwise became dark like the dark parent. This phenomenon, to which the purely descriptive term of change of dominance applied, called for an explanation in terms of developmental events. The flash of insight which came to the thirty-three-year-old man in 1911 combined *Beobachtung und Reflexion* on sex aberrant adults and color types of caterpillars in a general theory of genic action. For many years the flame then lighted was to burn bright and passionately in Goldschmidt's mind and to illuminate the thinking of a generation of biologists.

The early Mendelians had recognized the genic constitution of organisms and the Drosophila workers had uncovered in detail the chromosomal mechanism of heredity. These great discoveries were of classical character, solidly based on a multitude of well-established facts. By themselves, however, they would not have been sufficient to place genetics in the central position within biology which it was to assume. Beyond the "static" analysis of gene transmission an insight was needed into the dynamic role of genes in cellular physiology, biochemistry, and development. Goldschmidt recognized this aspect and outlined in brilliant generalizations a physiological theory of genetics. Necessarily the prophetic nature of this achievement—romantic in the sense of Ostwald's classification of great scientists and their discoveries—transcended its factual basis. It was the audacity of the theoretician unimpeded by the still scanty data which gave a new focus to biological thought.

Before describing in some detail the essentials of the Lymantria work and the deductions derived from it, external events in Goldschmidt's life need to be related. In 1911 the Kaiser Wilhelm Gesellschaft for the promotion of science was founded in Germany and began to organize its institutes, which were to become superlative centers of research. Boveri was offered the directorship of the Institut für Biologie and it was he who with extraordinary insight selected as leaders of its departments four men, all but one still in their thirties, and two of them destined to win Nobel Prizes: Spemann, Hartmann, Warburg, and Goldschmidt. On January 1, 1914, Goldschmidt was appointed a member of the institute then being built in Berlin-Dahlem. At the same time he was awarded a traveling fellowship in order to go to Japan and look for new races of Lymantria. In Tokyo, as guest of the Imperial University just before the outbreak of World War I, he bred the forms which he had collected, and was able to send the living material back to Germany where Seiler, his former student in Munich and now first assistant in Dahlem, carried the cultures through the tumultuous years until Goldschmidt's return, which was greatly delayed. When Goldschmidt arrived in San Francisco on his homeward journey from Japan, he found that the British blockade made his further travel to Germany impossible. After two months in the Zoology Department of the University of California he proceeded east to become a guest in Harrison's laboratory at Yale University for a period of years "during which I learned to love and admire my great host." Breeding work was done during summers at Harvard's Bussey Institution and at Woods Hole. His family was permitted to leave Germany and joined him late in 1915. Then, early in 1918, like many other Germans, he was interned in a civilian prisoner camp in Georgia and not released until after the war was ended.

In later years Goldschmidt would entertain his guests by vivid and good-humored accounts of his sojourn in a local jail, the transfer to the state prison, and his solitary march the length of lower Manhattan to Pennsylvania Station, handcuffed under guard of two soldiers with loaded guns!

Before leaving Munich, Goldschmidt had given some accounts of his Lymantria work on sex determination. He published additional preliminary papers on the same topic in American journals throughout the war years. In 1917 he wrote two book manuscripts in which he formulated in detail his new insights and generalized them broadly. When he finally took up work in the Kaiser Wilhelm Institut he quickly brought these up to date. In 1920 there appeared the *Mechanismus und Physiologie der Geschlechtsbestimmung*, a book which gave not only his own theories on the physiology of sex determination but also a comprehensive account of the accomplishments of the preceding twenty years on the mechanisms; the *Quantitative Grundlage von Vererbung und Artbildung;* and the first of the documentary "Untersuchungen über Intersexualität," an article of 199 pages. This was to be followed by five more papers under the same heading, the final one dated 1934.

The impact of the two books and of the supporting material was great, both in Germany and abroad. For Germans the years of the war, the defeat, the revolution, and the not-yet-ended inflation constituted a deep cleft which separated two eras. The desire for something more permanent than the catastrophic course of history made scientific achievements objects of consolation and hope. Within the field of biology the main sources of elation were Goldschmidt's syntheses and Morgan's *Material Basis of Inheritance*, soon made available in a German translation by Nachtsheim, another former student of Goldschmidt's.

The aberrant sexual types which Goldschmidt had produced in Lymantria crosses were at first considered under the term "gynandromorphism," a designation which was then used to describe a variety of abnormal forms in which female and male traits occurred side by side in the same individual. Soon, however, Goldschmidt recognized that his gynandromorphs differed from other gynandromorphic insects. Most of these were specimens in which the male and female parts had different genetic constitutions, such as having one X-chromosome in the cells of one part and two X-chromosomes in the other. Contrary to this situation, all of the cells of the Lymantria intergrades were genetically alike. In 1915

Goldschmidt coined the term "intersex" for them. The crosses in which the intersexes were produced showed that two genetic elements were involved, a maternally inherited female tendency F, and an X chromosomally inherited male tendency M, thus giving a normal female the formula FM and a normal male FMM. This scheme, according to which sex is determined by the relative "strength" of F and M—one M being less effective than F and two M's more effective—had been proposed also by Morgan, adjusted to the somewhat different situation regarding sex determination in flies as compared to moths.

In most organisms, where one meets with the normal sexes only, the scheme was not subject to test. In Lymantria the genetic analysis showed that different races had F's and M's of different effectiveness. Within each race the sex factors were so balanced that the relations $1M < F$ equals female and $2M > F$ equals male always were upheld, but hybrid combinations of F's and M's from different races could lead to FM intersexes whenever a "weak" F could not sufficiently overbalance a "strong" M, or to FMM intersexes whenever a strong F could partly outweigh two weak M's. Depending on the relative strength of F and M in hybrids between different races one could obtain slight, intermediate, or extreme intersexuality, including complete sex reversal. The experimental proof for these relations preceded by years Bridge's famous discovery of triploid intersexes in Drosophila. There, the quantities of the constant F and M factors can be read off the numbers of chromosomes of different sex forms and a direct demonstration can be provided that increasing quantity expresses itself in increased strength of the sex determiners. Goldschmidt himself had interpreted the racially different strengths of the Lymantria sex factors in terms of different quantities of F or M genic molecules.

It was surprising that the imbalances of the sex factors in the intersexes of Lymantria did not lead to intermediate development between male and female for the sex-determined or sex-controlled parts of the specimens. Rather, the intersexes are mosaics of male and female differentiated tissues. More surprising still, there seemed to exist two radically different kinds of intersexes. The quantitative balance theory of sex determination would lead one to expect that whether a given imbalance was on the basis of an FM or FMM genotype should make no difference, since it was not the *number* of M factors in the X-chromosomes but the relative effective strengths of the two sex antagonists which counted. Yet, the morphology of intersexual FM types appeared to differ in some basic way from that of FMM ones. Influenced by his observations on the changing color intensity of hybrid caterpillars, Goldschmidt modeled his thoughts on intersexuality into the "time law" theory. The sexually mosaic nature of intersexes and the occurrence of two different types or series was explained by the assumption that development of an intersex, from fertilization on, begins under the influence of one of the two sex factors and, at a turning point, changes over to development under the influence of the opposite sex factor. The spatially mosaic nature of intersexes he considered as the morphologic consequence of a mosaicism of sex tendencies in time, female FM intersexes having started as females and then switched to maleness, and male FMM intersexes having gone through the opposite sequence.

The theory of the time law led to a long series of morphologic and developmental investigations by Goldschmidt and his students and to searching critical analyses by other authors, among whom Kosminsky in Leningrad, Seiler in Zurich, and Baltzer in Bern were outstanding. Theoretically, "a sexual mosaic" could be produced by an alteration of determination in time or by simultaneous but alternative determination of different cells. The latter possibility is realized in the intersexes of the moth Solenobia, as has been proven by Seiler. The painstaking, factfinding work of this investigator, which led him to abandon the time law, stands in fascinating contrast to Goldschmidt's all-embracing theoretical convictions.

In the later 1940s and subsequently, Goldschmidt and Seiler had repeated literary exchanges concerning the interpretation of the facts in Solenobia. Goldschmidt was unwilling to concede that Seiler's observations were at variance with the demands of the time law. Similarly, he had attempted earlier to fit into his original concepts some specific findings by Kosminsky on Lymantria which were not easy to understand in terms of the time law.

It seems not possible today to gain a final view of the Lymantria situation. In contrast to the single, albeit widely variable type of intersexes in Solenobia, in Lymantria a major difficulty is presented by the occurrence of the two above-mentioned kinds of intersexes, male and female. In reality there are even more kinds, since female intersexuality in which sex genes of the Japanese Gifu race are involved and certain intergrades obtained by Kosminsky differ morphologically from all other types of female intersexuality and in their coarse mosaicism resemble typical male intersexuality. The phenomenon of Gifu intersexuality remained basically ununderstood by Goldschmidt, and even the morphology of male intersexuality had aspects not immediately derivable from his theory. Perhaps the two, or more, different kinds of intersexes in Lymantria are less different in their development than has been assumed for so many decades; this is an important area for renewed study. That the existence of two developmentally opposite types of intersexuality is actually contrary to Goldschmidt's own tenets was mentioned above. This was early pointed out by F. Lenz but not considered worthy of comment by Goldschmidt. The endeavors of other authors to devise different interpretations than his own were dissected mercilessly. With respect to a specific critique he concluded: "This attempt, which only resulted in a poorer version of my own ideas, has confirmed my conviction that there just does not exist any other explanation than the one found by me" (translation). His adversary's reaction to this was understandable: *Roma locuta est*—Rome has spoken.

Goldschmidt's general theory of genic determination of development was based on his interpretation of the action of the sex factors. All genes, he postulated, are enzymes and different alleles of specific genes are different quantities of a specific enzyme. The function of these gene-enzymes in development is a typically chemical one: to produce "hormones" of differentiation, in quantities which are proportional to the quantities of the genes themselves. Development, then, is based on the tuned, harmonious chemical reaction system of the totality of genes: "The mass law of reaction velocities is one of the basic laws of genetics." He illustrated his theory with schemes of curves, in which the abscissa represents time of development and the ordinate the amount of gene product. To account for the time law of intersexuality he fitted the M and F curves so as to cross each other at the appropriate assumed turning points. He applied the theory to a variety of phenomena observed by himself and others in different organisms—the patterns of lepidopteran wings and their temperature dependence, the polymorphism of natural populations and the head shape of Daphnia—and derived a theory of evolution which was based on selection of different multiple alleles causing different reaction velocities. He recognized that any quantitative shift in one gene would be physiologically feasible only if other genes and the reactions dependent on them were shifted correspondingly so as to produce a new physiological balance. This, he pointed out, accounted for the observation that geographic races, in contrast to local mutants, differ from one another by numerous traits, in Lymantria as well as in birds and mammals. If one disregards the specific interpretation of allelism as variation in gene quantity Goldschmidt's views of evolution are seen to be very similar to those of the later Neo-Darwinians.

Incessantly, Goldschmidt elaborated his general views, defended them against criticism, and emphasized their significance. If Darwin had his protecting bulldog, Huxley, and his prophet, Haeckel, Goldschmidt was his own bulldog and prophet. In 1927 appeared his book *Physiologische Theorie der Vererbung* which tried many new forms of curves to account for the development of the Lymantria sex types and of caterpillar coloration, for dosage effects, deficiencies, multiple factors, mutation, and other genic phenomena in Drosophila and in general. It was a brilliant performance, but psychologically made partially ineffective by harsh attacks on the Drosophila geneticists whose discoveries he praised highly but whose refusal to accept a quantitative theory of allelism and mutation and to go beyond "the narrowest interpretation of the factor theory" he disdained. Ten years later, one further presentation appeared, in English, *Physiological Genetics*. This book not only furnished another outline of Goldschmidt's dynamic developmental views but was actually a comprehensive review and critical analysis of all known important facts on genic action. As such it served for many years as a basic text in this area of genetics. An important change in the author's viewpoint from earlier years concerned the nature of the gene. He abandoned the idea of alleles being different quantities of the same molecule in favor of looking at a chromosome as a single gigantic macromolecule in which local "steric changes [account for] deviations from wild type which may be described as mutations, even as point mutations, though no actual wild-type allelomorph and therefore no gene exists." We shall return to this new concept.

Earlier than *Physiological Genetics*, Goldschmidt had written a monograph, *Die sexuellen Zwischenstufen* (1931), a remarkably complete survey of the great variety of intersexuality in the whole animal kingdom. Although more than thirty years have passed, during which new material has accumulated and some of Goldschmidt's interpretations have had to be abandoned because of later work, it is still a very useful book.

Twice in the 1920s Goldschmidt had again traveled to the Far East to collect additional material of Lymantria. For two years he was professor at the Imperial University in Tokyo where he continued his large-scale breeding work. Otherwise the Dahlem Institut served as an ideal place for research with its ample resources and the freedom of its staff from any other obligations than "to lay ostrich eggs," as Baltzer once wrote in a poem sent on the occasion of Goldschmidt's fiftieth birthday. Apart from intersexuality, geographic variation as a general phenomenon formed the core of Goldschmidt's activity during those years. From 1924 to 1933 appeared the six voluminous "Untersuchungen zur Genetik der geographischen Variation." Here the sex races, the differential characters of the life cycle—length of larval development, number of larval instars, nature of diapause, growth and size, color and marking of larvae and moths, cell-, egg-, and chromosome sizes—were all analyzed with respect to their genetics, physiology, and ecologic-geographical distribution. Goldschmidt particularly emphasized the adaptive nature of many racial traits in the sense that length of diapause, for instance, has to be related to the seasonal cycle of the respective geographic region. The significance of this work has been characterized by Mayr in his *Animal Species and Evolution* (1963; p. 298):

"The present generation of evolutionists can hardly appreciate the enormous impact of the demonstration by Schmidt (1918) for the fish Zoarces, by Goldschmidt (1912–1932) for the gypsy moth (*Lymantria dispar*) and by Sumner (1915–1930) for the deer mouse Peromyscus that the slight differences between geographic races have a genetic basis."

In 1934, after twenty-five years of work on Lymantria, Goldschmidt published a final comprehensive monograph and regarded his analysis "as finished as far as my own person is concerned." The facts which he had found and the views he had formulated would continue to play important roles in his future research.

The study of geographic variability had originally been initiated with the thought that its genetic analysis would provide a key to the problems of speciation. When, however, all the material had been assembled, Goldschmidt announced that his views had changed fundamentally. The many races of *Lymantria dispar*, he emphasized, still belong to one single species. The small mutations of the Neo-Darwinians would explain geographic diversity within a species but, he argued, could not serve as bridges over the fundamental gap which separates one species from another. This skeptical view, first pronounced at the Sixth International Congress of Genetics in Ithaca (1932), was gradually replaced by a positive theory which would account for speciation and even the sudden origin of much greater newness in evolution. Goldschmidt invoked the occurrence of "macromutations," causing fundamental changes in the structure of mutated individuals and thus abruptly leading across the unbridgeable gaps. He recognized that from the point of view of the old species the macromutants would appear to be monsters, but he showed that some of these might be "hopeful monsters" which had an evolutionary future. When he was appointed Silliman Lecturer at Yale University, he decided to formulate a comprehensive treatment of his concepts of evolution and chose as a topic *The Material Basis of Evolution* (1940). The first half of the book is given to microevolution, the origin of intraspecific diversity. Its conclusion is: "Subspecies are actually, therefore, neither incipient species nor models of the origin of species. They are more or less diversified blind alleys within the species." He thus was led to the concept of macroevolution which dominates the second half of the book. The brilliant main part of this section is entitled "Evolution and the Potentialities of Development." Here it is demonstrated, by reference to experimental changes in development, to hormonal effects resulting in vertebrate metamorphoses, to the alternation of larval and adult forms in insects, to the diversity of the sexes in the same species, and many other examples, that "the developmental system of a species is capable of being changed suddenly so that a new type may emerge without slow accumulation of new steps." It is further shown that mutations exist which affect early development and thus are liable to have large consequences. From these facts it is concluded that true evolution, macroevolution, can be and actually is based on genetic changes which are in effect unlike those of micromutations. By these changes development is recast in a fundamentally new pattern, that of new species.

In this fascinating book the knowledge of a lifetime of zoological experience, genetic experimentation, and all-embracing reading in biology is assembled in order to support its thesis. Goldschmidt knew well that his theory was contrary to the imposing structure of the New Systematics in which the work of two decades by leading investigators on theoretical and experimental population genetics as well as taxonomy had culminated. It was no surprise to him that his theses were rejected by the great majority of Neo-Darwinian evolutionists. Furthermore, his subsidiary assumption that the required macromutations were "systemic mutations," by which he understood basic repatternings of chromosomal macromolecules, was an easy target for criticism. The concept of the unified action of the "chromosome-as-a-whole" did not appear acceptable. Actually, as we shall see, the concept originated in part independent of evolutionary considerations and was not really essential to them. Developmentally it was conceivable that a macro-effect of a genic change could be the result of some micro-changes in the chromosome as well as of the hypothetical macro-changes. Notwithstanding the fact that the evidence and the doctrine of the Neo-Darwinians have withstood the impact of Goldschmidt's erudition and interpretative initiative, *The Material Basis of Evolution* contains elements which may well enter future thoughts on evolution. It is significant that two years after the author's death a reprint of the book, which had been out of print, has made it available again.

Long before the Silliman lectures appeared, the course of world events had once more deeply affected Goldschmidt's personal fate. The Nazi revolution ultimately forced him to leave Germany. For a few years he still worked quietly in the Kaiser Wilhelm Institut, which under Hartmann and von Wettstein remained an island of decency in distressing surroundings. In 1936 he accepted a professorship in the Department of Zoology of the University of California at Berkeley. "This turned out to be one of the most happy events of my life," he wrote later, "crowned by my becoming an American citizen in 1942." The happiness had to be bought at a high price. The Goldschmidt of pre-Hitler days was one of the imposing figures of German intellectual life; the Goldschmidt of the United States between the wars was one of the large number of professors whose prestige was limited. What weighed heavier, his status as a scientist was much in doubt among many of his colleagues. The attacks against their opinions which he had waged for so long and their critical attitude to his own pronouncements did not provide a background for ready acceptance except where the person, not the scientist, was involved. And it did not make matters easier when the embittered, psychologically and physically heartsick man chose the time of his arrival in the country of Morgan's Theory of the Gene to announce in a funereal voice: "The theory of the gene is—dead!"

This statement was the outcome of fully justified doubts, mainly arising from the phenomenon of position effect, the fact that the effect of a gene in many cases is influenced by its chromosomal neighbors. Goldschmidt felt that the concept of materially and physiologically separate genic elements in the chromosome was in need of revision. He rejected the old "beads on a string" analogy and tried to replace it by a concept which considered chromosomes as single super-macromolecules in which changes at any site would affect the action of the whole. In a way Goldschmidt's theory of evolution as affected by his older concepts of the genes as enzymes and of the quantitative nature of differences among alleles had prepared him for his new views. If ordinary mutations would not alter the nature of the gene enzymes but only change their quantity, it seemed difficult to him to believe that such mutations could result in the large-scale repatterning of development which he regarded as necessary for macroevolution. By discarding the idea of the individually separate genes of formal genetics, he abandoned at the same time his own idea of genes as separate enzymes. The concept of the genic nature of the chromosome-as-a-whole permitted both localized minor changes available for microevolution and over-all repatterning of the chromosome necessary for macroevolution.

For many years he found no willing ears, but gradually the climate of opinion changed. His own probings into the gene concept met with those of others, e.g., Barbara McClintock, E. B. Lewis, and later the microbial geneticists. It was a sign of the times that in 1950 he was invited by the Genetics Society of America to give the main address at the Jubilee Meeting in celebration of fifty years of genetics. In the following year he was asked to open the Cold Spring Harbor Symposium on Chromosomes and Genes. When the Ninth International Congress of Genetics was organized in Italy, he was elected President. He took the occasion of the presidential address to discuss "Different Philosophies of Genetics," his own "physiological or dynamic" and the "statistical or static," with one of his main unnamed targets, R. A. Fisher, as Vice-President, smilingly sharing the rostrum.

Already before the work on Lymantria was terminated, Goldschmidt began to experiment with Drosophila. From 1934 on, this organism was his exclusive tool. He began by studying in genetically wild-type individuals the effect of larval and pupal heat treatments on the phenotype of the adult, and discovered that he could produce a great variety of abnormalities resembling those produced under normal conditions by mutant genes. The induced changes were not heritable. He and his students in Berkeley, and workers in many other laboratories, have deepened our understanding of these "phenocopies," his suggestive term which has gained wide usage. Two experimental papers in this area, by Goldschmidt and Piternick, appeared shortly before his death. Other Drosophila studies were concerned with homoeotic mutants, with complex, poorly understood mutation phenomena, with sex determination, and with a plethora of other items. Much of this activity centered on the problems of evolution. The homoeotic mutants served to demonstrate that fundamental morphological changes can be the outcome of genetic changes in single loci or groups of loci. If antennae or wings can become leglike, or if the two forelegs of Drosophila can fuse to form an unpaired structure resembling the labium of the Orthopterans, are these not models for macroevolutionary events? A different area was explored in the laborious inquiries into presumed mass-mutational or "dependent, successive, and conjugated spontaneous mutation" phenomena, which extended over more than a decade. Here was an experimental, even if not fully convincing, attempt to prove the existence of repatternings of chromosomes by means of deficiencies, inversions, and translocations as a basis of evolutionarily significant mutations. Some of these works were documented in voluminous publications so crowded with details that they were read carefully by few only, but at least one, the monograph by Goldschmidt, Hannah, and Piternick on podoptera, a leglike replacement of the wing region, has found an important place in Lerner's considerations of *Genetic Homoeostasis*.

After the Congress in Italy Goldschmidt suffered a severe heart attack. When he was able to travel back to Berkeley he was confined to his home for several months. This gave the seventy-six-year-old man the opportunity to compose *Theoretical Genetics*, a closely reasoned book of 563 pages! It is an amazing work, to use an adjective of which he was fond. "Theoretical Genetics comprises the problems connected with the following questions: (1) What is the nature of the genetic material? (2) How does the genetic material act in controlling specific development? (3) How do the nature and action of the genetic material account for evolution?" The treatment of these topics is highly personal. Goldschmidt repeats, revises, and expands the pronouncements of decades and relishes showing how he had been right before. With an unbelievable grasp of the literature he expounds, criticizes, and warns of erroneous narrowness. Some of the then newest discoveries are dealt with, such as those of Visconti and Delbrück, Pauling and Cori, Watson and Crick. If the book is one-sided, and if large parts of it do not stand the test of time, it is still an achievement which hardly anyone else could have matched. Its merit greatly surpasses its limitations. Many parts remain highly relevant to current work, others are valuable in giving insight into the ever-questioning and constructive thinking of a remarkable scientist.

The range of Goldschmidt's biological activities was much wider than can be described in a limited memoir. A few further aspects will at least be listed. During his sojourn in Ross Harrison's laboratory Goldschmidt did pioneer work in tissue culture, being able to follow spermatogenesis of the Cecropia moth in vitro. Later, in collaboration with Katsuki, he made a comprehensive analysis of a case of inherited gynandromorphism in the silkworm. Other work dealt with the structure of salivary gland chromosomes in Drosophila, maternal effects in inheritance, and heterochromatin. In human genetics Goldschmidt described a fascinating case of multiple race crossings on the Bonin Islands and, in a curiously naïve and highly revealing paper, a pedigree of personality traits in the "M Family," obviously his own. To these activities should be added his critical reviews of unusual types of sex determination, evolutionary genetics, and mimetic polymorphism.

His influence as a teacher and lecturer would deserve a special section. In his university functions in the early Munich period and in Berkeley, he was clear and critical, deeply conscientious, and sympathetic to the students, with highest standards for himself and without dogmatism. He had an important personal influence on the development of genetics in Japan, as Komai testified in a historical survey. Many of his students became outstanding investigators themselves. Before scientific and general audiences on most continents he was a much admired speaker. His textbook of genetics, which appeared in five editions from 1911 to 1928, was a masterpiece, full of new thoughts, broad, and didactically skillful. His several popular science books were translated into many languages. "This my work in helping to educate the masses is very dear to my heart especially in view of the havoc wrought by the journalistic popularizers of science." In Germany he was active and prominent in many committees and societies and his editorial work was unusually comprehensive. Once he wrote an article on "Research and Politics" (1949) which was reprinted in Portuguese, German, and French. A book on his travel in outlying areas of pre-World War II Japan was composed during his second, long stay in that country.

Of the many external distinctions which came Goldschmidt's way the honorary doctoral degrees from Kiel (1928), Madrid (1935), and Berlin (1953) and foreign and honorary memberships in many academies and learned societies may be mentioned. He was nearly seventy years old when he was elected to membership in the National Academy of Sciences in 1947.

The springs from which the stream of Goldschmidt's activities was fed were manifold. Our motives are complex and often unknown to ourselves and to others. Goldschmidt was a naturalist who enjoyed the phenomena of life. Although his main activities were in the laboratory, he spent repeated periods at marine stations, visited the coral reefs of the South Seas, and watched the glowworms in the caves of New Zealand. If his theoretical work often seemed to lack a sufficient basis of factual contributions, a general judgment that he was primarily a theoretician can easily be shown to be incomplete. The many plates and drawings of morphological detail, the hundreds of tables of experimental data are evidence of a biologist's realism. If, in spite of this great amount of specific work, the foundations of his hypotheses frequently were not established solidly, it was due to the breadth of his interests and the width of his generalizations, which no single individual could anchor securely. Positive and negative attributes were curiously mixed in this great scientist. The sweep and penetration of his theoretical insights, the manifoldness of his objects of study, the courage to abandon seemingly established notions, including some of his own, stood side by side with the lack of rigor with which he could use imperfect observations when they fitted into his deductions, or sometimes disregard perfect ones when they failed to do so.

He served Nature but he also tried to dominate her. He relished the mastery over mere facts which his quick mind and his astonishing memory seemed to make possible. As a young man he aspired to prominence in the world of science, and as an adult, when he had attained it he carried it with deep conviction. Yet in spire of his assured bearing it is likely that he felt insecure. His oversensitivity to or, sometimes, his brushing away of criticism, his exaggerated literary reactions to it, his principle not to solicit comments on his writings before they went to press, and particularly the absolute tone of many of his dicta are signs of hidden doubts. It would, however, be very wrong to judge the man only from his technical writings and his often forbidding appearance in the laboratory. He gained lasting friendships based on mutual loyalty of many men such as the zoologists Bütschli, Richard Hertwig, Methodi Popoff, Ross Harrison, the painter Henrik Moor, the conductors Karl Muck and Ernst Kunwald, and his technician-in-chief Michael Aigner. When he had to leave the Kaiser Wilhelm Institut sadness and tears were on the faces of most, from the cleaning lady to his colleagues. At the end of his last lecture in Berkeley before retirement, his students in a general class surprised him with the gift of an ancient Chinese figure, an unheard-of act in the large, necessarily rather impersonal university. In his nontechnical publications there is evidence of fairness, of devotion, and of humor, witness his *Ascaris eine Einführung in die Wissenschaft vom Leben* (1922), which has been enjoyed by tens of thousands of readers in its original version and in translations into five different languages. Witness also the touching letter of the seventy-six-year-old Goldschmidt to the ninety-four-year-old Karl Jordan, the lively *Portraits from Memory* (1956) in which he sketches his recollections of zoologists and the atmosphere of German universities in their classical period, and his sensitive obituaries and biographical memoirs on Bütschli, Popoff, Kofoid, Federley, and others. He was warm and solicitous to the young people around him. He could keep separate his disagreement in matters of science from appreciation of the opponent as an individual. He was a relaxed and in a deep sense charming host in his and Mrs. Goldschmidt's treasure-filled home in which their guests experienced a unity of beauty and humanity. His interests were not exhausted by intense occupation with his work. From youth to late middle age he was active and excelled in various sports. He played the violin and viola and once wrote: "Music is still to me what religion is to the believer." He became an expert in objects of art, particularly those of the Far East. With the pleasure of an amateur and connoisseur he collected all through his life select pieces of pottery, statues, prints, and paintings.

The first sentence of Goldschmidt's autobiography reads as follows: "I come from an old German-Jewish family." Many times in the book he returns to the experiences of anti-Semitism to which he had been subjected, as a child in the streets of Frankfurt, at the hands of teachers in school, and in the withholding of an officer's commission at the end of his military service, events which preceded by decades the disaster which forced him to leave Hitler's Germany. The facts and consequences of his background had an important part in the shaping of his personality.

To his associates the emphasis on his Jewish origin in the autobiography was a surprise. Only very rarely had he spoken of it. There is no mention of the fact in his outline for the National Academy. His mode of life was anything but parochial. His wife came from Protestant ancestors and his children grew up in an atmosphere which led them in their turn to choose non-Jewish spouses. His earlier identification with German patriotism had been strong. Nobody could forget his description of that November night in 1918 when the factory whistles in Georgia carried to the German prisoners' camp the announcement of the end of the war and sobbing rose from the barracks where the men mourned the defeat of the fatherland.

The part which an individual's work plays in the development of his science varies greatly. There are the unique persons, whose often single achievements are the towering, long-persisting landmarks on the road of history. There are the many modest workers who contribute their few bricks or their bits of mortar to the unlimited edifice of knowledge. There are those whose discoveries and insights remain unnoticed in their lifetime and even—who knows—forever. There are those who march in the forefront of their contemporaries, leading the way in making new finds, warning of barren deserts, promising lands of fulfillment. And there are those who erect the scaffold on which others reach the spaces where they can fit their parts into the yet imaginary construction. The scaffold is abandoned some day and it disappears while the building stands. Goldschmidt partook of the traits of many of these men: contributor of permanent parts, some very large; preceptor and critic of his era; designer of frameworks for the future.

SEWALL WRIGHT, THE SCIENTIST AND THE MAN

*JAMES F. CROW**

Before I met Sewall Wright I regarded him with the adulation usually reserved for deities and rock groups. Now that I know him well the admiration persists undiminished, but the awe has been replaced by simple respect and affection. As a graduate student it had been my ambition to be a postdoctoral fellow with Wright. World War II precluded any possibility of this, but the opportunity was only postponed, for I have enjoyed the privilege of having him as a colleague since 1954.

Sewall Wright was born on December 21, 1889. He was an early bloomer. Before ever starting school he wrote a little pamphlet, in block letters, on subjects of interest to him. The title was "The Wonders of Nature." Some of the subjects were constellations, squashes, ants, dinosaurs, bees, marmosets, and a wren that could not be discouraged from trying to nest in the mailbox. One of them is entitled "The Gizzard of a Fowl," and I reproduce it here.

Have you ever examined the gizzard of a fowl? The gizzard of a fowl is a deep red colar with blu at the top. First on the outside is a very thick muscle. Under this is a white and fleecy layer. Holding very tight to the other. I expect you know that chickens eat sand. The next two layers are rough and rumply. These layers hold the sand. They grind the food. One night when we had company we had chicken-pie. Our Aunt Polly cut open the gizzard, and in it we found a lot of grain, and some corn.

Not only could he read before starting school, but he could do arithmetic. On his first day in class he volunteered the information that he knew how to extract cube roots. The teacher was so impressed that she

This article is based on a talk given at the University of Chicago at a symposium in honor of Sewall Wright's ninetieth birthday, December 1979. It is contribution 2487 from the Laboratory of Genetics, University of Wisconsin. The author is grateful to Dr. Wright for information provided in numerous conversations over many years and and to Dr. William Provine of the Department of History, Cornell University, for a draft of an early chapter of his forthcoming biography of Sewall Wright, the source of some of the material in this article.

*Department of Genetics, University of Wisconsin—Madison, Madison, Wisconsin 53706.

took him to the eighth-grade room to demonstrate this skill. The main thing that he seems to have achieved by this demonstration was instant unpopularity with the other students. From that day on he volunteered as little as possible in the classroom.

Wright was also adept physically. He did gymnastics and was a football quarterback in his high school days. This physical skill persisted. I recall his daring and agility in climbing a rock formation in his seventies. He still walks several miles every day.

If Wright was an early bloomer, he has also been a late bloomer. How many people have published their first full-length book at nearly 80? Once he got into the habit it seems he could not stop, and he wrote three more within the decade. His four-volume set, *Evolution and the Genetics of Populations,* is unique—already a classic [1].

He has retired two times. The first was from the University of Chicago in 1954 and the second from the University of Wisconsin in 1960. But the retirements were only time markers. His work continued unabated and it still goes on. His most recent paper was published in the September 1980 issue of *Evolution* [2].

Wright has had essentially all the honors for which population geneticists are ever considered: the Elliot Medal and the Kimball Award of the National Academy of Sciences, the Lewis Prize of the American Philosophical Society, the Weldon Memorial Medal from Oxford University, and the National Medal of Science from the United States Government. He has been president of the Genetics Society of America, the American Society of Zoologists, the Society of Naturalists, the Society for the Study of Evolution, and the 10th International Congress of Genetics. He has received nine honorary degrees. He is a member of the National Academy of Sciences, the American Academy of Arts and Sciences, and the American Philosophical Society, a fellow of the American Statistical Association, a foreign member of the Royal Society of London, an honorary member of the Royal Society of Edinburgh, and a foreign member of the Royal Danish Academy of Science and Letters. He is the only geneticist to be elected a fellow of the Econometric Society. His most recent honor is the Darwin Medal, given in December 1980 by the Royal Society of London for work "in the field in which Charles Darwin himself labored."

Sewall Wright's father, Philip G. Wright, taught for many years at the tiny Universalist Lombard College in Galesburg, Illinois. He was an economist by training, but among other things he taught mathematics, astronomy, surveying, and English composition. He also was director of the gymnasium. He had a printing press and produced the college bulletins. Later he wrote several books including one, *The Tariff on Animal and Vegetable Oils,* to which Sewall contributed an appendix.

Sewall Wright had two gifted brothers. Quincy became a distinguished authority on international law. Theodore was chief engineer at Curtis-Wright, a Civil Aviation commissioner, and vice-president for research and acting president at Cornell University. Quincy and Sewall often operated their father's printing press and, among other things, were the first to publish the poetry of Carl Sandburg, then a writing student of P. G. Wright at Lombard College.

Sewall Wright attended Lombard College and graduated in 1911. During the summer of 1909 he obtained work as a member of a surveying party for a future railroad construction in South Dakota. He was a good worker and his skill in some of the calculations made him so valuable that he was invited to continue. This he did, missing a year of college. It was a rich experience in the Old West tradition, with hardships and adventures. He still remembers several words of the Sioux language which he learned from Indian contacts during this period. These were the same local tribes that, 33 years before, had destroyed General Custer and his troops at Little Big Horn and, 19 years before, had been massacred at Wounded Knee. Wright's surveying was cut short by a severe attack of pleurisy. During his illness, while spending the time in a railroad car, he read a book on mathematics. By the end of the summer he had recovered and returned to school. Because of the lung infection he had trouble getting life insurance, something that he found increasingly amusing as he entered his tenth decade.

Wright never made specific use of the techniques of quaternions. Yet, it is likely that his later evolutionary thinking owes something to this reading during the summer of 1910. At that time he was especially interested in the physical applications of quaternions. Much later, in discussing his concept of a multidimensional adaptive surface in which the abscissas represent gene frequencies and the ordinate represents fitness, he used a potential field analogy to describe the tendency for a point representing a population to move upward in the general direction of the nearest peak. It is interesting that J. B. S. Haldane also studied quaternions during an enforced leisure. While on a hospital ship following his injury from a bomb explosion in World War I, Haldane read Helland and Tait's *Introduction to Quaternions*.

Wright's interest in biology was kindled by Wilhelmina Key, a teacher at Lombard who had studied at Wisconsin and had a Chicago Ph.D. Later she moved to Cold Spring Harbor and became an active researcher in the Eugenics Records Office. Wright's first understanding of genetics came in her course from reading R. C. Punnett's account of Mendelism in the great eleventh edition of the *Encyclopaedia Britannica*, to which Wright in turn contributed in later editions. After graduating from Lombard College, he spent a graduate year at the University of Illinois, where he had the good fortune to meet William Castle, who was there as a visitor and who agreed to take him as an assistant at Harvard.

After receiving his doctorate from Harvard in 1915 he went to the United States Department of Agriculture in Washington, D.C., and Beltsville, Maryland, as senior animal husbandman. He held this position for about 10 years, leaving for the University of Chicago in 1926. The retirement age at Chicago was 65, while at Wisconsin it was 70; so in 1954 he moved from Chicago to Madison and has been there since. This is the best bargain the University of Wisconsin has ever had, for Wright was paid only a small supplement to his Chicago retirement annuity. In return, Wisconsin received 25 years of work, and who knows how many more there will be? Now 91, he walks several miles to work and remains full of intellectual vigor and zest for new knowledge and experiences. He travels frequently to give talks and consult; he recently went to England to receive the Darwin Medal.

Wright's wife, Louise Lane Williams, was also a geneticist. She taught for 3 years at Smith College and taught the college's first course in genetics. Her charm and cheerful disposition persisted throughout her lifetime.

Wright's Scientific Work

Stimulated by the opportunity to honor Sewall Wright, I have been looking through the collection of his scientific writings, especially his early work—and what a feast it has been! The breadth of interest, the care and attention to experimental detail, and the quality of the mathematical analyses were to be expected. Yet I was frequently surprised to find how early he had said things that are now part of the conventional wisdom of genetics.

Wright's first published paper appeared in 1912 [3]. It was a detailed morphological study of a fish parasite, the trematode *Microphallus opacus*. This was work that he did during the year at the University of Illinois. His second paper, in 1914, was genetical [4]. It was a suggestion that a distinction could be made between allo- and autopolyploidy by the frequency of homozygosis for recessive genes.

In the next few years three of Wright's four main areas of concentration were clearly apparent: correlation and variance analysis, animal breeding, and mammalian physiological genetics. His interest in evolutionary problems came soon after. It is convenient to discuss his work under these headings.

STATISTICS

Wright's first statistical paper appeared in 1917 [5]. Here he pointed out an error in the use of the probable error by Raymond Pearl to test Mendelian ratios.

Also in 1917 Wright published a paper on the average correlation within subgroups of a population [6]. The title sounds commonplace enough, but in his analysis of guinea pig weights he used what would now be called analysis of variance and covariance to subdivide these into within- and between-strain components. This was long before such techniques were developed by Fisher for general statistical use.

Although he had used the idea earlier, Wright's first path diagram appeared in 1920 [7]. Here he showed how to separate the variance of coat color into genetic and environmental components. In a randomly mating population the genetic fraction was about 42 percent, and this disappeared with inbreeding, as expected. The paper also included a transformation to linearize the data on the amount of white spotting. This is now called the probit transformation and was described in more detail in 1926 [8].

His blockbuster, "Correlation and Causation," was published in 1921 [9]. This is the first full treatment of the method of path analysis. One of the simplest and most useful of Wright's devices is to diagram causal sequences so that paths of direct causation among measured or unmeasured variables are indicated by single-headed arrows and correlations between anterior, unanalyzed causes are represented by double-headed arrows. The diagram gives a clear, graphic picture of the assumed causal connections. Each causal step is associated with a "path coefficient," a partial regression coefficient standardized by being measured in units of the standard deviation. From such a diagram Wright found simple rules by which one can easily write all the appropriate equations. The path coefficients then measure the relative importance of the different causal paths. A virtue of the method is that it makes immediately evident whether there are a sufficient number of relationships among the unknowns to permit a solution.

Although Wright continued to use the method and make refinements, its essence is in the 1921 paper. The method was applied especially to problems of livestock breeding. It became the standard procedure for analyzing the effects of inbreeding, selection, repeatability, assortative mating, and other aspects of genetic improvement of livestock. Its wide popularity owes much to the influence of Jay L. Lush. Lush worked with Wright, wrote a very popular textbook, and spread the gospel throughout the animal breeding world by his international travels and through his large, cosmopolitan group of students.

In the first few years after developing the method, Wright applied it to a great variety of problems, including correlations between prey and predator, growth and transpiration in plants, respiratory physiology in humans, and corn and hog correlations. The last is a monumental work based on production and prices of corn and hogs during the relatively stable period from the end of the Civil War to the beginning of World

War I. There were 510 correlations to take into account; the calculation job alone, long before the day of modern computers, was enormous. Wright found, remarkably, that about 80 percent of the variance in hog production and prices could be accounted for by fluctuations in the corn crop and the various intercorrelations and time lags. It is interesting that he had trouble getting the paper published as a Bulletin of the U.S. Department of Agriculture; it was assumed that an animal husbandman should not be writing about economic questions. It took a change of administration and the intercession of Henry Wallace, son of the then Secretary of Agriculture, to get the bulletin published in 1925 [10].

At that time Wright did all his own calculations. He still does. He has consistently refused computing help from others. The reason is clear: he prefers to be close to the data, to the numbers themselves, and to the intermediate values in the calculations. Even in this day of powerful computing equipment he prefers to work with a hand-held or desk-top calculator. Wright's papers almost never have mathematical errors. This happens despite the fact that in conversation, seminars, and lectures he makes algebraic mistakes as others do. The reason that these never creep into his papers is his great reliance on numerical calculations; everything is checked and rechecked.

Another problem studied by Wright was the relative influence of heredity and environment in determining IQ [11]. This pioneering study, published in 1931, used data from Barbara Burks on the various correlations among children reared by their own parents and by foster parents. Wright concluded that heredity played a large role, but he was careful to point out the uncertainties about both the data and the assumptions. It is curious that the famous study of separated identical twins by Wright's Chicago colleagues, Newman, Holzinger, and Freeman, was so poorly analyzed from the genetic standpoint. Although Newman had an office on the same floor as Wright, he and his colleagues made no use of Wright's powerful methods of analysis or his advice.

One of the mysteries of the history of science from 1920 to 1960 is the virtual absence of any appreciable use of path analysis, except by Wright himself and by students of animal breeding. Although Wright had illustrated many diverse problems to which the method was applicable, none of these leads was followed. The method seems made to order for disentangling causal relations and assigning the relative importance to different causal paths. Yet scientists in general and biologists in particular made essentially no use of it. One reason is that most biologists of that period did not emphasize quantitative methods. A second may be that the experimental paradigm was the carefully controlled experiment in which only one or two factors were manipulated. There was no absence of correlation analysis; but other methods became more popular, perhaps because Wright gave no recipe for assigning standard errors to his coefficients. Furthermore, his method did not lend itself to sheer mechanical application as does, for example, the very neat layout devised by R. A. Fisher for analysis of variance and factorial analysis. Each case requires a carefully thought out path diagram. Yet, I suspect that none of these is the really important reason. It is, I think, simply that Wright made no effort to publicize or sell his method or to urge others to use it. Although he wrote reviews, these were all highly technical. He wrote no textbook with cookbook instructions. The recent popularity of the method in the social sciences has come from its use and advocacy by influential people and from clear, elementary expositions.

The method was ignored by econometricians, including Wright's Chicago contemporary, Henry Schultz. The interest by social scientists began about 1960. According to my colleague, Robert Hauser, the first to appreciate the power of Wright's methods were Tad Blalock, Otis Dudley Duncan, and William Hodge. The last two had been students of Wright's personal acquaintance, William Ogburn; but he, too, did not use Wright's methods. Arthur Goldberger, a Wisconsin econometrician who has done much to promote Wright's methods in his field, learned of them from Duncan.

In the past 15 years it has become de rigueur to use path analysis in many of the social sciences—econometrics, psychology, sociology, political science, and education. From 1969 to 1978 the annual number of citations to Wright's work in the Social Science Citation Index grew from 28 to 114, according to Hauser. Now, 60 years after its invention, path analysis is the "in" thing. More than once I have heard stories of Wisconsin social scientists who are surprised to discover that Wright is not only still active but moreover is at their own university.

In his early formulations Wright made no distinction between sample estimates and population parameters, nor was he greatly concerned with determining the precision of estimates. He did not give any systematic method for dealing with situations where the number of equations exceeds the number of parameters to be estimated. With modern statistical developments and high-speed computers Wright's methods have been made more rigorous and powerful. Yet his 1925 paper on corn and hog correlations is a remarkably complete analysis of voluminous and complex data, half a century ahead of its time.

MAMMALIAN GENETICS

Wright's interest in genetics centered on mammals. At Harvard he was associated with William Castle's studies on rats, guinea pigs, and rabbits. His work for the U.S. Department of Agriculture led him to give consid-

erable attention to farm livestock. His own experimental work emphasized guinea pigs, and this work continued throughout his Washington
and Chicago periods.

In 1917 and 1918 he published a series of papers, very sophisticated
for their time, on comparative coat color inheritance in mouse, rat,
guinea pig, rabbit, cattle, horse, swine, dog, cat, and human [12]. The
papers are remarkable for pointing out gene homologies in the different
species and for relating the results to then current knowledge of the
chemistry of pigmentation.

As early as 1918 he used path analysis to study the inheritance of size
factors and the correlation of body parts, using MacDowell's data from
rabbit strains of different sizes [13]. For example, the length of the
humerus was analyzed into components of general size, factors specific
to limbs, factors specific to the forelimb, factors specific to the homologous upper-limb parts, and factors affecting each part independently.
From this he could determine the relative importance of each component in determining the total variance.

Wright's early work on guinea pig teratology is still widely quoted. He
also studied such diverse topics as the inheritance of susceptibility to
tuberculosis, with Paul A. Lewis, and the genetic control of histocompatibility, with Leo Loeb. Here is an example from 1927 of how well he
understood C. C. Little's rules for graft rejection: "It is probably not so
much the presence of a single dose of the genes which is required as far
as the absence of reaction against a transplant is concerned as the lack of
a strange gene in the donor; while the presence of a strange gene in the
host is without significance" [14].

Wright's greatest interest was in coat color inheritance and, as I shall
discuss later, his studies on guinea pig pigments were greatly influential
in his views of evolution. By systematic matings he produced many combinations among the various color genes and was especially impressed by
the complexity of the interactions. He was characteristically thorough
and inventive in analyzing these interactions. Knowing his early statistical insights one is not surprised at the clever and original mathematical
procedures, but these analyses also made use of the most up-to-date
knowledge of enzyme activity then available. I still remember vividly the
impression that his 1941 paper on physiological genetics made on me
[15]. He explained dominance and epistasis elegantly and quantitatively
on the basis of enzyme kinetics and an assumed flux equilibrium. He also
devised a kinetic interpretation of H. J. Muller's concepts of hypomor-
phic and antimorphic genes based on interallelic interactions. Wright
even predicted the discovery of a new type, a mixomorph; this is a gene
whose product is high in substrate-combining ability but poor in conversion to the product. Such an allele was later found in Drosophila, the eye
color mutant *pearl*.

The history of genetics took a sudden turn in 1941 with the work of Beadle and Tatum, the beginning of wide interest in biochemical genetics and the forerunner of molecular genetics. Wright had planned to expand his 1941 review article into a book on physiological genetics, but he decided that the newer chemical approaches using microorganisms would so revolutionize the field as to make his type of analysis no longer the best approach, and gave up his book plans. It seems to me that his type of analysis is by no means obselete, and it is time to look again at the kinds of gene interactions that he studied, using the powerful techniques that molecular and cell biology provide.

ANIMAL BREEDING

Wright inherited a guinea pig colony when he first went to work for the United States Department of Agriculture. For many generations some of the population had been divided into highly inbred lines, and a full set of records was available. Wright continued the work during the difficult years of World War I when guinea pig food was hard to obtain. Furthermore the animals were housed in poor quarters and the hot Washington area summers were especially hard on them. It is remarkable that, with such poor control over environmental conditions, he was able to get such interpretable results. This also, I think, gave his work a greater pertinence to the actual problems encountered in livestock breeding. In 1920 he wrote a 67-page bulletin on animal breeding that reads remarkably well today [16]. In this he suggested combining inbred lines to produce vigorous hybrids while permitting effective genetic control over weakly heritable traits. The bulletin sold for 15 cents.

Wright's very detailed analysis of the effects of inbreeding and cross-breeding in guinea pigs could be used as a textbook today [17]. He clearly saw that the usual (but not invariable) decline in vigor with inbreeding, the fixing of different traits in different lines, the immediate recovery of vigor when inbred lines were crossed, and the quantitatively predictable decline of vigor with renewed inbreeding are all direct consequences of Mendelian inheritance and dominance. I might mention that Wright himself is a demonstration that inbreeding is not always deleterious; he is the product of a first-cousin marriage.

The year 1921 saw the publication of another of Wright's classics entitled "Systems of Matings" [18]. He analyzed mathematically a wide variety of mating systems involving inbreeding, assortative mating, and selection. It was a vivid illustration of the power of path analysis. This five-part series has been reprinted several times and reading it has become part of the initiation rites for future animal breeders.

In 1922 he proposed that the quantity which he had designated by the letter F, the correlation between the genetic values of alleles in uniting gametes, could be used as a general measure of inbreeding [19]. Earlier Raymond Pearl had used a function of the ratio of the actual number of ancestors to the maximum number as a measure of inbreeding. Wright's measure was clearly superior and has been standard ever since it was proposed. It is regularly included in elementary genetics courses.

Later Wright showed how his F-statistics could be generalized to measure various components of population structure. He first did this analyzing the history of Shorthorn cattle, for which there were especially complete pedigree records. He was able to show that the major cause of increased homozygosity was the small effective size of the population rather than consanguineous matings. He was convinced that breed improvement came not only from selection, but also from more or less random changes leading to occasional superior herds, from which individuals were exported to upgrade other herds. These observations played a large role in the later development of what he called the "shifting balance" theory of evolution.

In the late 1940s the Wrights spent a year in Edinburgh. One result is a paper published in 1951 in which Wright presented a thorough and systematic discussion of population structure, widely cited since [20]. The methods first used to analyze the history of livestock breeds were extended and generalized to analyze the evolution of natural populations.

During his Edinburgh stay he wrote a succinct and useful summary of the genetics of quantitative variability, to be part of a book on *Quantitative Inheritance,* edited by Reeve and Waddington of the Edinburgh Institute for Animal Genetics [21]. The manuscript was lost on the trip from Edinburgh to London and he had to rewrite it from scratch. It was a long and frustrating task, but Wright characteristically undertook it with good humor. He said he thought the manuscript was the better for having been rewritten.

Over the years Wright has given numerous talks to animal breeding groups and has written several articles on this subject. His influence permeates the whole field of quantitative livestock improvement and, to a lesser but important extent, plant breeding. He and R. A. Fisher laid the biometric foundations for the entire science. Wright's influence has been more indirect than direct, having been spread through the community of livestock breeders and experiment station workers through his intellectual descendants, especially Lush and A. B. Chapman and their students.

EVOLUTION

Wright stopped his guinea pig colony when he moved from Chicago to Wisconsin. He had done most of the work himself, keeping records and making matings. This, along with teaching and a heavy load of correspondence and manuscript review, had kept him very busy. Since his move to Madison he has been able to spend more of his time on his deepest interest, the neo-Darwinian theory of evolution. A lifetime of thought, along with a thorough review and interpretation of the literature, is to be found in his four-volume set published by the University of Chicago Press [1]. An enormous amount of work went into these books, for Wright not only reviewed the voluminous experimental and observational literature but interpreted it and fitted it into his general framework. Often this involved extensive recalculations from the raw data.

Although he is best known for his theoretical work in evolution, Wright has also made substantial contributions to experimental population genetics. In the 1930s A. H. Sturtevant and Th. Dobzhansky began to think about using Drosophila to study the genetics of natural populations. This was the beginning of a large number of papers, mostly by Dobzhansky and his students, on the analysis of genetic variation in *Drosophila* species, especially *D. pseudoobscura*. Wright collaborated with Dobzhansky for several years. The first paper, in 1941, was a landmark [22]. It showed how the analysis of concealed recessive lethal genes in natural populations could give some information about the dynamics of evolutionary processes. Dobzhansky always hoped for more than such a study was capable of revealing, and Wright had to point out several times that the number of measured variables and equations were not sufficient to estimate all the relevant parameters. Eventually Dobzhansky's requests for more analyses than Wright could find time for led to the termination of the collaboration.

From the beginning of his theoretical studies, Wright was especially interested in the effects of random gene frequency fluctuations in a finite population. R. A. Fisher had earlier supplied an ingenious method for solving the problem, although he made an algebraic error that Wright corrected [23]. Fisher believed that such chance processes were unimportant in the larger evolutionary picture, contributing only random noise to the otherwise steady process of selective improvement of fitness through natural selection. To Wright, the trial and error process of trying out different gene combinations among numerous local populations was the essence of evolutionary progress.

In the 1920s Wright began to be interested in finding mathematical expressions for gene frequency change under the influence of mutation, selection, migration, and random processes. He had virtually completed his manuscript by 1926 when he moved to Chicago, but it was not published until 1931. This paper, "Evolution in Mendelian Populations," is one of the most cited papers (more frequently cited than read, I fear) in the evolutionary literature [23]. Once again, it was not Wright himself who popularized his work. Rather it was Dobzhansky's influential book, *Genetics and the Origin of Species,* which brought Wright's views to the attention of biologists.

Wright continued to generalize his equations and to find more convincing derivations [23–25]. A large step forward was made when he received a reprint from the great Russian mathematican, A. Kolmogorov, who had formulated this kind of problem very generally in the form of two partial differential equations, now called the "forward" and "backward" equations. Wright quickly incorporated Kolmogorov's formulation into his writings [26]. This was the source of much more sophisticated mathematical treatment of the problem by modern workers, of whom G. Malécot in France, M. Kimura in Japan, and S. Karlin and T. Nagylaki in the United States have been leaders.

In reading Wright's papers one is especially impressed with what he was able to do with elementary methods. He had no advanced mathematics training. What he knew was self-taught. For the most part, though, Wright has not relied on the mathematical literature but has simply worked out his own solutions by a combination of inventiveness and persistence. The fact that so much of what he has done is original accounts in part for the difficulty that mathematicians have in reading his papers. For example, he devised his own unorthodox treatment of partial derivatives.

I have been struck by the fact that Wright's papers have become easier to understand as he developed successively more general derivations [23–26]. I think it was Poincaré who said that the most general formulations were often easier to comprehend than special cases.

Wright's unique contribution to the theory of evolution by natural selection is his shifting balance theory. As I mentioned earlier, this is a direct outgrowth of his studies of the breed structure of domestic livestock and the complex interactions often observed among the genes affecting coat color in guinea pigs. He views evolution as a process of moving from one harmonious combination of gene frequencies to another that is even more harmonious and better adapted to the environment. The difficulty is that it is often impossible to move from one superior gene combination to a better one without passing through gene frequencies that produce a preponderance of combinations inferior to either. In Wright's familiar metaphor, to cross the valley between two fitness peaks requires some process other than deterministic natural selection.

He visualized this as happening, at least partially, by the influence of random factors. If one local population happens to develop a superior combination of genes, then this population will grow faster than its neighbors, send out migrants, and eventually upgrade the whole species. In this way, by a combination of random gene frequency drift in local populations, selection between these local populations, and one-way migration from the best adapted, evolution can proceed from one harmonious gene combination to another that is even better.

In Wright's view, random gene fluctuations, whether due to small population size or variations in selection intensities, are not just random noise that impedes the process of natural selection but are the basis for evolutionary creativity. Such a process will be facilitated if the population is geographically structured.

Wright has consistently held this view since the 1920s and has reiterated it many times. Although he has made minor modifications in his views, his consistency and persistence in adhering to the view are indeed remarkable, as his 1931 statement [23] and his most recent article [2] illustrate. This latest paper is not only a clear statement of his views but a spirited defense of them against those who have misinterpreted him. He has also recently described how his studies with guinea pigs and breed histories of livestock led to his shifting balance theory of evolution [27].

Philosophy

Very few geneticists have written seriously about philosophy. Wright is an exception. He discovered that his Chicago colleague, Charles Hartshorne, shared a similar view about the philosophy of organism, and they became life-long friends.

The philosophy is in the tradition of Leibnitz. Wright's view is that there is no material basis for a mysterious "emergence" of new properties as systems become more complex. This being the case, one is forced to assume that such properties as consciousness must necessarily reside in the most elementary particles. He has developed this view in several papers; a good example is his presidential address to the American Society of Naturalists, published in 1953 under the title "Gene and Organism" [28].

Wright's philosophical view has attracted some attention among philosophers, and he has several times been invited to participate in national and international conferences. It has not received any significant attention from biologists. Wright found it amusing that his views of the biological organism got more notice from philosophers than from his fellow biologists. His teacher, Wilhelmina Key, wrote him after receiving a reprint of his presidential address that at last he had written a paper that she could understand.

A Few Reminiscences

Some aspects of Wright's character, as seen by one who knows him well, are best illustrated by anecdotes.

One of my early experiences with Wright occurred when he was a reviewer of my first paper in *Genetics*. It was a thorough, sympathetic, and helpful review. He noted that one result, which I had stated to be approximately correct, was in fact exactly correct. He showed me a much simpler way to derive this result, which I incorporated into the paper, improving it substantially. I have long felt indebted to Wright for correcting this error, although I suppose I can claim priority for the concept of being approximately wrong.

Improving other people's papers has for many years been a regular occupation for Wright. I recall another manuscript which he reviewed. As usual, he recalculated the data and in this case reached a conclusion quite the opposite from that of the author. The author changed the conclusions accordingly, and the paper was published in this much-improved form. Thoughout his life Wright has been very generous with his time—generous to a fault. He has spent countless hours helping others with their analyses and as a manuscript reviewer for several journals. He never simply offers casual criticisms or suggestions for improvement. Rather, he recalculates, rederives, and reanalyzes. The paper is usually greatly improved, but I cannot help believing that this has not been the best use of his time; his own work, necessarily put aside, was much more important.

In the early 1950s Newton Morton, then a student of mine, was involved with me in a series of experimental studies of Wright's concept of effective population number. We needed some mathematical help and arranged to travel to Chicago to consult him. We arrived at the appointed hour with a carefully prepared list of about a dozen questions. To each question, Wright said that he did not know the answer. That was all there was to the discussion, and it was all over in less than 15 minutes. We spent a short time looking at some of Wright's guinea pig pelts and then went home. On the train back Morton, who was not inclined then or now to understate his opinion, announced that if Wright was the greatest population geneticist the subject was in bad shape. A week later we understood. The mail brought a 14-page, hand-written letter with *everything*. All our questions were answered and the problems were completely worked out. We discovered, as others have, that Wright does not like to speculate or give snap judgments about complicated problems. He wants to work out the solutions for himself; then he will talk as much as you please.

Like many others who spend much time thinking about abstractions, Wright is absentminded. It is not a recent trait; he has been so as far back as his friends can remember. Anecdotes abound. I shall repeat only the most famous one, widely circulated among geneticists. It is said that Wright was once lecturing on guinea pig coat colors and was using live animals for illustration. He was writing a genotype on the blackboard with one hand while holding the squirming guinea pig in the other. Having made a mistake he absentmindedly used the guinea pig as an eraser. Unfortunately, the best evidence is that the story is not true. But it could be. Anyone who has seen Wright in action, talking and writing at breakneck speed and erasing the board with any convenient object at hand would find the story believable.

I should like to close with one more anecdote, one that is typical of Wright's selflessness, generosity, puritan conscience, and work ethic. While he was writing his four-volume set of books that I have now referred to several times, he received a stipend from the National Science Foundation. The amount was small and it did not change over several years. The NSF was very happy to make a cost-of-living adjustment. But, when I happily told Sewall about this, he refused the offer. He told me that his productivity was declining at a rate which he calculated to be about the same as the decrease in the value of the dollar and that he did not deserve an increase. He never accepted it.

REFERENCES

Sewall Wright's complete list of publications is much too large for inclusion here. The following titles are those specifically referred to in the text. Almost all of Wright's papers are by him only; when there is a coauthor, this is mentioned.

1. *Evolution and the Genetics of Populations*, vol. 1, *Genetic and Biometric Foundations*, vol. 2, *The Theory of Gene Frequencies*, vol. 3, *Experimental Results and Evolutionary Deductions*, vol. 4, *Variability Within and Among Natural Populations*. Chicago: Univ. Chicago Press, 1968, 1969, 1977, 1978.
2. Genic and organismic selection. *Evolution* 34:825–843, 1980.
3. Notes on the anatomy of the trematode, *Microphallus opacus*. *Trans. Am. Microsc. Soc.* 31:167–175, 1912.
4. Duplicate genes. *Am. Nat.* 48:638–639, 1914.
5. On the probable error of Mendelian class frequencies. *Am. Nat.* 51:373–375, 1917.
6. The average correlation within subgroups of a population. *J. Washington Acad. Sci.* 7:532–535, 1917.
7. The relative importance of heredity and environment in determining the piebald pattern of guinea-pigs. *Proc. Natl. Acad. Sci. USA* 6:320–332, 1920.
8. A frequency curve adapted to variation in percentage occurrence. *J. Am. Stat. Assoc.* 21:162–178, 1926.
9. Correlation and causation. *J. Agric. Res.* 20:557–585, 1921.
10. Corn and hog correlations. *U.S.D.A. Bulletin* 1300:1–60, 1925.
11. Statistical methods in biology. *J. Am. Stat. Assoc.* 26, suppl.:155–163, 1930.
12. Color inheritance in mammals. *J. Hered.* 8:224–235, 373–378, 426–430, 473–475, 476–480, 521–527, 561–564, 1917; ibid. 9:33–38, 87–90, 139–144, 227–240, 1918.
13. On the nature of size factors. *Genetics* 3:367–374, 1918.
14. Transplantation and individuality differentials in inbred families of guinea-pigs. *Am. J. Pathal.* 3:251–283, 1927 (with L. LOEB).
15. The physiology of the gene. *Physiol. Rev.* 21:487–527, 1941.

16. Principles of livestock breeding. *U.S.D.A. Bulletin* 905:1–67, 1920.
17. The effects of inbreeding and crossbreeding on guinea pigs. *U.S.D.A. Bulletin* 1090:1–63, 1922; ibid. 1121:1–60, 1922.
18. Systems of mating. *Genetics* 6:111–178, 1921.
19. Coefficients of inbreeding and relationship. *Am. Nat.* 56:330–338, 1922.
20. The genetical structure of populations. *Ann. Eugenics* 15:323–354, 1951.
21. The genetics of quantitative variability. In *Quantitative Inheritance,* edited by E. C. R. REEVE and C. H. WADDINGTON. London: Agric. Res. Council, 1952.
22. Genetics of natural populations. V. Relations between mutation rate and accumulation of lethals in a population of *Drosophila pseudoobscura. Genetics* 26:23–51, 1941 (with TH. DOBZHANSKY).
23. Evolution in Mendelian populations. *Genetics* 16:97–159, 1931.
24. Statistical genetics and evolution. *Bull. Am. Math. Soc.* 48:223–246, 1942.
25. Adaption and selection. In *Genetics, Paleontology, and Evolution,* edited by G. L. JEPSON, G. G. SIMPSON, and E. MAYR. Princeton, N.J.: Princeton Univ. Press, 1949.
26. The differential equation of the distribution of gene frequencies. *Proc. Natl. Acad. Sci. USA* 31:382–389, 1945.
27. The relation of livestock breeding to theories of evolution. *J. Animal Sci.* 46:1192–1200, 1977.
28. Gene and organism. *Am. Nat.* 87:5–18, 1953.

OSWALD THEODORE AVERY AND DNA

ALVIN F. COBURN, M.D.[*]

To one who had no scientific association with Avery's work it seems appropriate to place the following information on the record. I am motivated to make this report because repeatedly I note that many persons in high echelons of science are unaware that Oswald T. Avery envisaged the implications of the discovery of his "transforming factor." As early as 1943 Avery did indeed understand the significance of DNA in microbial genetics, the discovery of which culminated his extraordinarily creative life as a member of the Rockefeller Institute.

Avery practiced meticulously the sermon that he often preached to his assistants and to the many younger colleagues who came to him for guidance: "Apply your brakes when tempted to blow your own horn." His rigorous self-discipline, along with constant modesty and a deep humility —in the noblest sense of the word—made it impossible for Avery to go far beyond the "facts" in his published work. His high regard for the printed word deterred him from theorizing in print—only the "facts" were admissible—and his own wisdom certainly prevented him from pointing out the great significance of his discoveries, perhaps even to the many young men who had the great good fortune to work with "the Fess" (short for professor), as he was called.

Prior to 1940 O. T. Avery was only a name—a name which belonged with Theobald Smith, F. Gowland Hopkins, and Marie Curie, persons who had created new disciplines for mankind to explore. Avery had opened the doors to the world of immunochemistry, an achievement vividly brought home by Michael Heidelberger, who occupied the laboratory contiguous to mine. Whenever Fess Avery was mentioned, Heidelberger manifested reverence to such a degree that there was a distinct pause

[*] Visiting professor of pathology and research professor of orthopedic surgery, University of Virginia, Charlottesville, Virginia 22903.

in the conversation. For Heidelberger and the many younger scientists who had sought out Avery for advice on their research, the Fess was considered the mentor par excellence. It was well known that Avery examined every facet of a problem so thoroughly that the "consultation" rarely lasted less than two hours, and the research problem was frequently explored in depth for three to five hours. However, the Fess maintained a silence on his own work.

This creative and sympathetic interest in the research of others touched me personally in April 1942. Because I was ignorant of Avery's studies in progress, our first meeting was exceedingly painful to me. I had just been called on active duty by the U.S. Navy and was invited to attend an April meeting of the National Research Council in Washington, D.C. The subject to be discussed was streptococcal problems in the armed services, for already the navy was confronted with a high incidence of rheumatic fever in the recruit training camps at the Great Lakes Training Center. More than a dozen distinguished civilian biologists were assembled in the conference room. There were tedious harangues about the streptococcal menace. Each speaker seemed to concentrate on the periphery but declined to take an aggressive stance.

Finally, I was questioned. Nervously, I stated that without adequate preventive medicine or control measures, three things could reasonably be expected to happen: (1) There would be a high incidence and rapid spread of hemolytic streptococcal respiratory infections. (2) Certain strains would develop "mutants" or "sports" that would be highly infective (as contagious as postinfluenzal streptococcal pneumonia in World War I at army camps in Texas). (3) Perhaps one or more of the streptococcal mutants would be genetically resistant to sulfonamides.

There was dead silence when I had finished. Nobody spoke a word of agreement or approval; there seemed to be no interest in what was obviously a figment of my imagination. Presumably I had used words that were not acceptable in scientific circles—*genetic changes, bacterial variation, mutants, sports.*

Soon it was time to recess, and we paraded to the lower level for a standup snack lunch. Chagrined, depressed, and fearful of reprimand for my shocking verbal goof, I picked up a sandwich and a bottle of Coca Cola and slunk into a far corner to recover from my embarassment. Then, although I had thought that not one person at our conference table had believed a single word I had uttered, a short man with a soft voice joined me and said, "I was most interested in your remarks. Please tell me more." Thus, O. T. Avery, in his gentle way, opened my flood gates, and I expressed my concern to him in detail.

As I documented these statements with findings from our laboratory and referred to the recent report of Beadle and Tatum [1], Avery's attention increased. It had not occurred to me that the Fess had any interest in bacterial variation; so I was surprised when he invited me to join him for supper on the train back to New York. I well remember that during the course of our dinner conversation he emphasized how much he wished he were my age so as to be able to work on some of the biological problems that fascinated me. At that time, I knew only that Avery was *sympatico* with the young naval officer who believed that changes in the genes of beta hemolytic streptococcus might result in different biochemical mechanisms and increased infectivity for man.

One year passed. It was March 1943. Hemolytic streptococcal infections had become the navy's greatest stateside medical problem, and the beds of the hospital at the Rockefeller Institute were filled with the navy's enlisted men from New York's pier number 92. Practically all of these patients were infected with a single serological type, group A hemolytic streptococcus (type 19). Avery's sympathetic understanding back in 1942 had not been misplaced, and I felt that it would not be too much of a liberty for me to express my appreciation by inviting the Fess to spend a Sunday with us before the navy transferred me to Portsmouth, Virginia.

I met Avery at the North White Plains Station and drove him home. He suggested that we take a walk. I looked at his highly polished black shoes and his neatly pressed blue Sunday suit and hesitated. Nevertheless, he wanted to walk; so we set out for a two-hour hike through Conyer's orchard and the Round Hill Road back to our home. Except for this particular day the Fess always wanted to listen prior to expressing his opinion and before generously offering sound advice. However, on that March Sunday the dialogue was different; Avery wanted to talk; for once he was exuberant!

He said, "You are going away; I do not know when we shall meet again; I want to give you a little information that some day may be helpful to you in your work." For the only time in our many dialogues I had no idea what was on Avery's mind. He said that on the preceding day he had reported at a meeting of the trustees of the Rockefeller Institute his work on the "transforming factor." I was still completely in the dark until he described work done fifteen years previously by Fred Griffith of London. At last, I began to tune in on Avery's wavelength; Griffith's name and work were familiar to me.

It so happened that in 1931 Fred Griffith had extended to me a helping hand across the Atlantic Ocean in the serological identification of *Streptococcus pyogenes*. But Avery, I knew, was dedicated to the pneumococcus! What was the connection between these two bacteriologists? For two hours the Fees related his story, beginning with Griffith's 1928 publication on the changing of the serologic type of the pneumococcus, which sparked Avery's own research, and concluding with his report of the day before to his board of trustees on the chemistry of "the transforming factor."

Then came the climactic remark: the factor which transformed the genetically transmitted type of the pneumococcus was desoxyribonucleic acid! My first reaction, *not* expressed aloud, was: "So what? How is this going to help us win the war?" All that I could then do was to thank the Fess perfunctorily for telling me his most intimate secret. My thoughts were wandering to triumphant Rommel in North Africa and the Japanese fleet around the Coral Sea. Nevertheless, the fact that Avery had taken the trouble to brief me on a factor in genetic control of the pneumococcus had made its impression.

The incubation period for understanding the implication of the information that Avery had given me was approximately eight weeks (rather long for an infecting agent!). By mid-May I was living in a room in Virginia with windows blacked out at night, only yards away from the waves of the Atlantic Ocean. Weekly, German submarines were sinking oil tankers only a few miles off shore. The war prospects looked as black as the beach did on the morning after an oil tanker was sunk. My thoughts sought an escape from the grim reality ahead. It was under these circumstances that I recalled what the Fess had told me as a gift for any future association with streptococcal problems.

Then flashed through my mind this idea: perhaps this desoxyribonucleic acid will not only change bacteria but will also modify Charles L. Hoagland's culture of vaccinia virus. Perhaps it will change the genetically controlled biochemical mechanisms of all microorganisms! The excitement within me was so great that I immediately wrote the Fess. Two things had suddenly become as clear as crystal: first, Avery, through his decades of dedication to the study of the pneumococcus, had at long last in his hands a substance of enormous importance to microbial activity; second, I realized for the first time how deeply Avery's work was founded on the discovery made by Fred Griffith. It therefore seemed appropriate to send to Avery my picture of Griffith. Avery treasured this 1936 snapshot of the English bacteriologist with his dog, Bobby, which it has been said was the only existing picture of F. Griffith.

Two years later, in January 1946, returning "stateside" after World War II, I revisited the Rockefeller Institute and lunched with Dr. Charles Hoagland. I asked casually, "What's new in science since I left these shores?" He responded immediately. He did not mention his own important work or the work of his colleagues at the institute.

"The geneticists," he replied, "have caught on with enthusiasm to the significance of the Fess's discovery of desoxyribonucleic acid and are very excited about it." It was only then, through the catalyzing mind of Charles Hoagland, that I realized for the first time that the Fess had produced information that transcended the microbial world and might even be involved in the genetic problems of mankind. Avery had probably envisaged this when he "lectured" to me in March 1943. But always, as in the past, Avery kept his foot on the brakes and gave me only the facts which might be relevant to my own microbial research interests.

During the next decade (both in New York and in Nashville) Avery expressed to me his interests in the pursuit of certain studies with the collaboration of a geneticist. Later he explained why he preferred to leave the next step to others. This decision he had made prior to that crisp fall morning in 1952 when Dr. Roy C. Avery brought in the morning newspaper. The front page proclaimed that the Nobel Prize had been given for the discovery of streptomycin. O. T. Avery seemed pleased that an American bacteriologist was the recipient.

Recently, after the passing of nearly a quarter century, I received an inquiry from one of America's most distinguished geneticists, a member of the coterie of Nobel laureates spawned by Avery's identification of his transforming factor. The question posed was: Did Avery ever understand the implications to genetics of his own discovery of DNA? The answer was obviously "yes." Had the brilliant geneticist been confounded by Avery's modesty?

I replied that I knew from a conversation which I had held with O. T. Avery in 1943 that he was certainly aware at that early time of the significance of his transforming factor for microbial genetics, and probably also of its implications for human genetics. In any case, Avery was certainly aware of the broad implications for mankind before 1946, since by then the geneticists were expressing keen interest in his work on desoxyribonucleic acid. I only wished in replying to the inquiry that there were some way of documenting the substance of the conversation that Avery and I had had in March 1943. There were no letters; Avery budgeted his correspondence! He wrote to his family, and he was meticulously careful to send "bread and butter" thank-you notes, which his hostesses found to be masterpieces of gallantry. Until 1969 I was unaware that he had written anyone concerning the implications of his great discovery. It was logical to assume that my letter of May 1943 to the Fess, in which I recalled our March conversation, had no doubt long since been destroyed. At least this is what I presumed years after O. T. Avery's death.

Fortuitously I wrote to the Fess's brother, Dr. Roy C. Avery, in Nashville and made a personal request. Could I have back the only available snapshot of Fred Griffith sent to the Fess in May 1943, the picture that he had framed and always kept on his desk? I heard nothing for months and concluded that the picture had been thrown away. In due course, however, Dr. Avery located the picture cached in a trunk. When the frame was removed for mailing, he found that an extraneous object had been placed as a backing for the picture—my letter of May 25, 1943 to the Fess, which revealed my excitement upon catching on to the significance of what Fess had told me that March Sunday morning in 1943!

To document this narrative I submit: (1) A snapshot of O. T. Avery taken after dinner on a Sunday in March 1943, the day after he had announced informally to his board of trustees that he had discovered that his "transforming factor" was desoxyribonucleic acid. (2) My letter to Avery in 1943, found nearly a quarter century later incarcerated by the Fess in the frame of the picture of his peer Fred Griffith of the British Ministry of Health, a rare biologist whom he had never met but to whom he must have always felt indebted.

Since drafting this exposition I have been informed that the substance of

Fig. 1.—O. T. Avery with a young Coburn, at our home March 1943

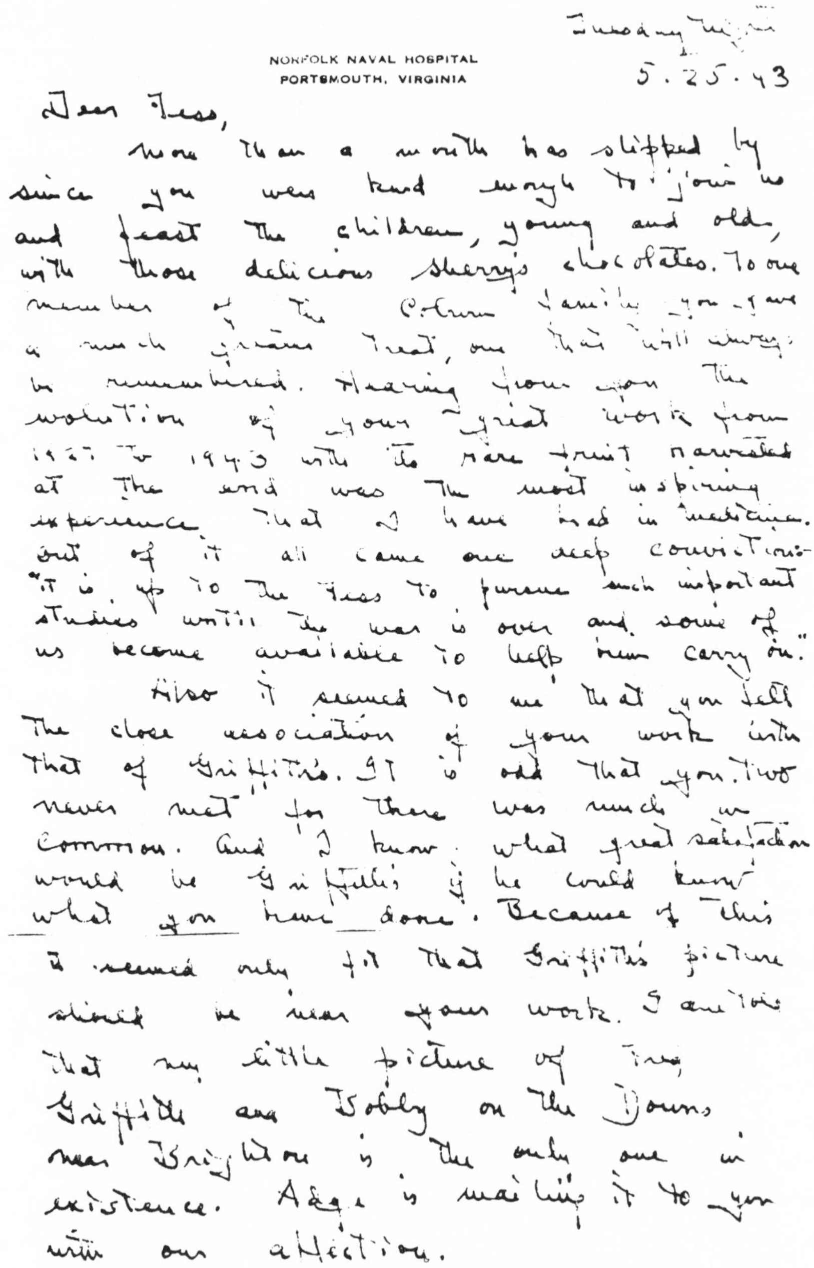

NORFOLK NAVAL HOSPITAL
PORTSMOUTH, VIRGINIA

Tuesday [night?]
5.25.43

Dear Fess,

More than a month has slipped by since you were kind enough to join us and feast the children, young and old, with those delicious sherry's chocolates. To one member of the Cohnn family you are a much greater treat, one that will always be remembered. Hearing from you the evolution of your great work from 1937 to 1943 with its rare fruit narrated at the end was the most inspiring experience that I have had in medicine. Out of it all came one deep conviction: "It is up to the Fess to pursue such important studies until the war is over and some of us become available to help him carry on."

Also it seemed to me that you felt the close association of your work with that of Griffith's. It is odd that you two never met for there was much in common. And I know what great satisfaction would be Griffith's if he could know what you have done. Because of this it seemed only fit that Griffith's picture should be near your work. I am told that my little picture of Fred Griffith and Dobly on the Downs near Brighton is the only one in existence. Adge is mailing it to you with our affection.

FIG. 2.—Reproduction of my letter to Avery, May 25, 1943

Avery's March 1943 revelation to me was recorded in a letter of May 13, 1943 to his brother, Roy C. Avery. This was published in part in 1964 by Carsten Bresch [2, p. 130] as follows:

Die molekulare Grundlage der genetischen Information Avery beschrieb am 13.5.1943 diese Entdeckung—wohl die grosste der Genetik seit MENDEL—in einem Brief an seinen Bruder:[1] ". . . But at last perhaps we have it. The active substance is not digested by crystalline trypsin or chymotrypsin, it does not lose activity when treated with crystalline ribonuclease . . . polysaccharide can be removed. . . . Lipids can be extracted . . . without impairing biological activity. The extract can be deproteinized. . . . When extracts, treated and purified to this extent . . . are further fractionated by the dropwise addition of absolute ethyl alcohol an interesting thing occurs. When alcohol reaches a concentration of about 9/10 volume there separates out a fibrous substance which on stirring the mixture wraps itself about the glass rod like thread on a spool. . . . The fibrous material is . . . highly reactive and on elementary analysis conforms very closely to the theoretical values of pure desoxyribose nucleic acid (thymus type). (Who could have guessed it) . . . depolymerase capable of breaking down known authentic samples of desoxyribose nucleic acid has been found to destroy the activity of our substance—indirect evidence but suggestive that the transforming principle as isolated may belong to this class of chemical substance. . . . If we are right, and of course that is not yet proven, then it means that nucleic acids are not merely structurally important but functionally active substances in determining the biochemical activities and specific characteristics of cells. . . . But today it takes a lot of well documented evidence to convince anyone that the sodium salt of desoxyribose nucleic acid, protein free, could possibly be endowed with such biologically active and specific properties and that is the evidence we are now trying to get. It is lots of fun to blow bubbles but it is wiser to prick them yourself before someone else tries to."

[1] Original im Besitz von R. C. Avery, Vanderbilt University, Nashville, Tenn.

On November 1, 1943 Avery, McLeod, and McCarty submitted for publication their epic discovery [3]. In this report they cite as the origin of their work the findings of F. Griffith [4].

In conclusion, in March 1943 when concerned with streptococcal epidemics [5], I learned that the serologic type of the pneumococcus could be transformed by desoxyribonucleic acid. And so it came to pass: singleness of purpose, continuity of work, a laboratory haven protected by the constant vigilance of Rufus Cole, this summation at long last bore a rare fruit. Avery's devotion to the biology and chemistry of one bacterium over the decades produced a discovery—"wohl die grosste in der Genetik seit Mendel" [2]. It was the immunochemist O. T. Avery who, while making the most significant contribution to modern genetics, ushered in a new era for the human family. Although Avery may not have used the term *DNA*, he was certainly the first to demonstrate its role in bacterial genetics, and he was one of the first, if not the first, to envision its significance for all mankind. As a man of great wisdom he made no claims, and the enormous significance of his contribution to the future of science was overlooked by most of his distinguished peers.

I share the reverence held by many for the selflessness and conservatism that characterized the Fess and wish that I could have presented this personal information anonymously, but there seemed no alternative to telling this story in the first person. In fact, I shall complete this exposition by citing an example of Avery's high degree of cautiousness as told me by Dr. Yale Kneeland of the Columbia-Presbyterian Medical Center. This occurred during a drive from Manhattan to Long Island across the windswept Triborough Bridge. Dr. Kneeland was at the wheel beside his passenger, Avery, who looked down at the dashboard and saw the indicator at the number 80. Avery inquired: "Don't you think that we are travelling a bit too fast, Dr. Kneeland?" The latter then also looked down, saw what was upsetting the ever-cautious Fess, and replied: "Dr. Avery, according to the speedometer we are going a bit less than 40 miles an hour. That's the radio indicator that you are looking at!"

REFERENCES

1. G. W. Beadle and E. L. Tatum. Nat. Acad. Sci. (U. S.), Proc., 27:499, 1941.
2. C. Bresch. Klassische und molekulare Genetik. Berlin: Springer-Verlag, 1964.
3. O. T. Avery, C. M. McLeod, and M. McCarty. J. Exp. Med., 79:137, 1944.
4. F. Griffith. J. Hyg., 27:113, 1928.
5. A. F. Coburn and D. C. Young. The epidemiology of hemolytic streptococcus. Baltimore: Williams & Wilkins, 1949.

INTERDISCIPLINARY GENETICS: INTRODUCTION

Muller investigated the nature of the gene in the years before the identification of the genetic material and the formulation of the double helix. A Nobel Prize for demonstrating the induction of mutations by X-rays came after the start of the atomic age. With the insight of hindsight, Muller reviews the rapid progress in identifying the central role of genetic nucleic acids in the origin of life on this planet and presumably any other planet in this or other solar systems.

By 1960, the base composition of the DNA of enteric and other bacterial species and their bacteriophages had been determined. Bacterial and viral geneticists had demonstrated "sex" in these life forms by identifying recombinant progeny. Lanni correlates DNA base composition and the results of genetic studies by assuming that crossability between bacterial species or viral "species" indicates genetic homology. This assumption had already been validated in eukaryotic species and only needed extension to the double helix of prokaryotes to detect homologous sequences for genes in bacteria and viruses.

Caspari relates the role of genetics in embryology, not yet known as developmental biology, from the vantage of 1960. The phenogenetic approach to understanding the impact of a mutation on the normal course of development is designed to identify the spatial or temporal fork in a primary pathway. As a pioneer in developmental biology, Caspari was aware of the emerging significance of biochemical genetics in understanding the molecular basis of development and the complexity of secondary biochemical reactions and interactions in development. It should be noted that the Jacob-Monod paper on genetic regulatory mechanisms in the synthesis of proteins appeared in 1961.

The reality of biological evolution is supported by evidence from each biological discipline as it matures in its time. Davis adds compelling evidence from population genetics, biochemistry, and molecular genetics. Assuming a common origin of life and then understandable evolutionary processes, Davis marshalls evidence from comparative biochemistry to demonstrate the foundations of common cellular functions and structures and from molecular genetics to show the universality of the genetic code and the mechanisms of genic and chromosomal mutation. The amino acid sequences of certain polypeptides found throughout major groups and the nucleotide sequences of the corresponding cistrons have yielded an emerging hybrid discipline that might be termed paleogenetics.

Lowenstein and Zihlman discuss human evolution and molecular biology by going directly to the source: the nucleotide sequences and the number of nucleotide substitutions in specific cistrons. The concept of the molecular clock, the rate of amino acid substitutions in certain polypeptides, is examined in terms of the paleontological record to determine the time of divergence in evolutionary lines. The correlations suggest a new and valuable approach to understanding the processes of human evolution. Unfortunately, the prehuman forms have not left detectable fossilized DNA or protein, leaving only other primates for study.

GENETIC NUCLEIC ACID: KEY MATERIAL
IN THE ORIGIN OF LIFE

H. J. MULLER*

I. *How Do Living Things Differ from Lifeless Ones?*

The differences between living and lifeless things that are big enough to be seen are for the most part so far-reaching and manifold that even primitive man, the world over, despite his prevalent limitations in coming to generalizations, has tried to divide all nature into these two categories. The category of the living, containing himself, he tends to endow with the main attributes supposedly characteristic of himself, including especially the possession of a sentient spirit, capable of acting upon matter yet also, so he thinks, of existing apart from it. Sometimes the distinction between living and lifeless becomes blurred when these spirits are thought of as coming to dwell not only in plants, animals, and human beings but also in sticks and stones, the earth, the wind, and heavenly bodies: a widespread doctrine known as animism. Yet despite this extension, his cleavage between living and non-living usually remains for him a fundamental one. And even scientists find it useful and sensible today, both on operational and theoretical grounds, still to make this distinction, although with rare exceptions they no longer think of it as based in any inherent spirit or vital force.

* Department of Zoölogy, Indiana University, Bloomington, Indiana, Contribution No. 708.

This article represents a lecture given at a Symposium on the Origin and Nature of Living Matter, held at the Albert Einstein Medical College on June 8, 1959, in connection with the graduation of its first medical class.

Ever since our range of view has been increased so as to include objects of microscopic and submicroscopic sizes, it has remained evident that, with only a limited set of possible exceptions, living things form a natural group, for all their enormous diversity, that is set apart in deep-seated and remarkable ways from lifeless things. When now we seek to define the difference between the living, or animate, and the non-living, or inanimate, in some exact yet general way, we find that some of the most outstanding characteristics by which we recognize many animate bodies to be such—for instance, motion, contractability, conduction of the impulses to react, sensitivity to light, sound, touch, etc., specific powers of digestion or conversion, even adjustment to such features of the environment as heat or drought—are present only in some types of organisms or in particular stages and under given conditions. Under certain circumstances some organisms—even human cells—can remain quite dormant. Even most of the peculiar chemical substances such as carbohydrates, fats, proteins, and vitamins found, in many variant forms, in all except a few submicroscopic organisms do not actually define things as living. For organisms that are no longer living may still possess these materials, although doubtless with alterations in some components and in the manner in which they are fitted together that have so far eluded us.

Still, all living things larger than viruses do have all these materials, and at some stage of their existence have a turnover of them, or metabolism, whereby they are manufactured from more or less different incoming materials and broken down into outgoing waste, so that the organism is constituted of materials that are subject to flux. And in all cases, at some stage or stages, the income exceeds the outgo, so that the organism grows while maintaining in the main its own specific composition. And this growth is so ordered as to cause the individual cells to form two cells, that is, to undergo a binary replication. And when the individual consists of many cells, there is a stage when individual cells separated from the parent, or cells formed by the combination of two that have separated from the parent, redivide and redifferentiate to form a new multicelled individual. The balanced flux has sometimes been compared with a flame or a whirlpool, although it is inordinately more complicated. But nothing like the orderly growth of such a complicated organization, leading to its regular binary replication, is known outside of living matter.

Now, as we all know, living things differ from lifeless ones in that the changes worked in them by impinging environmental agents are commonly of such a nature as to protect them from those agents if the circumstances have harmful potentialities, or to allow them to take advantage of the circumstances in furtherance of their own maintenance and reproduction if there are such possibilities. Skin rubbed tends to thicken; muscles used, to increase; food encountered, to arouse digestive secretions; and so on; quite the opposite to ways in which inanimate bodies react, with the exception of some mechanisms contrived by man. Occasionally, to be sure, sacrifices are made, as when a skink commits autotomy on its tail and, leaving it in your hand, escapes to grow a new tail; or when the mammary gland cells undergo autolysis to feed the young; or when a worker bee, like a kamikazi, undertakes the suicidal act of stinging you, losing its own life but thereby helping to save the hive. Yet in all such cases it is evident that the reaction to the situation—if that situation is of a type frequently encountered by the organism's ancestors—is such as to tend to be useful for the continuance or the multiplication of the organism or its relatives. That is what we mean when we say that the organism's reactions are *adaptive*. Moreover, adaptive reactions are the result of adaptive structures. From species to species these structures vary enormously, yet always they are adaptive, from the point of view of that type of organism. And by adaptive we mean *only* that they are conducive to the multiplication of that type and therefore, usually but not always, to the survival of that individual.

It was the existence throughout living nature of this appearance of design for a purpose, in such a multitude of forms—this intricate adaptation of all features and functions to the seeming goal of survival and reproduction—that made the great majority of thinking men before Darwin so sure that living things had been consciously planned. But Darwin showed that, starting with organisms of the simplest sort known, if only they had the capabilities, which all organisms do, of reproducing themselves, of varying, and of handing down to some of their offspring some of these variations, they would through natural selection or the survival of the fittest diverge from one another, step by small step, in different lines and become in the course of geological time more and more complexly and diversely adapted to the varied potentialities for existence that nature afforded (1). He showed, moreover, that all these now so manifold types represent different stages in one vast branching structure of evolution, which he called

"The Tree of Life." Living things, then, do form a natural group, all interrelated with one another no matter how different today, and form in a sense a category of their own apart from the lifeless things out of which they grow.

A little consideration shows clearly why it is that the adaptations they have developed have all seemed to be directed toward the same ultimate end: survival and multiplication of their kind. It is of course because those variations that have survived and multiplied have each time tended to be just the ones which were themselves conducive to that survival and multiplication. Moreover, survival without multiplication would not have been enough, even if variation had occurred. For survival without multiplication for one form is an impossibility when everything else is exerting the pressure of multiplication on it and thus tending to crowd it out. Besides this, multiplication is also essential to allow sufficient range of variation, inasmuch as a thousand or more variants must arise that are duds and have to be sacrificed for every single accidentally helpful innovation that succeeds and spreads out into the places left by those that die out.

But lifeless things can also undergo changes; that is, they can vary, and the impact of outer forces will cause them to vary. The reason why, despite this, they have not undergone evolution likewise is now evident. It is because they are unable to multiply, or to replicate themselves, as you may choose to call it, and in this process to pass on their changes to a self-increasing group of replicas. Thus multiplication or replication that transmits variations constitutes the essential mechanism of the process of biological evolution (2–9). And, this having been the case, multiplication has become at the same time the unconscious goal toward which are directed all the adaptations evolved in the course of ages of this multiplication process itself. The means thus becomes its own end, in the ever-repeated cycle of living.

If then we would wish to have deeper knowledge of the basis of the distinction between inanimate and animate matter that allowed the latter to assume its diverse complexities, and if we would seek to know the manner in which the transition took place, if indeed it did, from the inanimate to the animate, we should look into the problem of what it is in living matter that gives it this unique property of reproducing itself, and of reproducing itself in just such a way that its variations become incorporated into its replicas. Here now is where we must turn from Darwin—not yet to the biochemists, but first to the students of heredity, reproduction, and cell structure of the nineteenth century, and to their twentieth century descendants, the geneticists.

II. *Demonstration of a Specific Genetic Material*

Shortly after aniline dyes and high-power magnification with the aid of condensers and oil immersion lenses were developed in the seventies, the chromosomes and their maneuvers in mitosis and in fertilization were observed. Thereupon a group of four students of cellular phenomena, Weismann (10), Oscar Hertwig (11), Strasburger (12), and Kölliker (13), on the basis of observations on chromosome constancy and on the seeming identity of the sets of chromosomes contributed by the sperm and egg to the offspring, as contrasted with the dissimilarity of everything else visible in these germ cells of opposite sex, independently of one another arrived, in the early eighties, at the conception that the chromosomes form the basis of heredity and development. And Wilhelm Roux (14), reasoning from the longitudinal mode of division of these chromosomes, inferred, further, that they contain qualitatively different hereditary determiners, arranged in line.

Unbeknown to these or to most other workers of the time, Mendel (15) had meanwhile demonstrated the particulate nature of inheritance in peas, as shown by the fact that there are hereditary determiners there which preserve their separate identities, segregate accurately and cleanly from corresponding determiners which had been received from the other parent, and undergo recombination with non-corresponding determiners. Soon after Mendel's principles were rediscovered in 1900 it was realized (16–18) that the chromosomes undergo, visibly, just the type of maneuvers, in preparation for the production of mature germ cells, as had been deduced for the determiners of Mendel. And the fact that the chromosomes do contain these determiners, strung in lines much as Roux had first supposed, was proved conclusively by the work on the fruit fly Drosophila carried out at Columbia University in the half-decade 1910 to 1915 (19). The voluminous work on a multitude of organisms done since that time, as well as many further developments in the work on Drosophila itself, has served to verify and elucidate many details of the processes at work and to bring to light new ones. Most important among these are the exchanges between chromosomes, known as crossing over; the occasional occurrence of breaks of chromosomes followed by reattachment of the fragments in a new order, a phenomenon known as structural change; and the sudden sporadic transformation of a given determiner or "gene," as it came to be called (on Johannsen's [20] suggestion), into a new form, inheritable and replicable as such. This event, known as "gene mutation," turns out to provide the corporeal basis of that process of inheritable variation required by the Darwinian theory of evolution by natural selection.

Driesch (21) had already proposed before the turn of the century that the then postulated determiners of heredity are enzymes ("ferments" in the language of that day). During 1915 to 1917 Troland (22, 23) of Harvard, in his *Enzyme Theory of Life*, proposed that they are enzymes which not only carry out conversions useful in metabolism, heterocatalytically, but also convert food products into more enzyme molecules like themselves, autocatalytically: that is, they replicate themselves, and so, through their mutation and replication, they form the basis of living matter and of its evolution. As for the nature of the replication process, he postulated that to each part of the enzyme a free-floating part of an identical nature, present in the protoplasm, became attached, so that a pattern of parts like that in the original enzyme thereby became built up. He believed this process to be plausible on the ground that, in general, like molecules tend to take up positions adjacent to one another, as in crystals.

As I pointed out in treatments (2, 3) of 1921 to 1926 (the latter entitled "The Gene as the Basis of Life"), we do not know of any enzymes or other chemically defined organic substances having specifically acting auto-catalytic properties such as to enable them to construct replicas of themselves. Neither was there a general principle known that would result in pattern-copying; if there were, the basis of life would be easier to come by. Moreover, there was no evidence to show that the enzymes were not products of the hereditary determiners or genes, rather than these genes themselves, and they might even be products removed by several or many steps from the genes, just as many other known substances in the cell must be. However, the determiners or genes themselves must conduct, or at least guide, their own replication, so as to lead to the formation of more genes just like themselves, in such wise that even their own mutations become incorporated in the replicas. And this would probably take place by some kind of copying of pattern similar to that postulated by Troland for the enzymes, but requiring some distinctive chemical structure to make it possible. By virtue of this ability of theirs to replicate, these genes—or, if you prefer, genetic material—contained in the nuclear chromosomes and in whatever other portion of the cell manifests this property, such as the chloroplastids of plants, must form the basis of all the complexities of living matter that have arisen subsequent to their own appearance on the scene, in the whole course of biological evolution. That is, this genetic material must underlie all evolution based on mutation and selective multiplication.

In 1921 I had, however, left it an open question whether the genetic material, along with its building blocks, contained the main features of the mechanism by which it replicated itself, or whether on the other hand that mechanism chiefly lay in the manner of organization of the accompanying protoplasm, to which the genetic material only furnished guides or blueprints according to which the protoplasm manufactured it. The latter view would have presupposed that the protoplasm had originated first and had the capability of manufacturing not only the genetic material but also its own complex organization. This was the view which would naturally develop out of Haeckel's idea (24, 25) that in the transition from inanimate to animate matter proteins and other constituents of living matter gradually developed and through a kind of natural selection gained an ever greater ability to build up or anabolize, to counteract their continual decay or catabolism, and thereby grew and even, by falling apart into smaller pieces that again grew, came to reproduce and finally also to manufacture specific genetic materials that had differentiated in them. The Russian Oparin has since the early 1930's espoused this view and has followed the official Communist Party line by giving the specific genetic materials a back seat. It is in essence somewhat like the view proposed by the great mathematician von Neumann (26) when he devised in outline a complex machine capable of constructing, among other things, more machines like itself, and at the same time capable of copying blueprints embodying a variable design or code that was conferred upon the daughter-machine during its manufacture and that modified the non-replicational operations of the daughter-machine in diverse ways.

However, as I pointed out in my discussion of 1926, natural selection based on the differential multiplication of variant types cannot exist before there is material capable of replicating itself and its own variations, that is, before the origination of specifically genetic material or gene-material. It is an illusion to think that without having this unique replicational property any organization like protein or protoplasm could gradually become more and more anabolic, in the sense of producing a structure having exactly its own complexities. For there is no way by which any variation in it, no matter how conducive to its growth in general, could come to produce more of its own distinctive features and thus to share in its growth and multiplication. On the other hand, one is stretching chance too far to postulate that a complex finished mechanism like von Neumann's, capable of reproducing both itself and specific blueprints, could arise by the accidental coming together, into the precise arrangement necessary, of a host of preformed parts.

We are thereby reduced to accepting the proposition that the genetic material contains its own mechanism of replication, and a relatively simple one at that, whereby it is able to multiply, to vary or mutate, and then to replicate its own mutations, even without the aid of any organized protoplasm. It would thus be subject at once to the Darwinian process of natural selection, that is, to the differential multiplication of those mutants whose changes enabled them to multiply faster or more securely or in otherwise inhospitable situations. And thereby it would gradually acquire an increasing accumulation of such accessory chemical substances, manufactured by it or under its influence, as were conducive to its further multiplication. In other words, it would step-by-step build up a set of surrounding materials, or protoplasm, that served its own end of self-reproduction. This genetic material, then, would have constituted the basis of life from its very beginning. And its distinctive secret, which made life and all of its further evolution possible, would lie in the chemical constitution that gave it the ability to replicate, to replicate even its own mutations, and to mutate to an unlimited extent, in such ways as allowed it to engage in the manufacture and organization of complex accessory organic substances, without itself becoming changed or used up in the process. Such material, it was pointed out, is to be found in the chromosomes and in some of the plastids of animals and plants, and perhaps least encumbered with other material in some of the viruses.

III. *Chemistry of the Genetic Material*

What then is this genetic material, and what is its chemical secret? Researches of the nineteenth century by Miescher (28), Altmann (29), and Kossel (30) in Germany, carried further by P. A. Levene (31) at the

Rockefeller Institute, showed that the chromosomes are composed mainly of nucleoprotein, that is, of a combination of protein, chiefly of relatively simple types, with nucleic acid, and that the latter consists of nucleotides, compounds of a purine or pyrimidine base, a 5-carbon sugar, and phosphoric acid, strung together to form larger or smaller chains or groups of chains. Did the genetic material consist of the protein, as seemed most likely because of its known potentialities for complexity, diversity, specificity, lability, versatility, and ability to act enzymatically and immunologically; or did it consist of the nucleic acid, or of a combination of the two? Or was it perhaps comprised in the 5 per cent or more of other material, of largely unknown composition, with which the chromosomal nucleoprotein always seemed to be accompanied?

A partial answer did not come until 1934, when Stanley (32) and his co-workers were able to obtain active tobacco mosaic virus in a crystal-line-like form, free from any substances but protein and nucleic acid. Since it could, when introduced into host cells, cause its own replication, and even the replication of its variant types—that is, its mutants—it fulfilled the definition of genetic material. It happened that its nucleic acid was of the type containing ribose, not deoxyribose, but this RNA managed as well for it as the DNA that is in the chromosomes of higher forms manages for them. Thus the accessory material, that other than protein and nucleic acid, was ruled out.

Some ten years later the bacteriologists Avery, MacCleod, and McCarty (33), working on the material which, extracted from one variety of bacterium and applied to another, is able to confer on the latter genetic properties of the first type, found that they could obtain it as nearly pure nucleic acid, with no or very little protein. Although they thought of this material as merely the means of inducing a transformation, it became probable on genetic grounds that it consisted of genetic material of the first variety, which became *transferred into* the second, and this view has since received considerable support. From this the conclusion would follow that the genetic material consists of just the nucleic acid. Following this, in Hershey and Chase's (34) brilliant work on labelling the protein and nucleic acid of bacteriophages with radioactive sulfur and phosphorus, respectively, it was shown that when these infect their host bacteria, only the nucleic acid and not the protein, or hardly any of the latter, enters. Finally, in work on the tobacco mosaic virus, Fraenkel-Conrat (35) and others of

Stanley's group, and independently Gierer and Schramm (36), have shown conclusively that the nucleic acid if stripped of its protein before infection can nevertheless succeed in replicating itself, as well as in causing the production *de novo* of protein like that removed. Moreover, it can after stripping be given a cloak of protein from a different variety, yet on replication it will cause the production of its own distinctive kind of protein. Thus the composition of the genetic material becomes reduced to that of the nucleic acid alone, RNA or DNA, as the case may be.

The crowning feat of this quest, but by no means the last, has been the epoch-making study carried out by the young American geneticist Watson in combination with the English biochemist Crick (37) on the constitution of the DNA of chromosomes, as judged from the X-ray diffraction pictures of chromosome structure obtained by Wilkins, Franklin, and others of a British group (38, 39), on the one hand, and from the results of biochemists studying nucleotides obtained from chromosomes, on the other hand. As all geneticists are now keenly aware, they showed that the only structure that will adequately fit the requirements is an immensely long (relative to its width) double, helically coiled chain of nucleotides, running in parallel but in opposite directions. The nucleotide in any given position may be of any of the four types, containing either a purine (adenine or guanine) or a pyrimidine (thymine or cystosine). These bases project inward from each chain toward a complementary base across from it on the other chain, the two bases being attached across the chain by two or three hydrogen bonds for each pair, adenine always finding its complementary base thymine in the right place across from it and guanine its complement cytosine. It is the exact sequence of these nucleotides in line in the chain that must determine what chemical influence a given segment or region of the chain exerts in forming its protoplasmic products.

Now the greatest beauty of this construction is that it permits of the exact replication of the double chain, so as to result in the emergence of two double chains, each one just like the original double chain. This process takes place, on this theory, when the two component chains constituting the double chain separate. For then each base, deprived of its former complementary base in the other chain, becomes able to fasten down across from itself in the vacated position by means of hydrogen bonds, as before, a free nucleotide gathered from the medium. This happens when a nucleotide comes along which is of the same type as the one that was formerly in that position: namely, of the type with a base complementary to that in the nucleotide already present. Thus after the binding together of all these newly acquired nucleotides in line chainwise, in the positions in which they now find themselves, a double chain emerges in the place of each previously separated single chain. One of the members of each double chain is an old one and the other a complementary-type newly formed one, and both of these double chains are identical in their arrangement of nucleotide-pairs with the original or mother double chain before its division.

The connecting up longitudinally into a chain of the newly attached, originally free, nucleotides is doubtless facilitated by the high-energy phosphate bonds which the free nucleotides had been able to carry in a triphosphate arrangement. It is also this potentiality which (as the biochemists have known for some twenty years) enables the free nucleotides to serve as vehicles of energy transfer of many kinds leading to endothermic reactions involving other syntheses and to the performance of work. And it may be that this ability can still be called upon in some way in the case of the chained nucleotides after the replication of the chains is finished (5).

That the replication of genetic nucleic acid does in fact involve the separation of the original double chain and the building of a new single chain next to each original component chain, as this theory demands, has recently been demonstrated in the elegant studies of Levinthal (40) and of Meselson and Stahl (41). In these experiments the old chains of phages were labelled by means of distinctive radioactive and stable isotopes, respectively. It was then found that in replications subsequent to the first separation of the marked double chains these individual marked chains were inherited as units, retaining their marked atoms intact. Similar findings involving tritium labelling in chromosomes of the bean *Vicia* were made by Taylor (42).

This work demonstrates among other things that, as had been suspected on considerations of genetics, the genetic material, unlike protoplasmic constituents, is not subject to flux: that is, the original atoms within it remain there permanently, without turnover. The turnover or metabolism occurring in the other material of a cell represents, we might say, the fire that the genetic material keeps going outside itself, to get that other material to work for it, in the service of its own distinctive goal: its own survival and replication. Thus metabolism is not the essence of life but a kind of upper-level expression of life after the genetic material has succeeded in making for itself a workshop of protoplasm.

To clinch the point that the nucleic acid chain, taken together with its building blocks, the free nucleotides, contains most if not all of the mechanism necessary for its own replication, it has recently been shown by the biochemist Kornberg and his associates (43, 44) that a chain of as few as three nucleotides, when placed in vitro in a medium containing also free nucleotides and just one enzyme that catalyzes the reaction, gets to work and puts the free nucleotides together into more chains. Moreover, these chains have their different types of nucleotides in the relative frequencies anticipated for replication products derived by the Watson-Crick complementation process. The three nucleotides together, then, might be said to constitute a protogene. Results with ribonucleic acid (RNA) somewhat similar to Kornberg's, which were with deoxyribonucleic acid (DNA), have been obtained by Ochoa and his associates (45).

The groups of nucleotides that serve any given primary function in cellular organisms or even in viruses, and that we ordinarily have in mind when we use the term "genes," have been shown by the beautiful genetic analyses of Benzer (46–48) on phage, taken in connection with the work of others on other organisms, to be composed usually of thousands—indeed, in most organisms tens or hundreds of thousands—of exactly strung nucleotides. Yet the possibility of biological evolution must already have been present when the first few nucleotides, uniting chainwise, found themselves in a position to replicate and to mutate into new and ever longer nucleotide sequences.

Benzer's remarkable studies of the past five years involve the mapping of the positions of mutations in the phage chromosomes through the comparison of the frequencies with which they undergo recombination with one another in biparental inheritance. They are the same in principle—but oh how far removed in technique!—as the chromosome mapping carried out by these means in Drosophila some fifty years ago, but the frequency of formation of new combinations is about 1,000 times higher so that they allow mapping on a thousand-fold smaller scale in these tiny organisms. By these means Benzer has been able to obtain evidence that usually a mutation involves the substitution of just one nucleotide, lying at a given point in a chain of some 200,000 nucleotides. Thereby, through a chemical upset, a different one of the four possible types of nucleotides becomes substituted in the place of the one that was originally present at that point. In consequence, the long group of nucleotides constituting a given gene produces a different effect on the organism.

Benzer has been able to promote the occurrence of such substitutions by offering the living phage incorrect nucleotides, analogues of those normally present, and these abnormal nucleotides on further replication under normal conditions are likely to choose from the medium complements that would not have matched the original nucleotide and that thus initiate a new but again stable self-replicating type. Other means also have been found of inducing nucleotide substitutions. Most recently Freese, a physicist starting his biology as a student of Delbrück, then of Benzer and of Watson, has been getting evidence (49) that different agents promote different types of substitutions, so that three main types can be distinguished, comprising the three conceivable kinds of substitutions. Moreover, the chemical steps involved are beginning to be unravelled (44). Any such agent, however, whether "chemical" or, like radiation, "physical," produces its results broadcast, affecting according to the accident of its point of impact any one of thousands of different nucleotide sites, and the nature of the effect produced by the given mutation on the organism depends more on what site has been changed than on just which of the three possible changes has occurred. Thus we are still almost as far as ever from producing the kind of mutational effect we wish to.

IV. *The Evolution of Protoplasm*

If now we grant that with the appearance of the first successful chains of a few nucleotides in a medium containing their precursors, free nucleotides of triphosphate type, the basis of life came into being and the trigger was thereby pulled for the extraordinary chain-reaction known as biological evolution, with all its subsequent miracles of intricate adaptation, we are still at a primeval stage, an enormous way from the very first protoplasmic or even protein-containing forms. And although we call this early material living, because of its key properties of replication, mutation, and the selectively differential rate of replication of diverse types, nevertheless the ability to anabolize and catabolize and maintain a balanced flux, and the innumerable adaptive reactions that we usually think of in connection with living matter, have not yet appeared. Certainly most of these reactions are mediated, on a second or higher level of operations than those of the genetic nucleic acid, by proteins primarily, assisted by carbohydrates, lipids, prosthetic groups, sterols, and a whole retinue of accessory substances, some of all these being in highly organized supra-molecular configurations. And each of these substances must have undergone a long evolution, in its genetic basis, to issue in the particular type of molecule, highly adapted for its functions, that exists today. Let us recognize, then, that the road from protogene to protoplasm was an almost interminably long one, compounded of millions of steps, selected through their usefulness in survival and further multiplication.

We cannot speculate here on even the general outlines of the succession of stages that must have been passed through in the course of this progression. But in all probability the first great series of steps consisted mainly in the development of the ability of the nucleic acid chains to get hold of given amino acids or peptides of types already present in the medium and fit them together into patterns that were somehow useful to the nucleic acid core by supplying it protection, strength, stability, and finally enzymes that enabled it to undertake ever more far-reaching chemical operations. Even the tobacco mosaic virus constructs or readapts a protective protein coat, as well as some other proteins, at least one of which serves it as an enzyme when in the host cell. The problem of the manner in which the nucleotides, through their own pattern of sequence, manage to determine the amino-acid sequences of a polypeptide chain has not yet been cracked; Gamow's (50) brave attempt at it has proved too simple to accord with known facts of the respective sequences. In the already historic symposium on "The Structure and Function of the Genetic Elements" held at Brookhaven, Long Island, June 1–3, 1959, a whole series of remarkable studies were reported by Crick (51), Brenner (52), M. B. Hoagland (53), G. L. Brown (54), and others in which the problem of investigating the way in which the nucleic acids construct the proteins was dealt with. Brenner has already succeeded in mapping on the nucleic acid chain of a phage of *Escherichia coli* several sites, not very distant from one another, each of which has to do with a different peptide that enters into the composition of a particular protein that protects the nucleic-acid-containing head of the phage. But in between these is another group of sites, having to do with osmotic properties. Moreover, Hoagland and Brown obtained evidence indicating that even for the placement of a single amino acid a highly complex structure may be necessary. Yet the road is now open for investigation, and a few years, or even weeks, may see much progress in the analysis.

The path from the construction of protein to groups of interacting labile and enzymatic proteins and then to protoplasm, with its highly organized structure, again represents an enormous evolutionary distance, regarding the stages of which we will not venture here to speculate. We can only say at the moment that the primitive genetic nucleic acid had in its time to utilize those types of organic molecules that already existed in formed condition in the medium about it as the result of the prior operation of inanimate physicochemical processes. Certainly amino acids, pyrrols, carbohydrates, fatty acids, alcohols, some prosthetic groups, etc. must already have been present in a medium in which even nucleotides were preformed. But as these materials were confiscated by the replicating nucleotide chains, the supply must have diminished to the dribble that continued to be produced by inanimate agencies. Thus there was a high premium on means of converting other materials, related to the ones needed, into the latter, and mutations in the nucleic acid chains that gave their proteins enzymatic properties of a type helpful in making these conversions were preserved.

As N. H. Horowitz (55) has spelled out with illustrations, ever more extended chains of conversion were thereby developed, which finally enabled some living things to synthesize all their organic substances, literally from the ground up. At the same time, proceeding catabolically in an almost opposite direction, enzyme systems were developed that allowed materials of diverse kinds to be broken down ever further and more effectively, both for obtaining energy for syntheses and the performance of work and for the garnering of building blocks with which to reconstitute the fragmented materials into the distinctive macro-molecules of the given organism. Along with all this, hosts of inter-conversions, used in protoplasmic operations and in the building of fine structures—operations that we are hardly aware of as yet—must also have been developed. Yet all this still allowed only a tiny stream of replications prior to the time when the problem of effectively converting the energy of visible light into that of organic compounds was overcome. Only after that could the stream of turnover of materials achieved by living things on earth become rapid, continuous, and strong.

The main line, now fully protoplasmic, in which this revolutionary development occurred, seems to have been ancestral not only to the algae but to all the higher organisms, both plant and animal, although the latter sacrificed their chlorophyll. For the similarities in bacterial, plant, and animal protoplasm—their enzyme systems, genetics, and other features—is too far-reaching to be explained on the ground of parallel evolution. But the whole course of molar morphological and physiological evolution from the first fully photosynthetic organism on to all the higher forms, the story on which today's biology courses concentrate almost all of their attention, is probably no longer nor more intricate, if as much so, as that of the plodding biochemical development that preceded it. And all of the development, both before and after that critical point, let us again emphasize, depended on mutations in the genetic nucleic acid.

V. *The Preconditions for Life*

Our discussion of the origination of living matter has thus far been focused on defending the thesis that this event should be identified with the formation of nucleotide chains, or (to use our biological synonym) "protogenes." We have taken for granted the prior origination, through natural physicochemical operations, of these nucleotides themselves, as well as of many and varied other types of organic molecules which the primitive nucleotide chains first used ready-made and later manufactured. Yet, as has so often been urged, we do not see these organic substances arising from inorganic ones today in a state of nature without the intervention of already existing living things. Why should there be this seeming contradiction?

It was Darwin who long ago pointed out that in the earth's early history there was a vast amount of time for diverse organic substances gradually to accumulate through natural chemical reactions, but that today such accumulation would be prevented, and the operations masked, by the fact that organisms are already in existence that would take in and utilize such substances as fast as they were formed. Although I did not until recent years learn of this passage of Darwin, I have been regularly teaching the same thing to classes for some four decades or more, for it is a fairly obvious inference, although it did not attract widespread attention until a few years ago.

In the meantime, however, increasing evidence has accumulated from diverse quarters that there are a number of different ways in which organic substances inevitably become built up on any planet having atmospheric and aqueous envelopes, crust composition, temperature, and incident radiation like that prevailing on the primitive earth (56). For one thing, hydrides of diverse kinds, including ammonia, water, methane, and other hydrocarbons, must originally have been abundant during the anoxic period (as they are today on Jupiter) before the escape of most of the earth's atmospheric hydrogen, and these as well as some of the nitrides lend themselves readily to the formation of various organic molecules. Urey's student Miller (57, 58) at Chicago demonstrated the synthesis of various amino acids as well as other organic materials, as expected, when electric discharges were passed through gaseous mixtures of such substances.

But even in an atmosphere like that of today, minus its oxygen (which is doubtless, as the astronomer Russell (59) long ago pointed out, a product of photosynthesis by living matter and not a condition for life), the impinging short ultraviolet from the sun, not blanketed off by ozone derived from the oxygen, will result, as Urey has pointed out, in the formation of carbohydrates, amino acids, and other organic molecules from the water, carbon dioxide, and nitrogen. These will tend to rain down until some of them fall into positions under water or under ground where the larger molecules will be subject to a much lower rate of decay than of accumulation. And again, electric discharges, the radiation from radioactive materials and cosmic rays, and heat (as from volcanoes), acting upon materials near the earth's surface and in its waters, will supply the energy for syntheses. Moreover, as the materials are brought into contact with one another in diverse mixtures by the action of the currents in the water and air, ever higher and more diverse superstructures of organic chemical combination tend to be formed. Thus it is to be expected that at last, just *before* the appearance of life, the very ocean had become, in Haldane's (60, 61) vivid phraseology, a gigantic bowl of soup. Drop into this a nucleotide chain and it should eventually breed!

VI. *Life Elsewhere and Elsewhen*

According to calculations of Carl Sagan (62), it may even have happened on the moon, in its early days before it lost its atmosphere, that the incident ultraviolet, gamma, and cosmic radiation, acting on the outskirts of its atmosphere, produced a rain of organic materials, later covered and protected by an accumulation of meteoric and perhaps also volcanic dust.

This may even now be awaiting our arrival. Moreover, it is possible that before all water had disappeared this material succeeded, fleetingly, in recombining to give rise to a primitive and abortive life. Today however, in the absence of any effective atmospheric blanket, such material would, if brought to the surface, be baked by heat as well as distintegrated by prolonged burning with ultraviolet and ionizing radiation.

Mars seems to provide the only place in our solar system other than the earth that has had even halfway suitable conditions for the origination and continuation of life, if we may believe the recent reports of temperatures well above the boiling point on the surface of Venus. Although on Mars nearly all the free water is gone, a little precipitation still occurs, and there must be much bound water. As we have noted in connection with our earth, free oxygen is not required for life. And although Mars goes into a state of deep freeze at night, the daytime temperatures over much of its surface are suitable.

As for more direct evidence of Martian life, there are of course the familiar observations of seasonal changes in the darker, lower, and therefore moister areas, from brownish in the winter to a grayish, possibly greenish hue in the summer, suggestive of a masked chlorophyll such as we find among lichens at high altitudes on earth. To this phenomenon three recent findings are now to be added. The first consists of observations of changes in the polarization of light reflected from the dark areas that show them to be covered in winter with minute objects, less than a millimeter in diameter, that grow somewhat larger in the summer. Second, we must note the apparent ability of these particles to remove, destroy, or, more likely, overgrow, precipitates of dust that have been left on these areas by winds. Third, spectroscope studies of these same regions indicate the presence of those distinctive carbon-hydrogen bonds which on earth are found only in materials produced by living things.

We may judge, *very* conservatively, that "we" will get to Mars within the next quarter-century, provided that "we" do not have a nuclear war. Surely, considering the inhospitality of Martian conditions, we will find that life has not evolved nearly so far there, nor become nearly so abundant or diverse, as on the earth. Will this alien life, like ours, have its basis in some kind of genetic nucleic acids? Or are there other compounds capable of replicating their variations, and similarly versatile? If so, why have we found no evidence of them here? With our present limitations of knowledge and imagination, the most plausible view is that even on Mars the basic material of life will consist of a nucleic acid chain. Beyond that, however, anyone would be rash to predict the chemistry, morphology, and physiology of Martian life on its upper levels, corresponding to those of our protoplasm. But in any case, it seems very unlikely that the biochemistry of Martian life-forms would in general be very similar to earth's. Their meat would be our poison and vice versa. All this should provide one of the most fascinating fields for exploration by the future biologists, biochemists, and geneticists who are now children in school.

Let us not forget here that our solar system is but one speck in many billions in our own galaxy and that there are many billions of galaxies. Modern indirect astronomical evidence indicates that a considerable proportion of stars, perhaps 10 per cent (63), have their families of planets. A fair proportion of these planets must be of a size, temperature, and physical make-up similar to our earth's. It is true that, on the recent theory according to which only third- and later generation stars provide enough of the moderately heavy elements needed for the construction of life as known to us, there may be relatively few stars thus supplied which are nearly as old as our own sun (64). Yet even with this limitation, it remains likely that life has already originated on an enormous (in an absolute sense) number of planets in that part of the universe whose light is capable of reaching us. Since its origination, much of this life must have undergone a branching Darwinian evolution. And in the course of this evolution, on the older of these worlds, higher forms must have appeared.

These forms would doubtless be more or less analogous to ours in many ways, just as the structure of an octopus is analogous to that of a man. And, as shown by this illustration, intelligence is polyphyletic. However, there are too many critical accidents and concatenations of circumstances determining the direction of evolutionary developments for us to admit the possibility that on any of these other worlds plants or animals have evolved that could be classified in our earthly phyla, much less in any of our classes, orders, or genera (65). What a comparative study all this would make, and how enlightening, provided that the contacts did not somehow result in our destruction!

Leaving aside these speculations on life elsewhere, let us, in closing, descend again to earth and remind ourselves that in pursuing this investigation into life's basis and origins, we have at last reached the point, through work like that of Kornberg, where we are actually beginning to synthesize in the laboratory those genetic chains which, according to the theory based on a combination of so-called classical genetics with the most modern revolutionary developments, constitute·the most primitive form of life. In so doing we have arrived, although by a different pathway, at that act of origination of animate material which, short of the coming of man himself, constituted the most epochal event of all earth's history. What will we progress to from this in solving life's problems further and in contriving, by artificial means, living things like or better than those which from time immemorial we have always regarded as being beyond the capability of mere man to create?

VII. *Addendum*

Since the completion of the above article a considerable number of studies on biochemical genetics have appeared that further strengthen the evidence for the conclusion that it is the genetic nucleic acid—the gene material—and its activities which lie at the foundation of all other materials and operations in living things. However, in the Spring, 1961, issue of PERSPECTIVES (4:177–198), Winston Salser has advocated a version of the older view, espoused by Haeckel and recently in more detail by Oparin and other Lysenkoists, according to which the protoplasm originated first, developed anabolic and catabolic reactions, and later gave rise to genetic nucleic acid as an aid in the control and transmission of minutiae of its operations. According to the argument presented in our present article this view would put the cart before the horse. For anything that can undergo biological evolution must be possessed of the faculty of replicating both itself and an unlimited variety of mutations of itself that would be capable of exerting widely different chemical influences, and there is no good evidence that this faculty exists in any other material of living things than the genetic nucleic acid.

It is true that, once the complications of protoplasm have been evolved, a number of different mechanisms, ultimately gene-dependent, can come into existence whereby a given product can in its cellular environment react "autocatalytically" by serving as a primer in the construction of more of the same product. This happens, as Salser points out, in the case of adaptive enzyme formation, as when a given permease results in the entry of a substrate which stimulates the production of both that permease and the adaptive enzyme which acts on the substrate. Similarly, as Sonneborn (66) has found, the prior existence of a given pattern of ectoplasmic materials in a Paramecium or the instigation of the production of a given type of ectoplasmic antigen in it may result in the priming of the reactions whereby just that pattern or that antigen becomes reproduced, out of a number of other alternatives, when the cell as a whole reproduces. It is to be expected that evolution took much advantage of the possibility of having adaptive protoplasmic mechanisms of this kind.

However, these mechanisms are dependent on the prior existence of elaborate gene-products and gene-dependent protoplasmic operations (67). And although these may be switched on and off from one to another, and can even present a fair repertoire of alternatives, nevertheless they do not, like the genetic nucleic acid, have in themselves unlimited possibilities of further inheritable variation and evolution. Instead, their further evolution results from the occurrence of advantageous mutations in the genetic nucleic acid that underlies their existence. For this reason the present writer in an earlier paper (68) designated these autocatalytic mechanisms as "epigenic."

REFERENCES

1. C. DARWIN. The origin of species by natural selection. London: J. Murray, 1859.
2. H. J. MULLER. Amer. Nat., 56:32, 1922.
3. ———. Proc. Int. Cong. Plant Sci., 1:897, 1929.
4. ———. Sci. Mon., 29:481, 1929.
5. ———. Proc. Roy. Soc. B, 134:1, 1947.
6. ———. Proc. Amer. Philos. Soc., 93:459, 1949.
7. ———. Science, 121:1, 1955.
8. ———. Bull. Amer. Math. Soc., 64:137, 1958.
9. ———. Accad. Naz. dei Lincei, Quad. No. 47:15, 1960.
10. A. Weismann. Über die Vererbung. Jena: G. Fischer, 1883.
11. O. HERTWIG. Jenaische Ztschr., 18:276, 1885.
12. E. STRASBURGER. Neue Untersuchungen über den Befruchtungsvorgang bei den Phanerogamen, als Grundlage für eine Theorie der Zeugung. Jena: G. Fischer, 1884.
13. A. VON KÖLLICKER. Z. wiss. Zool., 42:1, 1885.
14. W. ROUX. Über die Bedeutung der Kernteilungsfiguren. Leipzig: Engelmann, 1883.
15. G. MENDEL. Verh. d. Naturf. Vereins in Brünn, 1865, 4:1, 1866.
16. C. CORRENS. Bot. Zeitung, 60:65, 1902.
17. W. S. SUTTON. Biol. Bull., 4:24, 1902.
18. E. B. WILSON. J. Exp. Zool., 2:371, 1905.

19. T. H. MORGAN, A. H. STURTEVANT, H. J. MULLER, and C. B. BRIDGES. The mechanism of Mendelian heredity. New York: Holt & Co., 1915.

20. W. JOHANNSEN. Amer. Nat., **45**:129, 1911.

21. H. DRIESCH. Analytische Theorie der organischen Entwicklung. Leipzig: Engelmann, 1894.

22. L. T. TROLAND. Cleveland Med. J., **15**:377, 1916.

23. ———. Amer. Nat., **51**:321, 1917.

24. E. HAECKEL. Generellen Morphologie, **1**:167, 1866.

25. ———. Die Welträtsel. Leipzig: Alfred Kröner, 1908.

26. J. G. KEMENY. Sci. Amer., **192**:58, 1955 (citation of von Neumann's machine on pp. 64–67).

27. A. OPARIN. The origin of life. New York: Macmillan, 1938. New York: Academic Press, 1957.

28. F. MIESCHER. Die Histochemischen und Physiologischen Arbeiten von F. Meischer. Leipzig: Vogel, 1898.

29. R. ALTMANN. Arch. Anat. und Physiol., Physiol. Abt., **23**:524, 1889.

30. A. KOSSEL. Arch. Anat. und Physiol., Physiol. Abt., **27**:223, 1893.

31. P. A. LEVENE and L. W. BASS. Nucleic acids. New York: The Chemical Catalogue Co., Inc., 1931.

32. W. S. STANLEY. Science, **81**:644, 1935.

33. O. T. AVERY, C. M. McLEOD, and M. McCARTY. J. Exp. Med., **79**:137, 1944.

34. A. D. HERSHEY and M. CHASE. J. Gen. Physiol., **36**:39, 1952.

35. H. FRAENKEL-CONRAT. J. Amer. Chem. Soc., **78**:882, 1956.

36. A. GIERER and G. SCHRAMM. Nature, **177**:702, 1956.

37. J. D. WATSON and F. H. C. CRICK. Nature, **171**:737, 964, 1953.

38. R. E. FRANKLIN and R. G. GOSLING. Nature, **171**:740, 1953.

39. M. H. F. WILLIAMS, A. R. STOKES, and H. R. WILSON. Nature, **171**:738, 1953.

40. C. LEVINTHAL. Proc. Nat. Acad. Sci. USA, **42**:314, 1956.

41. M. MESELSON and F. W. STAHL. Proc. Nat. Acad. Sci. USA, **44**:671, 1958.

42. J. H. TAYLOR, R. S. WOODS, and W. L. Hughes. Proc. Nat. Acad. Sci. USA, **43**:122, 1957.

43. A. KORNBERG. *In:* D. McELROY and B. GLASS (eds.), The chemical basis of heredity, p. 579. Baltimore: Johns Hopkins Press, 1957.

44. I. R. LEHMAN, S. B. ZIMMERMAN, J. ADLER, M. J. BESSMAN, E. S. SIMMS, and A. KORNBERG. Proc. Nat. Acad. Sci. USA, **44**:1191, 1958.

45. S. OCHOA and L. A. HEPPEL. *In:* The chemical basis of heredity, p. 615.

46. S. BENZER. Proc. Nat. Acad. Sci. USA, **41**:344, 1955.

47. ———. *In:* The chemical basis of heredity, p. 70.

48. S. BENZER and E. FREESE. Proc. Nat. Acad. Sci. USA, **44**:112, 1958.

49. E. FREESE. Proc. Nat. Acad. Sci. USA, **45**:622, 1959.

50. G. GAMOW. Nature, **173**:318, 1953.

51. F. H. C. CRICK. Brookhaven Symposia Biol., **12**:35, 1959.

52. S. BRENNER and L. BARNETT. Brookhaven Symposia Biol., **12**:86, 1959.

53. M. B. HOAGLAND. Brookhaven Symposia Biol., **12**:40, 1959.
54. G. L. BROWN, A. V. W. BROWN, and J. GORDON. Brookhaven Symposia Biol., **12**: 47, 1959.
55. N. H. HOROWITZ. Proc. Nat. Acad. Sci. USA, **31**:153, 1945.
56. H. C. UREY. The planets. New Haven: Yale University Press, 1952.
57. S. L. MILLER. Science, **117**:528, 1953.
58. S. L. MILLER and H. C. UREY. Science, **130**:245, 1959.
59. H. N. RUSSELL, R. S. DUGAN, and J. Q. STEWART. Astronomy. Boston: Ginn, 1927.
60. J. B. S. HALDANE. The causes of evolution. London: Longmans, Green, 1930.
61. ———. *In:* New Biology No. 16, p. 12. London: Penguin, 1954.
62. C. SAGAN. Proc. Nat. Acad. Sci. USA, **46**:393, 1960.
63. G. P. KUIPER. J. Roy. Astron. Soc. Canad., **50**:57, 105, 158, 1956.
64. M. BURBIDGE and G. BURBIDGE. Science, **128**:387, 1958.
65. H. J. MULLER. Amer. Biol. Teacher, **23**:65, 1961.
66. T. M. SONNEBORN. *In:* L. C. DUNN (ed.), Genetics in the 20th century, p. 291. New York: Macmillan, 1951.
67. J. LEDERBERG. Stanford Med. Bull., **17**:120, 1959.
68. H. J. MULLER. *In:* Genetics in the 20th century, p. 77.

GENETIC SIGNIFICANCE OF MICROBIAL DNA COMPOSITION

FRANK LANNI, Ph.D.*

Now that deoxyribonucleic acid (DNA) in some microorganisms and ribonucleic acid (RNA) in others are widely accepted as the material carriers of genetic information, and now that both the functional genes and the nucleic acid macromolecules are conceived as linear structures minutely differentiated along their lengths, attention has focused sharply on details of microstructure (1–5). The root idea, which is still only a brilliant hypothesis, is that an intimate correspondence exists between the sequentially ordered mutational sites that make up a functional gene and the similarly ordered nucleotides, or nucleotide pairs, that make up the carrier molecule of nucleic acid or the relevant segment of such a molecule. There is a justified hope that experimental validation or invalidation of the idea will soon be forthcoming.

There is also a danger. The intense and necessary preoccupation with the fine structure of genes and molecules can easily divert attention from significant relations at higher structural levels. In fact, so great is the devotion to fine structure that one can sense in some quarters a skepticism about the value of a panoramic view.

This essay proposes first that near-identity in overall DNA composition is a highly correlative and heuristic feature of genetically interacting microbes (6). The later part of the discussion deals with somewhat more speculative aspects of the genetic significance of DNA composition. Occasional reference is made to microbes with RNA genomes.

* Department of Bacteriology, Emory University, Atlanta 22, Georgia. The author's research is supported in part by grant E-857 from the National Institute of Allergy and Infectious Diseases, U.S. Public Health Service.

The author wishes to express his appreciation to numerous colleagues, at Emory University and elsewhere, for their valuable suggestions and criticisms, and especially to Dr. John M. Reiner for his inspiration and guidance.

I. *Correlation between DNA Composition and Genetic Homology*

Associated closely with the root idea is a second familiar idea: namely, that genotypic similarities and differences between two organisms reflect similarities and differences in the nucleotide fine structure of their DNA. A testable consequence should be that organisms which show extensive *phenotypic* similarity—as judged conveniently for bacteria by their close association in a determinative scheme such as Bergey's—ought to resemble one another in DNA base composition. The monumental studies of bacterial DNA by Lee, Wahl, and Barbu (7) and later studies by Spirin and colleagues (8) have largely confirmed this expectation. There are, however, many notable exceptions, as might be expected from the notoriously arbitrary character of phenotypic classifications. To propose DNA composition itself as a taxonomic criterion may be valuable for taxonomy but would merely beg some of the questions that we have in mind.

Accordingly, we look for a more reliable sign of actual genetic homology. We find it easily in the ability of two microorganisms to undergo specific genomic interactions (matings), as evidenced by genetic recombination and certain other phenomena. In current thinking, these interactions involve specific molecular pairings, which themselves imply considerable similarity in the nucleotide fine structure of the genomes. If, moreover, the homologous portions of two genomes were to encompass a large number of nucleotides, it would be reasonable to expect a similarity in overall base composition of the appropriate nucleic acids. For DNA genomes, it seems immaterial for our purpose whether the genomes mate directly or whether the actual pairing occurs between accurately coded RNA intermediaries (9).

A survey of the literature reveals a striking correlation. Taken in pairs, all microorganisms whose genomes are known or reasonably suspected to be at least partially homologous (through direct or indirect signs of interaction), and whose DNA composition is known or may be confidently surmised, are effectively indistinguishable as to quantitative DNA base composition. Conversely, microorganisms for which there is no positive sign or reasonable suspicion of genetic homology may, and frequently do, differ appreciably in base composition.

The relevant data are gathered in Table 1, which lists DNA analyses for the well-known T phages of *Escherichia coli* B, several temperate phages or their virulent mutants, and selected bacteria. The tabulation neglects trace constituents in the DNA of several of the listed organisms.

In the following discussion the characteristic measure AT/GC— (adenine + thymine)/(guanine + cytosine)—is used for brief specification of DNA composition. For convenience, a DNA is called *equimolar* if AT/GC falls between 0.87 and 1.10. Such a definition, which may not be totally arbitrary (see below), minimizes distraction by the inevitable disagreements in analyses of similar material by different workers.

TABLE 1

BASE COMPOSITION OF DNA FROM SELECTED MICROBIAL SOURCES

DNA SOURCE	MOLAR PROPORTIONS OF BASES*				AT/GC†	REFER- ENCE
	Adenine	Thymine	Guanine	Cytosine		
Coliphages:						
T2	1.30	1.30	0.73	0.67‡	1.86	(10)
T4	1.29	1.32	0.73	0.65‡	1.89	(10)
T6	1.30	1.30	0.73	0.67‡	1.86	(10)
T5	1.21	1.23	0.78	0.78	1.56	(10)
T1	1.08	1.00	0.92	1.00	1.08	(11)
T3	0.90	1.12	0.94	1.04	1.02	(12)
T7	1.03	1.06	0.98	0.93	1.09	(13)
λv	0.85	1.14	0.92	1.08	1.00	(14)
Salmonella phages:						
A1	0.94	1.33	0.75	0.98	1.31	(14)
P22 (virulent)	"equimolar"					(15)
Enteric bacteria:						
Escherichia coli B	1.02	0.98	0.98	1.02	1.00	(16)
Escherichia coli K12	1.04	0.96	0.99	1.01	1.00	(17)
Salmonella typhosa	0.94	0.94	1.07	1.06	0.88	(8)
Salmonella paratyphi A	0.99	1.01	1.00	1.00	1.00	(7)
Salmonella typhimurium	0.98	1.01	1.02	0.99	0.99	(7)
Shigella dysenteriae	0.94	0.92	1.07	1.07	0.87	(8)
Shigella flexneri	1.00	1.03	0.99	0.98	1.03	(7)
Other bacteria:						
Diplococcus pneumoniae III	1.19	1.26	0.82	0.72	1.59	(18)
Streptococcus pyogenes	1.34	1.32	0.66	0.68	1.97	(7)

* Calculated to a total of 4.00.
† AT/GC = (adenine + thymine)/(guanine + cytosine).
‡ Refers to 5-hydroxymethylcytosine instead of cytosine.

The following subsections deal briefly with a variety of genomic interactions (1–5). In each case the proposed correlation is challenged maximally by comparison of organisms whose ancestral relation is unknown.

A. GENETIC RECOMBINATION IN PHAGE

The production of genetic hybrids between two organisms is at present the surest sign of genetic homology. In phage, the phenomenon is best illustrated by recombination between the independently isolated members of the T-even group: T2, T4, and T6 (19, 20), which do not differ significantly in DNA base ratios. There are no examples of recombination between phages known to differ appreciably in base composition.

B. PHAGE-BACTERIUM HOMOLOGY: LYSOGENY

In certain bacteria which survive infection by so-called *temperate* phages, the phage genome attaches in highly stable fashion to a specific site on the bacterial genome, and thereafter the two genomes reproduce synchronously until accident or artifice intervenes. The attached phage genome is called *prophage*, and the bacteria are said to be *lysogenic*. The specific association of prophage with the host genome is widely interpreted as a sign of genetic homology. In one case, that of the temperate phage λ and its host *E. coli* K12, recombination between the two genomes has been demonstrated (21, 22).

The relevant DNA analyses leave much to be desired. Phage λv, a virulent (non-lysogenizing) mutant of λ, appears to contain equimolar DNA, like that of K12; but the briefly reported base ratios for the phage deviate somewhat from certain well established equalities (T/A = C/G = 1) for ordinary duplex DNA (23). The analysis for phage A1, a temperate phage of *Salmonella typhimurium*, is even more distressing in this respect. Our caution in accepting this analysis seems justified by the more recent report (15) that a virulent mutant of the closely related phage P22 has "equimolar" DNA, like that of the host. While further analyses of coliphages and salmonella phages would clearly be valuable, a more searching test of the correlation would be furnished by data for a series of temperate phages whose hosts represent a wider variety of DNA composition.

C. GENETIC RECOMBINATION IN BACTERIA FOLLOWING CONJUGATION

The most relevant examples are the successful crosses between *E. coli* and *Shigella dysenteriae*, *flexneri*, and *boydii* (24), *S. typhimurium* (25), and *S. typhosa* (26). All these bacteria, or their relatives within the same genus, possess equimolar DNA. There are no examples of recombination between bacteria known to differ in DNA base composition.

D. TRANSDUCTION

The term is used in its limited sense to refer to phage-mediated transfer of genetic markers from one bacterium to another. The phage vector is usually (not necessarily) temperate for the recipient bacterium, and it may be temperate also for the donor strain.

As regards phage-bacterium homology, the temperate phages P22 (27) and λ (28), whose DNA composition has already been discussed, are both successful transducers. In particular, transduction by λ appears definitely to involve recombination of λ prophage with the host genome (21, 22). As regards bacterium-bacterium (donor-recipient) homology, the most cogent examples are the intergeneric transductions between strains of *E. coli* and *Sh. dysenteriae* (29, 30), all presumably with equimolar DNA. Analysis of the DNA of the phage vector, P1, has not been reported.

E. TRANSFORMATION OF BACTERIA BY HETEROLOGOUS DNA

Intergeneric transformations by free DNA have been achieved between a Type II pneumococcus and three strains of viridans streptococci, including strain D (31). Complete DNA analyses are available only for a Type III pneumococcus (Table 1) and for several non-viridans streptococci (typical example in Table 1). Fortunately, the minor reservations left by these comparisons have been dealt with by Dr. Julius Marmur, who has kindly informed us of unpublished results. After confirming the transformation between the strain D streptococcus and the Type II pneumococcus, Marmur was able to establish a near-identity in the DNA base composition of the original strains by indirect but highly sensitive comparisons of two independent kinds: band positions in a cesium chloride density gradient (32) and the optical characteristics of thermal denaturation (33).

F. GENETIC RECOMBINATION IN THE INFLUENZA VIRUSES

These viruses, with genomes currently believed to consist of RNA, are grouped into several serologically distinct types, each represented by many strains that cross-react serologically. Genetic recombination has been reported among strains of type A and among strains of type B but not between representatives of A and B (34, 35). RNA base analyses (36) showed that strains of a given type, A or B, did not differ significantly among themselves but differed significantly from strains of the alternate type. The ratio of (adenine + uracil)/(guanine + cytosine) fell between 1.22 and 1.28 for five A strains and between 1.38 and 1.43 for three B strains.

II. *Application of the Correlation*

Some of our friends have suggested that we set up a consulting bureau for microbial matrimony, to advise about the nucleic acid credentials of prospective mates. Actually, a widely circulated and frequently updated chart of microbial base compositions could have considerable utility, not only as a guide in match-making, but also in helping to extend the factual area over which the correlation might be tested.

In this section we indicate some other sorts of application, beginning with the situation that provoked the entire analysis.

A. HOMOLOGY (?) OF T PHAGES WITH *E. coli*

Stent (9) has recently reviewed a large body of circumstantial evidence from cytological, genetic, and radiobiological studies suggesting that phages T1, T3, and T7, but not T2, T4, T5, or T6, possess some genetic homology with their usual host, *E. coli*. Reference to Table 1 shows that phages T1, T3, and T7, and not the others, have an equimolar DNA like that of *E. coli*. The correlation thus provides independent evidence compatible with the proposal. *Serratia marcescens* (AT/GC = 0.72 [7], 0.69 [38]) can be infected by T3 and T7 (37) but differs from these phages in base composition. It might be instructive, therefore, to repeat with this host some of the experiments that suggested homology of T3 and T7 with *E. coli*.

B. ANOMALOUS INHERITANCE OF COLICINOGENIC DETERMINANTS

Colicins are specific antibiotic proteins elaborated by certain enteric bacteria. The hereditary colicin-producing ability of a given strain can be transferred to nonproducing strains by a process of cell conjugation. Intergeneric transfers have been made not only from one equimolar enteric bacterium to another, but also from one of these, *E. coli*, to *Klebsiella pneumoniae* (39). If we assume that the DNA of the latter resembles that of its very close taxonomic relative *Aerobacter aerogenes* (AT/GC = 0.75 [7], confirmed in several laboratories), we are led to think either that our correlation stands in jeopardy or that there might be something peculiar about the hereditary factors controlling colicin production. The facts support the second alternative. Unlike the classical genetic determinants of bacteria, the colicinogenic determinants may be present or absent in a given strain and, when present, may be either autonomous or integrated with the bacterial genome. These properties have been crystallized in the concept "episome" (40), which embraces also the genomes of temperate phages and certain other determinants.

A provocative fact is that the integrated colicinogenic determinants, whose chemical nature is as yet unknown, are found associated with a specific region of the bacterial chromosome (41), just as are the prophages in lysogenic bacteria (42). We consider this feature below.

C. CHEMICAL IMPLICATIONS OF "MEROMIXIS"

The three chief modes of gene transfer in bacteria (conjugation with one-way transfer, transformation, transduction) typically involve transfer of only a portion of the donor genome to the recipient. The term "meromixis" has been invented for these unequal matings (43) and might be extended to other specific interactions such as that between the relatively small prophage and the bacterial chromosome in lysogeny. In each case of unequal mating we imagine that the smaller genetic interactant is homologous with a part, not all, of the larger. Therefore, in our previous discussion of intergeneric transfers in bacteria, the signs of genetic homology were strictly limited to portions, sometimes quite small, of the donor and recipient genomes; yet the respective *unfractionated* DNA preparations were indistinguishable in base composition. The situation is similar in lysogeny, where a single host cell contains roughly one hundred times as much DNA as a single phage particle and perhaps a comparably greater number of functional genes.

These considerations long ago suggested to us that unfractionated microbial DNA preparations might prove to be highly homogeneous in base composition at the level of ordinary DNA molecules; and, in fact, recent ingenious analyses (32, 33, 44) have amply verified this guess. We disclaim here any priority to the idea; our purpose is merely to illustrate the diversified utility of our correlation. Turned around, the experimental data on homogeneity justify the use of analyses of unfractionated microbial DNA preparations in further attempts to extend and apply the correlation.

Nearly homogeneous as they may be, bacterial DNA preparations nevertheless display measurable compositional heterogeneity. As an example, *E. coli* DNA fractions have been obtained with AT/GC ranging from 0.86 to 1.19 (16). Integration of the same colicinogenic determinants with each of two bacterial genomes which differ in average base composition, as discussed above, may perhaps be referable to heterogeneity of such degree, but other explanations may be imagined. For lysogenic systems, an interesting possibility is that the DNA base compositions of a set of temperate coliphages might correlate systematically with the map positions of the respective prophages on the bacterial chromosome.

III. *Some Speculations*

Granted the correlation, we can easily derive it from the currently conjectured dependence of gene traits on DNA structure, and we might even proffer the correlation as confirmation of this dependence. Given two genetically interacting organisms, such as an escherichia and a salmonella, to which we might ascribe a common ancestry but which now differ in many gene-controlled traits, we can readily accommodate the existing genetic differences in DNA of analytically identical composition. If, as generally surmised, each functional gene contains hundreds of nucleotides, and if most ("point") mutations affect only a few nucleotides (45), then the combined effect of hundreds of minute mutations scattered over hundreds of functional genes in the genome could be analytically inappreciable. In bulk, the two genomes would be identical in fine structure, hence in overall base composition. Further, a tendency to compositional divergence via point mutations or more extensive changes such as deletions would be countered, at least partly, by three processes: (*a*) cancellation of prior compositional changes by further mutations ("statistical buffering"); (*b*) convergent natural selection in a similar habitat (for *Escherichia* and *Salmonella* in the vertebrate intestine); and (*c*) interclonal transfer of genetic material. It remains to be learned whether these processes are *quantitatively* sufficient to harmonize the existing taxonomic differentiation with the existing analytical identity in DNA composition.

By similar arguments, we might attempt to explain the near-homogeneity of a microbial DNA preparation with respect to the base composition of its individual DNA molecules. Neglecting how it arose, we might imagine a primordial microbial ancestor whose entire genome consisted of a single typical DNA molecule with a few thousand nucleotides. During millennia of survival, it could have evolved to the current genomes, equivalent in some cases to hundreds of such molecules, by incremental molecular additions followed by gene-functional differentiation via mutation. Again, statistical buffering, convergent natural selection, and interclonal transfer could minimize compositional divergence in the molecular population.

But these arguments have a specious tone, and in conclusion to this essay we shall summarize evidence suggesting the existence of more determinate controls over DNA composition. In particular, we shall challenge the sufficiency of any theory of DNA structure and replication that is attentive merely to the individual nucleotides and neglects their mutual relations in a chain (see, for example, the theory of Watson and Crick [46]).

To begin, Figure 1, a chart of the frequency distribution of microbial DNA compositions, suggests the following deviations from a random distribution:

1. The distribution is severely constrained within two symmetrical limits, which may well lie at 25 and 75 per cent GC (guanine + cytosine). It is not difficult to imagine how compositional imbalances greater than these would critically reduce the ability of DNA to code macromolecular specificities.

2. The individual GC values appear to coalesce into more or less distinct clusters, of which we wish at present to discern five. To explain these discrete characteristics, we might suppose that the clusters evolved independently from distinct primordial ancestors. Alternatively, if we admit the possibility of large mutational changes in base composition, we might suppose that certain compositions are excluded a priori or are disadvantageous

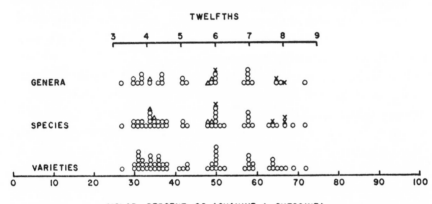

FIG. 1.—Frequency distribution of microbial DNA base compositions. The parameter GC, rather than AT/GC, is used to facilitate observations of symmetry. All of the 58 published analyses of bacterial DNA by Lee and colleagues (7) were plotted as "varieties" after rounding off to the nearest whole-number per cent. Averages for varieties of a given species were plotted as "species" to give what is regarded as the most significant portion of the chart. Averages for species of a given genus were plotted as "genera" merely to show the taxonomic range of the analyses; in a few instances species of one genus deviated widely and were represented separately with negligible effect on the plot. Bacteria not dealt with by Lee (symbols ×) and bacteriophages (symbols Δ) were included at the species and generic levels whenever analyses of the same species were available from two or more laboratories. This uniform rule for inclusion avoids tedious discussion of the reliability of several excluded analyses but does not affect the main features of the chart. For T-even phages, GC means guanine + 5-hydroxymethylcytosine.

for survival. Supporting the notion of inter-cluster migration are these considerations. First, each of the five clusters contains both gram-positive and gram-negative bacterial representatives. Second, some bacterial genera are represented in more than one cluster. Third, Spirin and colleagues (47) have lately reported several independent migrations via mutation from our central cluster (their values 52, 53, and 53 per cent GC) to a cluster on the left (42 and 43 per cent GC) and to one on the right (65, 66, 67, and 67 per cent GC). These provocative results need prompt confirmation.

3. The cluster means are distributed symmetrically around the center, which is itself a cluster mean. Further, the means show a remarkable fit to a set of duodecimal fractions from 4/12 to 8/12 inclusive; this explains our temptation to set the limits of the distribution at 3/12 and 9/12. For the present, we leave the contemplation of these numbers to a contemporary Pythagoras, who might notice with aesthetic thrill that the number 12 is the product of a magic number 4 (bases) by the magic number 3 (elements in the popular triplet codes).[1]

The pronounced symmetry of the distribution speaks strongly for a symmetry in the structural and/or functional significance of two sets of bases: A and T versus G and C. For ordinary two-strand DNA, the same base affiliations have already been recognized experimentally (23) from the well established molar relations $T/A = C/G = 1$, and theoretically in the familiar Watson-Crick structure (46), which locates A opposite T and G opposite C in the two complementary strands of ordinary duplex DNA. An implication of this structure is that the GC fraction of each of the individual strands is identical with that of the duplex. Hence, Figure 1 may be regarded as a chart of single-strand compositions, and the symmetry must then represent a new feature of DNA which cannot be inferred directly from the Watson-Crick model.

The foregoing considerations engender a suspicion that single polynucleotide strands are more regular in composition than ordinarily imagined. In support of the suspicion, nucleic acid preparations which consist

---

[1] When he read this passage, Pythagoras frowned, said something about eschewing mysticism, read further, and offered the following more concrete proposal. Think of the four bases as forming two groups, with G and C in Group I and A and T in Group II. Take a single DNA strand, number bases along the strand, and suppose that all the bases whose numerical positions differ by an integral multiple of 12 belong to the same group, either I or II. You will now have 12 position-classes, each uniform as to group and each containing 1/12 of all the bases. Do you follow me? Now let there be mutations that switch an entire position-class from one group to the other, and you will have your symmetrically distributed clusters arranged according to your so-called duodecimal fractions.

partly or exclusively of single strands almost invariably prove to contain equal proportions (within 5 per cent) of purines and pyrimidines (DNA) or of 6-amino groups and 6-keto groups (RNA), even though they may deviate considerably from all other simple compositional rules for duplex molecules. This is true for the single-strand DNA of phage ϕX-174 (48), whose GC content of 43 per cent happens, incidentally, to fall neatly in one of our clusters; for certain compositionally anomalous fractions of calf thymus DNA (49), which Dr. Aaron Bendich has kindly informed us appear to consist at least partly of single strands; and for many kinds of microbial and animal RNA (8, 50). It may be of interest that a GC content of about 42 per cent seems to be the lower limit for RNA from a variety of sources.

The tobacco mosaic viruses (RNA genomes) offer an especially striking challenge to theoretical interpretation. Two mutationally related strains, the ordinary TMV and M, are indistinguishable in RNA base ratios (51, 52) but differ profoundly in base sequence (53). Extensive rearrangement of the bases thus seems to have occurred within fixed rules governing the over-all composition of single strands.

These facts and those considered earlier call for some new ideas about nucleic acid structure and synthesis. Guided by recent evidence of extensive intra-chain pairing in RNA (54), we suggest as one idea that large portions of a given polynucleotide strand are actually complementary to one another (cf. 55) according to rules that produce a near-equality between, say, purines and pyrimidines but not necessarily between any two of the four bases. A mutual complementarity of this kind might even embrace the functionally distinct molecules in the DNA (or RNA) population from a given microbial source.

As a second idea, we suggest that the over-all composition of microbial DNA molecules is under specific control by mutable genetic factors that set general species-specific rules of composition within which the individuality (specific fine structure) of the ordinary genes must then be modulated. The concept of "species-specific master genes for base composition" readily accommodates a variety of observations: (*a*) the preliminary evidence (32), not previously mentioned, of *intra*molecular homogeneity in DNA base composition; (*b*) the compositional similarity of the individual DNA molecules in the population from a given microorganism; (*c*) the analytical identity in the DNA base composition of genetically homologous organisms that have had appreciable opportunity to evolve independently; (*d*) the discrete clusters noted in Figure 1; (*e*) the suggestive taxonomic and experimental evidence of inter-cluster migration, supposedly via mutation; (*f*) the analytical identity in the RNA base composition of strains of tobacco mosaic virus that differ profoundly in base sequence.

Attractive as these ideas are, further discussion at this point seems quite premature. We prefer to await experimental developments.

IV. *Addendum*

Some significant papers were missed during the earlier survey and some others have appeared recently.

G. R. Wyatt (56) reported DNA base analyses for a heterogeneous collection of 11 insect viruses. Expressed as per cent of GC and grouped, the results were 58.8; 51.5, 51.2, 47.6; 42.7, 42.7, 42.5, 42.2; 37.5, 37.4, 34.8. These values confirm in a remarkable way the discrete clusters depicted here in Figure 1. Wyatt commented: "The scale of AT:GC ratios appears to be a discontinuous, stepwise distribution, with groups of viruses having similar values."

R. Rolfe and M. Meselson (57) and N. Sueoka (58) have published definitive papers on the homogeneity of microbial DNA preparations. P.-Y. Cheng (59) has suggested that cells contain a mechanism for controlling the length of nucleic acid chains, whether DNA or RNA. L. F. Cavalieri, B. H. Rosenberg, and J. F. Deutsch (60) have obtained evidence that "the unaggregated, unit DNA molecule of *E. coli* is actually a din.er composed of two double helices, laterally bonded together." K. K. Reddi (61, 62, 63) has extended his earlier studies of structural differences in the RNA of tobacco mosaic virus strains and has suggested that "there is a definite regularity in the arrangement of both the purine and pyrimidine nucleotides." R. Markham (64) had earlier discerned a compositional regularity in the RNA of turnip yellow mosaic virus.

In connection with an extensive comparative survey of amino-acid sequences of proteins and polypeptides, F. Šorm wrote: ". . . it may be anticipated that the composition of the proteins and nucleic acids, too, will prove to be governed by a number of rules of limited complexity" (65). Evidence of *intra*molecular sequence patterns in proteins and polypeptides has been discussed by several workers, notably by Šorm and collaborators

(66, 67) and by D. Schwartz (68, 69, 70). A current survey in our own laboratory seems to indicate an extraordinary degree of intramolecular patternization in proteins, formally consistent with our main ideas regarding compositional regularities in nucleic acids. These results will be described elsewhere.

REFERENCES

1. Cold Spring Harbor Symp. Quant. Biol., Vol. 21, 1956.
2. W. D. McElroy and B. Glass (eds.). The chemical basis of heredity. Baltimore: Johns Hopkins Press, 1957.
3. Symp. Soc. Exper. Biol., Vol. 12, 1958.
4. Cold Spring Harbor Symp. Quant. Biol., Vol. 23, 1958.
5. Brookhaven Symp. Biol., Vol. 12, 1959.
6. F. Lanni. Bact. Proc., p. 45, 1959.
7. K. Y. Lee, R. Wahl, and E. Barbu. Ann. Inst. Pasteur, 91:212, 1956.
8. A. S. Spirin, A. N. Belozersky, N. V. Shugayeva, and V. F. Vanyushin. Biochemistry, 22:699, 1957; translated from Biokhimiya, 22:744, 1957. The same data are given by Belozersky and Spirin, Nature, 182:111, 1958.
9. G. S. Stent. Advances Virus Research, 5:95, 1958.
10. G. R. Wyatt and S. S. Cohen. Biochem. J., 55:774, 1953.
11. E. H. Creaser and A. Taussig. Virology, 4:200, 1957. Dr. Harold Amos has kindly supplied some exceptionally careful unpublished analyses for the DNA of phage T1. Three separate determinations gave average values close to 0.25 for the molar fraction of each of the four bases. The ratio AT/GC was 1.04, 1.00, and 1.00. The base recovery was 94, 102, and 97 per cent. Using C^{14}-labeled host DNA to monitor purification of the phage, Amos estimated that host DNA constituted less than 1 per cent of the DNA finally submitted to analysis.
12. The entries for phage T3 are averaged from practically identical data ascribed to Fraser and Nakamura in: F. W. Putnam: Advances Protein Chemistry, 8:175, 1953; and to Knight and Frazer in: C. A. Knight: Advances Virus Research, 2:153, 1954.
13. K. D. Lunan and R. L. Sinsheimer. Virology, 2:455, 1956.
14. A. Lwoff. Bact. Rev., 17:269, 1953 (quoting J. D. Smith and L. Siminovitch).
15. N. D. Zinder, J. Cell. & Comp. Physiol., 45 (Suppl. 2): 23, 1955 (quoting A. Garen and N. D. Zinder).
16. G. L. Brown and A. V. Brown. Symp. Soc. Exper. Biol., 12:6, 1958. The E. coli strain is not designated in this paper but was identified as strain B in a personal communication from Dr. G. L. Brown.
17. B. Gandleman, S. Zamenhof, and E. Chargaff. Biochim. et Biophys. Acta, 9:399, 1952.
18. M. M. Daly, V. G. Allfrey, and A. E. Mirsky. J. Gen. Physiol., 33:497, 1950.
19. M. Delbrück and W. T. Bailey, Jr. Cold Spring Harbor Symp. Quant. Biol., 11:33, 1946.

20. M. H. Adams. Bacteriophages, pp. 11, 331. New York: Interscience Publishers, Inc., 1959.

21. W. Arber, G. Kellenberger, and J. Weigle. Schweiz. Ztschr. allg. Path., 20:659, 1957.

22. A. Campbell. Virology, 4:366, 1957.

23. E. Chargaff. *In:* E. Chargaff and J. N. Davidson (eds.), The nucleic acids, 1:307. New York: Academic Press, Inc., 1955.

24. S. E. Luria and J. W. Burrous. J. Bact., 74:461, 1957.

25. L. S. Baron, W. F. Carey, W. M. Spilman, and A. Abrams. Bact. Proc., p. 46, 1958.

26. L. S. Baron, W. M. Spilman, and W. F. Carey. Bact. Proc., p. 29, 1959.

27. N. D. Zinder. J. Cell. & Comp. Physiol., 45 (Suppl. 2): 23, 1955.

28. M. L. Morse, E. M. Lederberg, and J. Lederberg. Genetics, 41:142 and 758, 1956.

29. E. S. Lennox. Virology, 1:190, 1955.

30. J. N. Adams and S. E. Luria. Proc. Nat. Acad. Sc., 44:590, 1958.

31. R. M. Bracco, M. R. Krauss, A. S. Roe, and C. M. MacLeod. J. Exper. Med., 106: 247, 1957.

32. N. Sueoka, J. Marmur, and P. Doty. Nature, 183:1429, 1959.

33. J. Marmur and P. Doty. Nature, 183:1427, 1959.

34. F. M. Burnet. Principles of animal virology, p. 414. New York: Academic Press, Inc., 1955.

35. G. K. Hirst, T. Gotlieb, and A. Granoff. *In:* G. E. W. Wolstenholme and E. C. P. Millar (eds.), Ciba Foundation symposium on the nature of viruses, p. 191. Boston: Little, Brown & Co., 1957.

36. G. L. Ada and B. T. Perry. J. Gen. Microbiol., 14:623, 1956.

37. M. H. Adams and E. Wade. J. Bact., 68:320, 1954.

38. S. Zamenhof, G. Brawerman, and E. Chargaff. Biochim. et Biophys. Acta, 9:402, 1952.

39. Y. Hamon. Compt. Rend. Acad. Sc., 242:2064, 1956.

40. F. Jacob and E. L. Wollman. Compt. Rend. Acad. Sc., 247:154, 1958.

41. L. Alfoldi, F. Jacob, E. L. Wollman, and R. Maze. Compt. Rend. Acad. Sc., 246:3531, 1958.

42. F. Jacob and E. L. Wollman. *In:* McElroy and Glass (eds.), *op. cit.,* p. 468.

43. E. L. Wollman, F. Jacob, and W. Hayes. Cold Spring Harbor Symp. Quant. Biol., 21:141, 1956.

44. M. Meselson and F. W. Stahl. Proc. Nat. Acad. Sc., 44:671, 1958.

45. S. Benzer. *In:* McElroy and Glass (eds.), *op. cit.,* p. 70.

46. J. D. Watson and F. H. C. Crick. Cold Spring Harbor Symp. Quant. Biol., 18:123, 1953.

47. A. S. Spirin, A. N. Belozerskii, D. G. Kudlai, A. G. Skavronskaia, and V. G. Mitereva. Biochemistry, 23:147, 1958; translated from Biokhimiya, 23:154, 1958.

48. R. L. Sinsheimer. J. Molecular Biol., 1:43, 1959.

49. A. Bendich, H. B. Pahl, G. C. Korngold, H. S. Rosenkranz, and J. R. Fresco. J. Am. Chem. Soc., 80:3949, 1958.

50. D. Elson and E. Chargaff. Nature, 173:1037, 1954.

51. C. A. Knight. J. Biol. Chem., 197:241, 1952.

52. W. D. Cooper and H. S. Loring. J. Biol. Chem., 211:505, 1954.

53. K. K. Reddi. Biochim. et Biophys. Acta, 25:528, 1957.

54. P. Doty, H. Boedtker, J. R. Fresco, R. Haselkorn, and M. Litt. Proc. Nat. Acad. Sc., 45:482, 1959.

55. J. R. Platt. Proc. Nat. Acad. Sc., 41:181, 1955.

56. G. R. Wyatt. J. Gen. Physiol., 36:201, 1952.

57. R. Rolfe and M. Meselson. Proc. Nat. Acad. Sc., 45:1039, 1959.

58. N. Sueoka. Proc. Nat. Acad. Sc., 45:1480, 1959.

59. P.-Y. Cheng. Proc. Nat. Acad. Sc., 45:1557, 1959.

60. L. F. Cavalieri, B. H. Rosenberg, and J. F. Deutsch. Biochem. Biophys. Research Comm., 1:124, 1959.

61. K. K. Reddi. Biochim. et Biophys. Acta, 32:386, 1959.

62. ———. Ibid., 36:132, 1959.

63. ———. Proc. Nat. Acad. Sc., 45:293, 1959.

64. R. Markham. Cold Spring Harbor Symp. Quant. Biol., 18:141, 1953.

65. F. Šorm. Coll. Czech. Chem. Commun., 24:3169, 1959.

66. F. Šorm, B. Keil, V. Holeyšovský, V. Knesslová, V. Kostka, P. Mäsiar, B. Meloun, O. Mikeš, V. Tomášek, and J. Vaněček. Coll. Czech. Chem. Commun., 22:1310, 1957.

67. F. Šorm and B. Keil. Coll. Czech. Chem. Commun., 23:1575, 1958.

68. D. Schwartz. Proc. Nat. Acad. Sc., 41:300, 1955.

69. ———. Nature, 181:769, 1958.

70. ———. Ibid., 183:464, 1959

GENIC CONTROL OF DEVELOPMENT

ERNST CASPARI, Ph.D.*

I. *Relation of Genetics and Embryology*

When genetics was born, at the beginning of this century, it was regarded as closely related to embryology. Some of the foremost workers in the field, particularly from the zoölogical side—Bateson and Morgan, for instance—had previously worked on embryological problems. It was apparently expected that the study of heredity would yield information that would lead to a deeper understanding of the processes of development.

Actually, genetics and embryology went in opposite directions. Genetics became an atomistic science which interprets biological phenomena in terms of the behavior of ultimate units, the genes. It therefore must ask questions about the physical and chemical nature of these units. The nature of the gene has been the core of genetics, and work on this problem has led to the great triumphs of the last decade that have revolutionized our concepts of biology. Embryology, on the other hand, was most profoundly influenced by the discovery of the phenomenon of induction by Spemann and his collaborators and has, during the last fifty years, been primarily concerned with the analysis of the mutual interaction of organ primordia and cell complexes, and more recently with the behavior of cells in cell complexes. The analysis has led to concepts like the "morphogenetic field" of Weiss and the "prepattern" of Stern, which refer to properties of cell complexes and cannot be described in terms of fundamental entities.

Genes are responsible for morphological differences between organisms. Since in higher organisms morphological differences are the result of differences in developmental processes, these processes must be under genetic control. The genes, on the other hand, are known to be active constituents

* Wesleyan University, Middletown, Connecticut; present address: Department of Biology, University of Rochester, Rochester, New York.

of the individual cells. The question of the genic control of development may be formulated by asking which of the cell properties that are under genic control are instrumental in the supracellular processes that constitute the development of a multicellular organism.

It must be admitted that we do not know so much about this problem as we do about many other aspects of genetics. We know at present a good deal about the nature of genes, and more material is added to our knowledge every year. We have a theory about genic action in the cell which is in good agreement with our picture of the structure of the genic material and supported by a formidable amount of experimental information, particularly from microorganisms. Finally, the behavior of genes in populations is well understood; it forms the basis of the modern theory of evolution. Developmental genetics stands intermediate between these levels of investigation. A fuller understanding of its problems would give us a consistent representation of the organismic world in all its aspects, in which the problem of identical reproduction would be central, and in which the other biological problems could be understood as problems of the properties, effects, and history of the materials responsible for identical reproduction, the genes.

II. "*Phenogenetic*" *Approach*

The problem of the genic control of development was first clearly formulated by Valentin Haecker in 1918 (1) and designated by a special name, "phenogenetics." He also supplied a method of attacking the problem: the mutant phenotype was to be followed back in development to the point where it becomes indistinguishable from the normal. At this point, it is assumed, normal and mutant phenotype first differ from each other, and all later differences are to be regarded as secondary consequences of this first abnormality.

This method, which may be called "phenogenetic" in the narrow sense, has been extensively used in a number of organisms, particularly the mouse, the chicken, and the fly Drosophila. The extensive work of Grüneberg (2) with a large number of mutants in the mouse may be taken as an example. From the adult phenotype, the mutant characters are traced back in development step by step. If several mutant characters are dependent on one gene, each one of them is traced back and finally linked up with an earlier appearing deviation from the normal development. In this way, a pedigree of phenotypic effects of one gene may be obtained, the earliest observed deviation from normal development being supposed the cause of the later appearing ones. As a generalization, it is postulated that the primary action of a gene in development is unitary, insofar as it appears at a particular time of development and in a particular type of cell. All later effects of the same gene are to be interpreted as developmental consequences of this primary effect.

This "phenogenetic" method is a morphological method. It has, therefore, the limitation of morphological methods in general—that, though it can lead to a great amount of insight and to a certain amount of generalization, it cannot lead to further analysis in causal terms. Sooner or later it has to be supplemented by experimental procedures just as the concepts of embryology, though derived from the observation of normal development, are based fundamentally on particular types of experiments: defect, isolation, and transplantation.

III. *Gene* a *in Ephestia*

It appeared to me that to analyze the action of genes in development, an approach similar to that used in experimental embryology should be employed. I decided in 1931 that the gene *a* in the meal moth Ephestia would be a favorable object for this study. The mutant gene *a* is a simple recessive gene which causes the eyes to be red—the eyes of normal moths (allele *a⁺*) are black. Study of the anatomy and development of *aa* animals showed that the effects of the gene *a* are not restricted to its influence on the eye pigment. The adult testis of *aa* Ephestia is colorless or only very weakly colored, whereas that of a^+a^+ Ephestia is vividly colored, red or brown, depending on another pair of genes. The brain shows a corresponding difference in pigmentation. Furthermore, the skin of *aa* larvae is colorless, while that of the wild type is pink or flesh colored. And, finally, the eyes of the larvae, which are much simpler than the large compound eyes of the adult, have less pigment in *aa* than in a^+a^+ animals. All these characters are multiple (pleiotropic) effects of the same gene.

The alleles a^+ and *a* determine, therefore, the amount of a pigment in certain organs. In a^+ animals the pigments are formed at different times in development: the pigment of the larval skin and eyes arises in the embryo; the testes start to become pigmented during the last larval instar; while the difference in eye color appears only in the pupa. The earlier phenotypic differences are apparently not the causes of the later ones; but similar processes, all controlled by the gene a^+, seem to occur—formation of pigmented substances in different organs at different times.

The relation between these different effects was analyzed by transplantation experiments. Testes from larval males which had not yet become pigmented were transplanted into the body cavity of larvae (males and females) of the other strain. The aa testes which were transplanted into a^+a^+ hosts became pigmented, obviously under the influence of the host. The sex of the host made no difference. On the other hand, a^+a^+ testes transplanted into aa animals formed pigment as though they had remained in the original normal animal; but more than that, transplanted a^+a^+ testes had an influence on the aa host: its own testes developed pigment, and its eyes became black. In other words, the a^+a^+ testes supplied something, probably a diffusible substance, to the aa host which was necessary for the formation of the pigment in testes and eyes (3).

Chemical identification of this substance proceeded in a circuitous way. Beadle and Ephrussi (4) and their collaborators applied a similar experimental method to Drosophila, where many more mutant genes affecting eye color are known than in Ephestia. They found that at least two of these mutant genes, v and cn, exert their effects through diffusible substances; that the diffusible substance lacking in v Drosophila (v^+ substance) is probably identical with that necessary for pigment formation in a Ephestia; and that the substance missing in cn is a reaction product of the v^+ substance. In a further series of investigations, it was found that if the medium on which v Drosphila are reared is supplied with the amino acid tryptophan, and if this medium becomes contaminated with certain bacteria, v flies develop the missing pigments (5, 6). It was concluded from these observations that the diffusible substance is probably derived from tryptophan and that certain bacteria are able to convert tryptophan to the substance in question. Two ways were now used to identify the substance. Butenandt, Weidel, and Becker (7) synthesized a number of derivatives of tryptophan, injected them into aa Ephestia pupae, and looked for the formation of black eye pigment. Beadle, Tatum, and Haagen-Smit (8) grew the bacteria able to produce the substance in the presence of tryptophan and isolated the active substance from the culture medium. Both groups of workers arrived at the same conclusion: the substance lacking in aa Ephestia and vv Drosophila is kynurenin. This substance had been observed earlier by Kotake (9,) who had found that the liver of mammals contains an enzyme which catalyzes the oxidation of tryptophan to kynurenin. The conclusion from these experiments is that the mutant genes *a* in Ephestia and *v* in Drosophila inhibit the oxidation of tryptophan to kynurenin. As a consequence, the formation of the pigment which is derived from kynurenin cannot be carried out.

IV. *Enzyme Theory of Gene Action*

On the basis of these observations on the genetic control of eye pigments in insects, Beadle and Tatum developed their famous theory of gene action (10, 11): a particular gene controls the formation of a particular enzyme; mutations of the gene lead to inactivation of the enzyme in question, thereby blocking the metabolic step controlled by the enzyme. They have supported this theory by a large number of brilliant investigations with the mold Neurospora and other microorganisms, and the theory forms at present the basis of our understanding of the action of genes in the cell.

The validation of the gene-enzyme theory for the gene *a* in Ephestia ran into a good deal of technical difficulty, and it took almost twenty years until the enzyme controlling the oxidation of tryptophan to kynurenin could be demonstrated in the moth. The biochemical methods available at the time were not sufficiently sensitive. A large number of investigators, including myself, tried to find the enzyme in homogenates of flies and moths, always with negative or inconclusive results. An enzyme present in only one or two organs is so diluted in ground-up whole animals that this is not a good starting material for enzyme studies. In 1958, Egelhaaf (12) succeeded in developing a method which permitted the demonstration of the enzyme in Ephestia. He incubated tryptophan with ground-up isolated organs of Ephestia and subjected the reaction mixture to paper chromatography; the formation of kynurenin could thus be demonstrated and measured by its characteristic blue fluorescence. Egelhaaf was able to show that the enzyme is present in the larval testis and in the adult ovary of a^+a^+ animals, but he could not demonstrate it in other organs. In agreement with the theory, no enzyme activity has been found in organs from *aa* animals. The a^+a hybrid ovaries have a lower enzyme activity than the a^+a^+ strain (13).

This last finding throws some light on the genetic phenomenon of dominance. At the level of enzyme activity, the a^+a hybrid is intermediate between the two pure strains. But the enzyme activity of the hybrid is sufficient to supply enough kynurenin for full pigmentation, so that in this respect the hybrid is indistinguishable from pure a^+a^+.

The primary action of the gene a^+ may now be regarded as the control of the enzyme necessary for the oxidation of tryptophan to kynurenin. Gene a either inhibits formation of this enzyme or causes a modified, inactive enzyme molecule to be formed—examples of both alternatives are known to occur in microorganisms. From the developmental point of view, the question of the determination of the pattern of manifestation has to be restated. The pigment appears only in certain organs—e.g., larval skin, adult testis, and larval and adult eyes—and it appears in each organ at certain definite times of development. The pair of alleles a^+ and a controls the formation of a precursor of the pigment, kynurenin. Certain organs acquire, at certain times of development, the competence to react in the presence of kynurenin by forming pigment. This ability to react is probably, in turn, under genetic control, but the a locus is apparently not involved in its determination. The pattern of pigment appearance is, therefore, not an indication that the gene "acts" in those cells which form the pigment and "does not act" in those which do not. The pattern of gene manifestation indicated by its enzyme activity is different from the pigment pattern. There again, certain organs (larval testis, adult ovary) form the enzyme, others do not, or possibly they produce it in lesser amounts. To restate the question, therefore: Is the gene "active" only in those cells which produce the enzyme and not in the others? If we regard enzyme formation as the only primary activity of the gene, then the gene would by definition not be active in those cells which do not form the enzyme. It becomes obvious that the question of activity of the gene depends on the specific gene effect we are considering. The results are better expressed by stating that the presence of the allele a^+ is a necessary condition for the formation of the enzyme but that the actual formation of the enzyme depends on the state of differentiation of the cell. Why in different cells with presumably the same genes an enzyme may or may not be formed remains one of the fundamental problems of developmental genetics.

V. *Formation of Cytoplasmic Structures*

When in the process of development, cells become different in their function and architecture, the specific structures which characterize particular types of differentiated cells—e.g., muscle fibers, secretion granules, etc.—are formed in the cytoplasm. The genic control of cellular differentiation as expressed in the cytoplasmic structures formed, is therefore, another aspect of the genic control of developmental processes.

The pigment granules which are formed in different organs of Ephestia under the influence of the gene a^+ may be taken as an example of such a cytoplasmic structure. These granules are not just inert precipitates of a metabolic waste product, but they carry out enzymatic functions. Isolated eye color granules of Ephestia and Drosophila are able to carry out a number of enzymatic activities, such as several steps of the citric acid cycle and the oxidation of tyrosine (14, 15). Consistent with this finding, the pigment granules have a complex composition. If the pigment is dissolved, the granules can still be seen under the phase microscope or after staining with basic dyes (16). The granules consist, then, of the pigment derived from kynurenin plus a "core" granule which has been shown to contain protein and ribonucleic acid (15, 17).

Hanser (16) proposed a hypothesis concerning the genic control of the core granule. She based her theory on findings with another recessive eye color mutant, *wa*. The *wawa* moths have white eyes, and the pigment derived from kynurenin is completely missing in all organs. Transplantation experiments demonstrated that *wawa* animals do form kynurenin, however (18). The chain leading to pigment formation is, therefore, interrupted in *wawa* animals at a later stage. *Hanser* found that in *wawa* eyes, the core granules are missing, and she concluded that the wild type allele *wa*$^+$ is necessary for the formation of the core granules and, therefore, in their absence, the pigment could not be deposited. If this theory were correct, it should be possible to find naked core granules in the colorless testes of *aa* animals—core granules are present in the eyes of *aa* animals since these eyes do contain pigment granules, though the pigment is chemically different from the kynurenin-derived pigment of a^+. Actually, no naked core granules could be found in colorless *aa* testes; then *aa* testes were implanted into a^+a^+ animals, and core granules appeared in the transplanted testes after 2 to 3 days and developed gradually into pigment granules (19). This finding does not disprove Hanser's hypothesis, since in *aa* testes the core granules might be present but too small to be visible. Nevertheless, it suggests another possible interpretation of the genic control of the formation of the pigment granule: the granule is composed of at least two components, the pigment and the constituents of the core granule. If a gene inhibits the formation of any one of these components, the others fail to organize into the structure. This interpretation seems to be supported by the appearance of core granules in the *aa* testis after it has been supplied with kynurenin.

The pattern of development of pigment in *a*+ animals may, therefore, be an expression of the appearance of one of the constituents of the core granules which makes the formation of pigment granules possible if kynurenin is supplied.

VI. *Secondary (Pleiotropic) Effects of Gene* a

It has been stated earlier that individual genes may affect more than one phenotypic character, thus producing multiple, pleiotropic effects. It is usually assumed that the primary action of the gene is unitary, probably consisting in the control of protein structure. It is therefore postulated that the different pleiotropic effects of a gene have to be understood in terms of secondary consequences of the primary gene effect. One of the goals of developmental genetics is, therefore, the unraveling of the causal interactions of the different pleiotropic effects of a gene. This attempt has not been uniformly successful, and the possibility that one gene may have more than one primary effect cannot be rigidly excluded. There is reason to believe, however, that in some cases secondary interactions may be so complex that they are difficult to analyze.

In agreement with the general theory of gene action, we accept that the alleles *a*+ and *a* control the enzyme transforming tryptophan to kynurenin. The fact that gene *a* inhibits the oxidation of tryptophan to kynurenin raises the question of the fate of the tryptophan which cannot be converted. Tryptophan is an essential amino acid in animals; it cannot be synthesized but must be ingested with the food. The *a*+*a*+ and *aa* larvae consume about the same amount of food and the same amount of tryptophan. Excess tryptophan in *aa* animals must, therefore, be disposed of in some way: it may be excreted or broken down by another pathway or stored. The first possibility was a priori unlikely, since insects do not excrete substances, except in gaseous form, during the pupal rest. Both *a*+*a*+ and *aa* animals were therefore analyzed for tryptophan. It turned out that *aa* moths contain significantly more tryptophan than the wild type. Since tryptophan is a constituent of proteins, it could conceivably be stored either as free tryptophan or in the proteins. Both assumptions turned out to be right. About 15 per cent of the excess tryptophan of *aa* animals is found in the non-protein fraction while the remaining 85 per cent is found in the proteins. Since there is no evidence that the total amount of proteins is increased in *aa* animals, it must be concluded that a protein or some proteins are qualitatively different in the two strains. This is not due to a direct action of genes on protein constitution but to the storage of excess tryptophan in *aa* proteins.

Other secondary biochemical effects of the gene *a* have been found. The pteridines are a group of chemical substances which are easily demonstrated by their intense fluorescence; their physiological importance is largely unknown. Kühn and Egelhaaf (22) have shown that the amounts of pteridines in different parts of the moth are different in a^+a^+ and *aa* animals. In this case, it could be shown that the difference in the pattern of pteridine distribution in the two strains is a secondary consequence of the lack of pigment in *aa*. If *aa* pupae are injected with kynurenin, they form the eye pigment and, at the same time, the pattern of pteridine distribution in the injected animals becomes normal.

It becomes clear that the consequences of a single biochemical block are more complex than might be at first presumed. Because an organism is an integrated system, a number of biochemical processes will be affected secondarily. The mutant organism will, therefore, differ from the wild type in quite a number of biochemical characters. Every one of these may be expected to affect developmental and physiological processes. In the case of the gene *a*, one more instance will be described.

It was observed accidentally that the testis of the late larva and early pupa contains a distinct yellow pigment which is not present in the early larva or in the adult. This pigment has been identified as the vitamin riboflavin; over 50 per cent of the total riboflavin of the moth accumulates in the testis in the last larval instar. Where does the riboflavin come from, and what becomes of it when it disappears in the late pupa? This question can be answered by dissecting animals at different stages of development and investigating their riboflavin content with the methods of paper chromatography. It was found that in the early last instar larval riboflavin is concentrated in certain hypodermis cells. In the late last instar it disappears from the skin, is briefly found in the blood, and is preferentially accumulated in the testis. It stays in the testis until about 6 to 8 days after pupation; at this time it disappears from the testis and is stored in the Malpighian tubules (23, 24).

This process of riboflavin accumulation in certain organs at different times of development can be observed in a number of different strains, and conclusions about the genic control of this process may be drawn. The whole process does not occur in the white-eyed strain, *wawa:* no riboflavin is stored in the testis nor in the skin nor in the Malpighian tubules; the main amount is constantly found in the blood. In other words, the organs cannot accumulate the substance. The effects of the mutant gene *a* are: (*a*) the amount of riboflavin stored in the pupal testis is increased in comparison to wild type; and (*b*) riboflavin is released from the pupal testis about two days earlier than in the wild type.

This last observation is of importance for the interpretation of the action of genes in development. We are dealing here with timing in development. Different processes—in this case, release and uptake of riboflavin—occur in different organs, and they are timed in such a way that two organ systems act in a complementary way at any one time. The *a* locus affects the timing of at least one of these processes. This effect cannot be regarded as the gene's primary effect since its action on the enzyme converting tryptophan to kynurenin has been identified as such. It must, therefore, be regarded as a secondary (pleiotropic) effect of the gene, even though the actual connection between the primary effect and this secondary effect is still unknown.

VII. *Pleiotropic Effects on Viability and Behavior*

Viability—the ability of the organism to survive from the egg stage to adulthood—is one of the most generalized characters of an organism; practically any change in the biochemical makeup of the organism may be expected to affect its chances to survive to the adult stage. Consonant with this expectation, it has been found that most gene mutations have an effect on the viability of the organism. The secondary effects of the mutant gene *a* have an adverse effect on viability.

More interesting is the fact that mutant gene *a* has definite effects on behavior. Dr. W. B. Cotter (25) has recently studied the mating behavior of isogenic a^+a^+ and aa strains and has found definite differences. If a female is brought into a jar with wild-type and red-eyed males, the red-eyed males approach it and start courting before the wild-type males.

However, they are less agile in carrying out the complex series of movements preceding copulation so that, on the average, wild-type males are more successful in copulation than *aa* males.

Nothing is known about how this difference in behavior is connected with the other actions of gene *a*. A suggestion may be made. The gene *a* is known to interfere with the metabolism of tryptophan. In mammals and man, it has been found by psychopharmacologists that a tryptophan derivative, serotonin, is preferentially present in particular parts of the brain and exerts, apparently, an important controlling influence on certain aspects of behavior. Kynurenin and serotonin are then produced from the same substance, tryptophan, by alternate pathways. It is conceivable that in *aa* animals where the pathway leading to kynurenin is blocked, the alternate reaction would proceed at a faster rate, resulting in a different level of the reaction product. If serotonin or a similar substance were active in insects, it would be understandable that a gene which controls the fate of tryptophan derivatives would have a pleiotropic effect on behavioral characters.

VIII. *Conclusions on Gene Action in Development*

Geneticists and biochemists generally agree that the genetic information is carried by the pattern of purine and pyrimidine bases in the chromosomal DNA. It is also generally assumed that in the cell this pattern is translated into the amino acid sequence of proteins. It is frequently proposed that mutant genes give rise to alterations of the amino acid sequence, with consequent changes in the function of the protein structure.

In the case of *a* Ephestia, this change of function consists in the inactivation of the enzyme converting tryptophan into kynurenin. This leads primarily to a lack of the pigment formed from kynurenin in certain organs (e.g., testis, eyes). However, the emphasis in our analysis has shifted from the primary effect to a large number of secondary effects. There is no particular difficulty in visualizing this situation; every organism is a complex integrated system. If there is a change in one component of an integrated system, secondary changes in a number of characters are to be expected. These changes will be more varied the more complex the system is. In a multicellular organism we expect to find more and different alterations than in unicellular organisms.

For the gene *a*, it has been shown that these secondary biochemical changes are quite complex and affect a number of substances in quantitative and qualitative ways. It is probable that more or less similar complex secondary biochemical changes accompany most mutations in multicellular organisms. There is no difficulty in principle in supposing that some of these changes may affect the intercellular interactions which are involved in embryonic processes.

The difficulty appears if we want to give specific interpretations of the genic control of these interactions, because we know very little about the forces acting in development in terms of cell physiology. An example would be the relation between an inductor and a competent reacting tissue. We know that an inductor releases something which determines the developmental fate of the reacting tissue. We also know that the reacting tissue will react only when it is in the competent state. The induction process may, then, be influenced either by an effect on the inductor or by an effect on the competence of the reacting tissue. We do not know what competence is physiologically; the ability of certain cells of Ephestia to form pigment, if kynurenin is supplied, may be taken as a model of developmental competence. We have seen earlier that, in this case, the presence of certain substances or structures in the cell might form the basis of competence. The ability to accumulate certain substances from the environment—exemplified by the storage of riboflavin in the testis sheath—might offer another possible mechanism for the explanation of competence. It should be stated that these two suggestions are not necessarily mutually exclusive. The gene *wa* certainly interferes both with storage of riboflavin in the cells and with the ability to form pigment granules in the presence of kynurenin. A common mechanism governing both the competence to form pigment and the accumulation of riboflavin, depending on the type of cell, could be imagined.

An important problem in the genic control of development is the control of timing. For an inductor system to function properly, the reacting system must become competent at the same time the inductor is supplied. Goldschmidt (26) regarded timing as the crucial problem of physiological genetics; on the basis of his observations on intersexes of Lymantria, he proposed the theory that the main action of genes in development consists in the control of developmental reaction velocities. Our observations on riboflavin storage in *a* Ephestia shed a certain amount of light on this matter, in that *a* induces a premature release of riboflavin from the testis sheath. Since there is good evidence that the primary action of these genes consists in the inactivation of an enzyme, the effect on the timing of the release of riboflavin must be regarded as a pleiotropic effect of the same gene.

We know of only one force that can closely coordinate the primary and secondary actions of genes so as to assure the exact time relations found in development: natural selection. It may be assumed that natural selection in the evolution of multicellular organisms has favored the establishment of coadapted gene complexes which insure coordinated timing of developmental reactions occurring in different tissues. It is consistent with our general theory of gene action to assume that these effects may frequently be secondary rather than primary effects of the genes involved.

In microorganisms, selective pressures will be mediated mainly through characters which are closely related to the primary action of the genes, i.e., the control of protein structure. The ability to synthesize substances and to resist harmful conditions of the environment may be assumed to play a major role in the evolution of microorganisms. In animals which live on organic materials synthesized by other organisms, ability to synthesize may not be of such paramount importance. Fraenkel and Blewett (27) have claimed that *Ephestia kühniella* can synthesize a certain amount of riboflavin, while the closely related species *E. elutella* cannot. In multicellular organisms it may be assumed that those characters of cells which influence their behavior in cell complexes and the timing of developmental processes are of great selective importance, since any strong abnormality in development will prevent the affected organisms from reproducing. Genes will be selected for their influence on developmental processes, and these may in many instances be pleiotropic effects which are some distance removed from the primary gene effects. Finally, it is well known that the selective value of genes and gene complexes in animals is strongly affected by their influence on characters of a very generalized kind: viability, fertility, and behavior.

Development cannot be understood in terms of our knowledge of primary gene action at the molecular level exclusively. We must understand the secondary consequences of a genic substitution. Finally, evolutionary considerations are necessary to understand the close co-ordination of different developmental processes, since selective pressure is the only force we know of which can fit together independent processes that are influenced by a large number of genes in such a way that normal development becomes possible.

REFERENCES

1. V. Haecker. Entwicklungsgeschichtliche Eigenschaftsanalyse (Phänogenetik). Jena: G. Fischer, 1918.
2. H. Grüneberg. The genetics of the mouse. The Hague: M. Nijhoff, 1952.
3. E. Caspari. Arch. Entw. Mech. Org., 130:353, 1933.
4. G. W. Beadle and B. Ephrussi. Genetics, 21:225, 1936.
5. Y. Khouvine, B. Ephrussi, and S. Chevais. Biol. Bull., 75:425, 1938.
6. E. L. Tatum. Proc. Natl. Acad. Sc., 25:490, 1939.
7. A. Butenandt, W. Weidel, and E. Becker. Die Naturwiss., 28:63, 1940.
8. E. L. Tatum and A. J. Haagen-Smit. J. Biol. Chem., 140:575, 1941.
9. Y. Kotake and T. Masayama. Hoppe-Seyl. Ztschr., 243:237, 1936.
10. G. W. Beadle and E. L. Tatum. Proc. Natl. Acad. Sc., 27:499, 1941.
11. G. W. Beadle. Physiol. Rev., 25:643, 1945.
12. A. Egelhaaf. Ztschr. Naturforsch., 13b:275, 1958.
13. A. Egelhaaf and E. Caspari. Unpublished results.
14. I. Ziegler-Günder and L. Jaenicke. Ztschr. f. Vererbungsl., 90:56, 1959.
15. B. M. Meissner. Unpublished results.
16. G. Hanser. Ztschr. ind. Abst. Vererbungsl., 82:74, 1948.
17. E. Caspari and J. Richards. Yearbook Carnegie Inst. of Wash., 47:183, 1948.
18. A. Kühn and V. Schwart. Biol. Zentralbl., 62:226, 1942.
19. E. Caspari and I. Blomstrand. Cold Spring Harbor Symp. Quant. Biol., 21:291, 1956.
20. E. Caspari. Genetics, 31:454, 1946.
21. A. Butenandt and W. Albrecht. Ztschr. Naturforsch., 7:287, 1952.
22. A. Kühn and A. Egelhaaf. Die Naturwiss., 42:634, 1955.
23. E. Caspari and I. Blomstrand. Genetics, 43:679, 1958.
24. E. Caspari. Proc. X. Internat. Congr. Genetics, 2:45, 1958.
25. W. B. Cotter. Unpublished results.
26. R. B. Goldschmidt. Physiological genetics. New York: McGraw-Hill Book Co., 1938.
27. G. Fraenkel and M. Blewett. J. Exper. Biol., 22:162, 1946.

MOLECULAR GENETICS AND THE FOUNDATIONS OF EVOLUTION

BERNARD D. DAVIS*

In the century since Darwin developed his theory, largely on the basis of comparative morphology and paleontology, genetics and comparative biochemistry have provided a great deal of further support. And in a dramatic further advance molecular genetics has now yielded a new, more direct kind of evidence for evolutionary continuity, extending from bacteria to man. Indeed, unless we assumed that continuity the study of molecular genetics in bacteria would not help us to understand human cells.

Yet various groups remain skeptical, for various reasons. Religious fundamentalists in the Judeo-Christian tradition object that the evolutionary view of man's origin destroys an indispensable basis for morality. Extreme egalitarians have difficulty with the implication that the genetic diversity within each species, on which natural selection depends, must include mental (as well as physical and biochemical) traits in man. Some literary people, following the line of Arthur Koestler, falsely ascribe to science the goal of discovering absolute truths, and they then criticize evolution for failing to meet that goal [1]. And while a distinguished philosopher of science, Karl Popper, accepted evolution as a fact, he questioned whether Darwin's theory (even in its modern, neo-Darwinian version) meets the criterion that he has proposed for distinguishing a scientific from a metaphysical theory: the ability to generate falsifiable (i.e., testable, refutable) predictions [2]. Popper has now conceded that

Much of this material, with more emphasis on its philosophical aspects, appears in a related paper: B. D. Davis, Molecular Genetics and the Falsifiability of Evolution, in E. Ullmann-Margolit, ed., *The Kaleidoscope of Science: Israel Colloquium Studies in History, Philosophy, and Sociology of Science*, Vol. 1. Atlantic Highlands, N.J.: Humanities Press, 1984.

*Bacterial Physiology Unit, Harvard Medical School, 25 Shattuck Street, Boston, Massachusetts 02115.

his criterion was too rigid [3, 4], but unfortunately creationists have continued to draw support from his original position [5].[1]

A recent poll of a representative sample of Americans [6] illustrates the extent of resistance to the theory of evolution: 44 percent of the respondents believed in the special creation of man occurring within the past 10,000 years, two other groups conceded a longer time scale or else accepted the theory of a *directed* evolutionary process, and only 9 percent accepted the scientific conclusion that our species has evolved by undirected natural selection. Though this result is discouraging it is not hard to understand. Scientific ideas on man's origin are relatively recent, while religious ideas carry the weight of long tradition, have much more emotional appeal, and offer a simpler basis for a moral consensus. No wonder so many people find these ancient, poetic myths about man's origin more credible and more satisfying.

Nevertheless, since the question of the origin of our species is a question of biology, only objective scientific inquiry, divorced from moral preferences, can provide an answer that corresponds to reality. And since nature has the last word on such questions it is hard to doubt that the scientific answer will ultimately prevail. But "ultimately" may be a long way off; for although liberal religion is primarily concerned with questions of value, and has given over to science its earlier function of also trying to explain the world of nature, that is not true of all religions. Meanwhile, the tensions between science and myths are likely to become worse, as advances in genetics, neurobiology, and sociobiology further contradict treasured preconceptions—political as well as religious—about human nature.

Evolution is thus central to our attitude toward reality and to our assumptions about human nature. It is therefore essential, for the future harmony of our society, to try to teach the subject more effectively. In most of its development evolutionary biology has depended on morphological homologies, both in the fossil record and among living species; but this approach has not revealed the continuum of transition forms between species that Darwin predicted. Moreover, while he expected further research in paleontology to fill in the gaps, we no longer entertain that hope. But now, at last, molecular genetics has provided a direct, radically different kind of evidence for such continuity.

So far, however, this powerful evidence has penetrated very little into the introductory teaching of evolution and into debates with the cre-

---

[1]Despite Popper's brilliance in analyzing the scientific method with physics as a model, his understanding of biology exhibits serious limitations, including a failure to recognize the significant elements of refutable prediction in Darwin's picture (e.g., correlation of age of fossils with their stage of evolution). I shall not deal here with these issues, which have been dissected well by a philosopher especially interested in biology [5].

ationists. In the recent spectacular legal victory of the American Civil Liberties Union against a creationist law in Arkansas, and also in recent books, the defenders of evolution have continued to focus almost entirely on the geological time scale and on the paleontological record, as in the Scopes Trial in 1923 [7].[2] It is surely time for our teaching to balance this approach, without decreasing our appreciation for Darwin's remarkable achievement. Not only does molecular genetics provide the most convincing evidence for evolutionary continuity, but this evidence should impress a public that is well aware of the power of this science in other areas. This article will therefore review some of the contributions of molecular genetics—as well as those of classical genetics, biochemistry, and microbiology—to evolutionary biology.

Darwin's Problem with Variation

Darwin's theory had two major components: variation and natural selection. The latter received virtually all the attention, because of its courageous philosophical and religious implications. But the basic theme, dramatized as "survival of the fittest," has been accused of being a mere tautology—and it would be, if it were concerned only with the obvious idea of differential individual survival. Darwin's great accomplishment was to link that idea to heredity, thus creating the much more consequential idea of net differential reproduction.

The really radical component of the theory, then, was the assumption of endless hereditary innovation, on which selection could act. But this was an ad hoc assumption. Wrestling all his life with this problem, Darwin came up with a mixed view: "hard" inherited variation, arising without direction by the environment, seemed likely to be the main source of novelty; but everyday observations seemed to point also to "soft" inheritance, responsive to use and disuse [8]. He therefore developed a logical but useless theory of "gemmules"—particles of inheritance in body cells that were responsive to use and disuse, and that were released to the germ cells and thus able to influence the next generation.

To be sure, Darwin did lean heavily on the implications of artificial selection for hereditary variation. But that process has a serious weakness as a model for natural selection. We now know that artificial selection depends largely (and it might conceivably have depended entirely) on genetic recombinations, arising from the range of genetic variation already existing within a species. Yet evolution of the enormously diverse living world from a common ancestor would require a much more thoroughgoing kind of genetic innovation. And since Darwin wrote *The*

---

[2]In a gratifying exception, a recent brochure by a committee of the National Academy of Sciences has given serious attention to the molecular evidence [7].

Origin of Species before the emergence of genetics, he had no evidence, or even a plausible mechanism, for explaining such innovation.

Darwin therefore could not proceed within the usual framework of science—that is, by means of a stepwise series of hypotheses, predictions, and confirmations. Instead, he had to make a large conceptual leap. His theory was thus in a sense premature. On the other hand, since some philosophers criticize the theory for its lack of testable predictions, one might also say that it made a grand prediction: a hereditary process that would reconcile the paradox of breeding true and yet creating novelty.

To appreciate how the development of genetics solved this problem, let us engage in a fantasy and pursue a hypothetical rearrangement of history, imagining that no one dared to propose the theory of Darwin and Wallace until it had a testable foundation in genetics. The two fields would then have arisen in a logical order.

A Hypothetical Scenario

The first step toward filling Darwin's big gap was DeVries's discovery of mutations, in 1900. Hereditary variation could then be seen to arise in two different ways: mutation provides the ultimate source of novelty, and the reassortment of genes in sexual reproduction enormously amplifies this variation. Nevertheless, for decades the actively growing field of genetics had little to contribute to evolution. Geneticists believed that species arose by giant mutations ("saltations"), rather than by the gradual changes invoked by Darwin, whereas evolutionists considered mutations to be exceptional monstrosities and assumed that the gradual steps of evolution must have arisen by some other mechanism.

Genetics and evolutionary biology were finally linked in the late 1920s, when the "particulate," Mendelian mode of inheritance, based on discontinuous traits, was shown to apply also to traits exhibiting continuous variation. Their apparent continuity arises from polygenic inheritance, in which many genes and their interactions with the environment contribute to the final numerical value of the trait. At the same time, Fisher, Haldane, Chetverikov, and Wright developed the quantitative science of population genetics [9]. This discipline provided a powerful new approach to evolution: measurement of gene frequencies in successive generations and identification of the factors that cause these frequencies to change.

One of these factors is genetic drift: a process in which the chance geographic separation of a small population limits its gene pool. As a result the isolated population retains some mutations that would have been diluted out by competing alleles in the major population, and these nonadaptive mutations influence the directions of further evolution toward valuable, adaptive traits. Recognition of genetic drift as a major process has greatly broadened the scope of evolutionary biology. Thus, although evolution as a whole depends on the selection of adaptive properties, not every step must be adaptive. This concept helps dispel the mystery of the survival of the intermediate steps in the evolution of a complex organ, such as the eye. Moreover, since genetic drift permits survival of individuals that might not have been the fittest in the original population, it removes evolution even farther from the tautology of "survival of the fittest."

Population genetics soon became the key discipline in evolution, and it changed profoundly our understanding of the nature of biological populations. Earlier biologists, under the influence of Aristotelian essentialism, had long characterized species (and races) in terms of an ideal type or essence, ignoring the presumably trivial individual deviations from the type. But the concept of uniform entities, though essential in physics and chemistry, is not appropriate for describing natural biological populations. In particular, genetic studies of biochemical traits (e.g., allelic forms of an enzyme) showed that a large fraction of genes exhibit polymorphism (multiple forms within a species). Each species thus conserves much more genetic diversity than meets the eye, and this diversity is now seen as a crucial rather than a trivial feature. As Mayr [10] has emphasized, this shift from a typological to a populational view has been one of the most important conceptual advances in biology.

Microevolution and Macroevolution

The development of population genetics revealed the true nature of geographic races, within both animal and plant species: these are populations whose prolonged reproductive separation has led to accumulation of significant statistical differences in their gene pools. Extension of this divergence would eventually create separate species by giving rise to reproductive incompatibility (because of incompatibility in behavior, genital fit, chromosomal organization, or perhaps histocompatibility antigens) [10, 11].

Population genetics also made possible the direct demonstration of microevolution, that is, evolution within a species (or occasionally yielding a closely related species) and in contemporary time. For example, in localities where industrialization darkened the tree trunks an originally light-colored species of moth became predominantly dark. Whereas such shifts had earlier been ascribed to physiological adaptation, the mechanism was now found to be genetic adaptation (i.e., selection for an initially rare genotype).

These developments converted the process of natural selection from a bold hypothesis into a mere description of fact, just as the discovery of capillaries did for Harvey's bold hypothesis of the circulation of the blood. The extrapolation to macroevolution, which creates the enormous diversity of the living world, was logically compelling, and it was supported by the more or less closely graded morphological homologies in the phylogenetic trees. Essentially all biologists have accepted the result of this synthesis of evolution with genetics—the so-called neo-Darwinian model of evolution—as the central organizing principle in their field [12, 13].

If we return to our hypothetical scenario, we see that evolution might have been logically postulated, without any big jumps, when heritable continuous variation was shown to involve mutable Mendelian genes. With the development of population genetics evolution would have become an inescapable inference.

The Unity of Biochemistry

Meanwhile, the study of comparative biochemistry provided impressive further support for macroevolutionary continuity, for although this field (like many novel developments in biology) was initiated simply to see what is there, it soon became apparent that the result fulfilled a firm evolutionary prediction: if all organisms have a common origin, then even after extensive divergence they might retain some common features.

Indeed, even by Darwin's time the microscope had shown a basic unity—all organisms are made of cells that share common structural features. But biochemical studies in the 1930s and 1940s revealed much more detailed conservation. The most distant organisms, ranging from bacteria to man, use the same building blocks for their proteins and nucleic acids and, with variations, very similar enzymatic catalysts, central metabolic pathways, and energy-transducing mechanisms. Subsequently, organisms of all kinds were found to use the same fundamental molecular mechanism for regulating the activity of genes (and of enzymes), that is, allosteric changes in the shape of a protein. Clearly, a great deal of molecular unity underlies the morphological diversity of the living world. Unfortunately this evidence has penetrated very little into the introductory teaching of evolution.

More recently, molecular genetics has demonstrated a particularly dramatic unity—the genetic code. If organisms had arisen independently they could perfectly well have used different codes to connect the 64 trinucleotide codons to the 20 amino acids; but if they arose by common descent any alteration in the code would be lethal, because it would change too many proteins at once. Hence the finding of the same genetic code in microbes, plants, and animals (except for minor variations in intracellular organelles) spectacularly confirms a strong evolutionary prediction.

The Impact of Bacterial Genetics

The discovery of biochemical unity between bacteria and higher organisms encouraged the search for unity also in their genetic mechanisms. The result was the development of bacterial genetics, which then laid the groundwork for molecular genetics.

The emergence of bacterial genetics was curiously delayed. Changes in the properties of cultures during serial cultivation were one of the earliest challenges of the infant science of bacteriology; yet study of bacterial variation did not become linked to the new science of genetics, begun in 1900, for at least four decades. There were several reasons. The bacterial cell was too small to permit any chromosomal organization to be visualized, and genetic recombination could not be demonstrated (because it is a very rare event, rather than the regular event seen in the reproduction of higher organisms). Moreover, bacteria multiply so rapidly (up to three generations per hour) that the progeny of a highly favored mutant could replace a parental population during overnight growth; and since this seemed to be too rapid to be caused by a Darwinian process, a Lamarckian, instructed change in inheritance was inferred. Bacteria were therefore regarded as virtually bags of enzymes having some vague, plastic mechanism of inheritance.

The barriers were not broken down until the 1940s, when Beadle and Tatum's discovery of a 1:1 relation between genes and enzymes, in the mold Neurospora, was soon extended to bacteria. Even more significantly, Avery's study of pneumococcal transformation founded both bacterial genetics (by demonstrating gene transfer between bacteria) and molecular genetics (by showing that the genetic material is DNA). Bacteria and their viruses then became the favored models for the study of molecular genetics, because of their relative simplicity and rapid growth and because very rare mutants or recombinants could be efficiently selected from huge populations.

Although the discovery of gene transfer and recombination in bacteria came as a surprise, in retrospect it is easy to recognize the evolutionary value of such an early emergence of these sophisticated processes. Prokaryotes clearly had to accumulate an enormous amount of genetic variation during the 3 billion years before they gave rise to eukaryotes. If its only source were successive mutations within a lineage, without any form of recombination, evolution would surely not yet have progressed beyond early prokaryotes.

In addition to providing a background for molecular genetics, the development of bacterial genetics provided a simple demonstration of natural selection: the experiment can be easily performed in the elementary biology laboratory, the overnight emergence of resistant bacterial strains in cultures containing an inhibitor or a bacteriophage. (Moreover, in human populations treated with antibiotics resistant variants may become predominant within a few months or years.)

Nine years after Avery's great discovery Watson and Crick launched the field of molecular genetics by elucidating the basic structure of DNA. The concepts and techniques that emerged have influenced virtually every branch of biology, and they have given rise to the rapidly growing daughter field of molecular evolution [14, 15]. I shall briefly describe some particularly pertinent developments.

Size of Steps in Evolution

As Watson and Crick noted, the structure of DNA not only explained the puzzle of gene replication: it also could explain mutations as inevitable occasional errors in base-pairing during replication. Further studies have revealed an extraordinary variety of additional molecular mechanisms of mutation, and the list is still growing. As a result, Darwin's thesis—that new organisms evolve by "numerous successive, slight modifications"—has acquired a precise, operational meaning. A "slight" modification now means a one-step hereditary change (a "mutation" in the broad sense).

In addition, the size of a mutation, originally described only in terms of phenotypic effect, can now also be defined in terms of information content, and the two measures do not necessarily vary in parallel. In particular, some single mutations can produce a large phenotypic effect—an obviously important process for evolution.

In the simplest mechanism, even a *single-base change* can have a broad effect, by affecting the formation of a number of products. Base substitutions may have this effect when they occur in DNA sequences that regulate several genes, code for a topoisomerase, or code for an enzyme that can modify other proteins. The single-base change of a frameshift can also cause a major change in the sequence of a protein.

Another source of broad effects is *DNA rearrangements*. Microscopically visible chromosomal rearrangements were discovered decades ago, but from McClintock's work on maize, and from more recent molecular studies in bacteria, it is clear that smaller rearrangements, invisible to the cytologist, are much more frequent. Moreover, they occur by highly evolved mechanisms, for they often involve mobile genetic elements (transposons), which have special terminal sequences and a gene coding for an enzyme that mobilizes these sequences. These transpositions not only can alter the quantitative expression of a gene, but they can create novel proteins by fusing parts of different genes: an obviously major evolutionary mechanism.

In eukaryotic cells *introns* (nontranslated intervening sequences between the translated parts of a gene) appear to serve to accelerate such recombinations of parts of different genes. Moreover, recombination also occurs in messenger RNA, and the resulting novel sequences can be copied into the DNA of the chromosome by reverse transcriptase. Since copying errors are much more frequent in RNA than in DNA (which has elaborate corrective mechanisms), it has been suggested that DNA guards genetic identity, whereas RNA promotes its modification [16].

In higher organisms *embryonic development*, including differentiation and morphogenesis, can enormously amplify the effects of a genetic change. (For example, a rare single-gene defect in man—polydactyly—results in hands with six fingers.) *Differentiation* is just beginning to be understood, for it depends on *selective gene regulation*, and several mechanisms have already been identified. These include binding of regulatory proteins, selective methylation of bases, shift of the DNA conformation to a left-handed twist (Z-DNA), rearrangements in DNA sequences, and selective gene amplification. *Morphogenesis* may prove to be much more obdurate, for it involves intricate patterns of product localization and feedback regulation in time and space rather than the linear sequences of events familiar to biochemists today.

A large phenotypic effect can also be achieved in one step by the introduction of *exogenous DNA* into a cell—and often into a chromosome—whether by a virus, by a plasmid, or as a naked DNA fragment. Though such "infectious heredity" was discovered in bacteria (where it causes rare, partial genetic recombination in otherwise asexual organisms), it has also been observed in the cells of higher organisms. Indeed, this ability of a virus to become integrated into a chromosome, and then to transfer some of the host DNA into other organisms, has intriguing evolutionary implications. Like bacteria, which were first studied as pathogens and then found to have a much broader role in the recycling of organic matter, viruses may also have developed their prominent role in pathogenesis only as a sidetrack—in this case, from an evolutionary rôle in mediating gene transfer [17, 18].

Molecular genetics is thus producing profound insights into the mechanisms of genetic variation, and the end is not in sight. Such studies leave little justification for the recent speculation (reviving an earlier saltationism) that species formation requires some unknown kinds of macromutations, or for the further claim that the neo-Darwinian picture requires radical revision [19, 20]. Moreover, these claims have been part of a broader argument, dramatized as "punctuated equilibrium" [19–21]. This proposition—that evolution is not gradual but alternates between long periods of stasis and short bursts of rapid change—has been widely criticized for claiming a spurious novelty, since Darwin explicitly recognized that "gradual" meant proceeding through a sequence of small steps but not necessarily proceeding at a constant rate. In fact, Gould has recently reduced his claims so drastically [22] that the scientific controversy has evaporated. Unfortunately, reverberations in the press continue to contribute to the antievolution movement by creating the false impression that the neo-Darwinian view is in disarray.

Nevertheless, it is true that the detailed mechanism of speciation is still a challenge. The theory of evolution demands gradual change; yet we fail to find a continuous series of transition forms, both among fossils and (with few exceptions) among living species. The obvious explanation is that the intermediates in the formation of a new species quickly move forward in evolution into the better-adapted new species (like a transitory intermediate in a chemical reaction), and so their numbers are too low to permit detection. Moreover, although the neo-Darwinian model requires that the genetic steps between species form a continuous series, it does not specify the size of the individual steps, either in terms of information content or in terms of phenotypic effect—and we have seen that either can be larger than earlier seemed possible.

Developmental biology will surely play a major role in elucidating the steps in the transitions between species, since the genetic requirements for effective development must limit the directions in which an organism can evolve. Indeed, with further advances in developmental biology, and with our growing ability to sequence and to manipulate genes, it may become possible to reproduce the transmutation of species before long.

Verified Predictions, and the Origin of Life

Since evolutionary theory has been faulted by some philosophers for its presumed lack of testable predictions, it is interesting to note that studies in molecular genetics have verified several major predictions (though these were usually recognized as such only retrospectively). For example, in order for increasingly complex organisms to evolve the genome would have to be able to expand. One mechanism of expansion, already noted, is the insertion of a block of exogenous DNA. Another is nonreciprocal recombination (unequal crossover) between homologous chromosomes, which yield a chromosome with two copies of a particular sequence: one of these can then maintain the original function, while successive mutations in the other can create a new function. Confirming this evolutionary mechanism, many families of proteins with closely related sequences have been found within an organism.

Another prediction arises from the wide range of genome sizes, which is 10^4-fold from the smallest viruses to vertebrates. A mutation rate (per unit length of DNA) that is appropriate for evolution in the former group would create an excessive frequency of mutations (mostly deleterious) in the latter, and so the rate should show a roughly inverse correlation with genome size. This prediction has been confirmed. Three mechanisms that alter the rate are known: variations in (1) the accuracy of the enzymes that replicate DNA, (2) the activity of the enzymes that correct errors in replication, and (3) the presence of mobile sequences. In addition, the mutation rate can be varied at specific sites: for example, methylating a base increases its mutability without changing its specificity of pairing. This ability to select for altered mutability—in the whole genome or in a specific part—no doubt increases the flexibility of evolution.

Finally, another prediction is that a single DNA sequence can theoretically be read in multiple ways, and evolution might be expected to take advantage of the economy that would result. Three different mechanisms have been observed: (1) reading not one but both strands of a DNA sequence, (2) reading a strand between different starting and stopping sites, and (3) reading it in different phases (created by the fact that coding occurs in continuous triplets).

Molecular genetics has also provided an increasingly detailed answer to an old objection to Darwinian evolution: that it lacks a plausible beginning. Earlier speculations, by Oparin and Haldane, had postulated that organic molecules would accumulate in a prebiotic "soup" and that, in a dramatic event, their spontaneous aggregation would eventually form the first cell. However, with advances in the study of molecular replication it became clear that nucleic acids, as repositories of information, had to precede proteins in prebiotic evolution: hence natural selection could appear even before the first cell. As Eigen [23] has proposed, the evolution of life (or, more precisely, the evolution of genetic information) would start with spontaneously polymerized, short nucleic acid sequences that slightly favored the formation of similar sequences; and through natural selection, at a molecular level, sequences would gradually emerge with increased precision of replication and hence increased capacity to transmit complex information. The recent finding of enzymatic activity in certain nucleic acid sequences [24, 25] dramatically supports this proposal.

Whatever the details of prebiotic evolution, it gave rise to cells very early: bacterial fossils have been found in rocks that were formed within the first billion years of the earth's 4.5-billion-year history. This speed suggests that life will inevitably evolve under the right chemical circumstances. The ultimate mystery is thus not the creation of life; it is the creation of a cosmos whose properties led to the evolution of life. And here we can again recognize different spheres for science and for religion: since science cannot provide an answer to the question of ultimate origins it has no conflict here with religious speculation.

Sequence Homology

Thus far we have seen that molecular genetics has supported the theory of evolution by propounding many mechanisms of variation, many testable predictions, and plausible mechanisms of prebiotic evolution. In an even greater contribution it has directly demonstrated, in two novel ways, the macroevolutionary linkage of present living organisms [26].

The first kind of direct evidence is based on a strong prediction: if evolution has occurred through the accumulation of mutations, then as organisms diverge in phenotype they should similarly diverge in DNA (and in protein) sequences. Several techniques have been developed for estimating similarity of sequence: immunological cross-reaction between proteins (initiated with hemoglobin more than 75 years ago!); formation of hybrid double strands by mixing DNA from two species under appropriate conditions; and direct sequence determination of proteins or of specific DNA fragments. The results have confirmed many of the classical branching taxonomic trees, which were based primarily on morphological criteria. For example, two closely related, very recently evolved species, man and the chimpanzee, have an average sequence homology of approximately 99 percent [27] compared with 70 percent homology between the mouse and the rat.

On the other hand, the parallel between genotypic and phenotypic properties becomes a bit weaker under some circumstances. Thus, when an organism encounters an entirely new territory (or a major environmental change within its territory) it can undergo *adaptive radiation*, rapidly creating a variety of new species to fill different ecological niches in that territory. On the other side of this coin, species that fill similar niches in different territories may exhibit *convergent evolution*, that is, they may develop similar structures even though these have been derived via very different routes.

Studies on molecular sequences have revealed striking, previously unrecognized examples of these classical evolutionary mechanisms. Thus, the widely varied songbirds found in Australia were earlier classified as relatives of various Eurasian and North American genera, which they closely resemble; but DNA hybridization now shows that they are much more closely related to each other than to any birds on other continents (reviewed in [28]). What is surprising is the extraordinary extent to which the Australian adaptive radiation, starting from an immigrant bird 65 million years ago, has produced phenotypes convergent with the various birds found in Eurasia. This finding emphasizes that even though chance variations make evolution possible, the environment closely guides its course.

Incidentally, this exception helps prove the rule. A skeptic might claim that the observed correlations between phenotype and DNA sequence simply reflect the nature of the material with which the Creator worked, rather than a predictable consequence of evolution. But he would then have to explain why the Creator developed different correlations on different continents.

The study of DNA sequences has opened up a wide variety of theoretical problems with implications for evolutionary theory. We have already seen that a remarkably small set of changes (as in the chimpanzee and man), presumably located primarily in genes regulating development, can cause large differences in morphology and behavior. Another discovery has been that of a "molecular clock," based on the surprisingly constant (and characteristic) rate at which various proteins fix neutral mutations (i.e., base substitutions that generate a synonymous codon or that code for a functionally equivalent amino acid). Moreover, studies of DNA in eukaryotes have revealed many unexpected properties. These include the presence of enormous amounts of highly repetitive, untranslated DNA; "pseudogenes" that are almost identical with active genes but are not expressed; introns interrupting the translated regions; reverse transcription from RNA to DNA; and localized recombination in somatic cells in the formation of antibodies (and very likely also in genes with other functions).

It is clear that selection acts not only on genes, individual organisms, and populations, but also on variable properties of the DNA itself. These include stability; the effect of the ratio of A-T to G-C base pairs on the tendency to "breathe" more readily in a given region (i.e., to expose the bases for external pairing); the presence of repeat sequences that permit internal pairing within a strand (thus forming a cloverleaf); the choice among synonymous codons; the presence of sequences that favor interaction with plasmids, viruses, and restriction enzymes; and the content of mobile sequences.

In particular, the presence of much apparently nonfunctional DNA presents a challenge. This category may include duplicated genes undergoing a shift from one function to another—or, more broadly, a reservoir of unused genes for evolution to "tinker" with [29]. Since the amount of this DNA is extraordinarily large, it has alternatively been suggested that much of it may be "selfish" or parasitic, accumulated simply because such opportunism is built into the mechanism of DNA replication [30, 31].

DNA Transfer across Species Barriers

In addition to DNA sequence homology, a second kind of direct molecular evidence for genetic continuity is the transfer of DNA between distant organisms—observed not only in the laboratory but also in nature. For example, the bacterium *Agrobacterium tumefaciens* initiates crown gall tumors in plants by integrating a tumor-inducing gene from a bacterial plasmid into a host cell chromosome [32]. In another example, a symbiotic luminescent bacterium has evidently taken up a gene from its host, the ponyfish, since the bacterium has the characteristic animal form of the enzyme superoxide dismutase (containing Cu or Zn) rather than the quite different form (containing Fe or Mn) found so far in all other bacteria [33].

In such lateral transfer the gene might be expected to persist more often when the donor and the recipient are closely related, but because the differences between native and foreign genes would then be much smaller the transfer would be harder to recognize. However, DNA sequencing may offer a method of detection. Thus, in a pair of sea urchin species long separated in evolution the gene for one histone seems to have been shared quite recently, since its sequence showed less than 1 percent as much divergence as that observed in other proteins, including other histones [34]. It therefore seems reasonable to infer a continual slow flow of bits of DNA between species.

Recognition of this flow of DNA is not entirely an esoteric matter: it is quite pertinent to the recent controversy over the hazards of recombinant bacteria. Most of the debate failed to recognize the virtual certainty that the human species has been continually exposed to natural recombinants, such as intestinal bacteria that had taken up DNA released from lysed adjacent host cells. Accordingly, the classes of recombinant bacteria being produced in the laboratory are neither as novel nor as threatening to our survival as was initially assumed. In addition, the successful spread of an organism depends, as we have seen, on a coherent genome rather than on a particular powerful gene; hence the introduction of random DNA of distant origin would almost always impair survival. Wider awareness of these considerations—among biologists as well as laymen—might have spared much unnecessary anxiety.

Conclusions

In summary, although Darwin presented massive evidence for natural selection in *The Origin of Species,* he lacked evidence—and even a plausible mechanism—for an equally crucial part of his theory: the assumption of continual, limitless hereditary innovation. Accordingly, his theory could not fully meet the criteria customary in the experimentally based sciences (although for that reason the intellectual achievement was all the greater, broadening as it did our concept of the scientific method). With the discovery of genes and mutations and then the extension of Mendelian genetics to polygenic traits, genetics provided the hereditary mechanisms required for evolution, and the later development of population genetics made it possible to demonstrate microevolution, based on chance variation plus selection. Moreover, the extrapolation from microevolution to macroevolution was a logical necessity, supported by the graded homologies of present living groups of organisms, by an expanded, precisely dated paleontological record, and finally by the demonstration of extensive biochemical features shared by all organisms.

Today, microbiology and molecular biology have provided major new perspectives in evolutionary biology. This field was concerned first with identifying the sequences and the kinetics of evolutionary change, but we are now analyzing the detailed underlying mechanisms. The results have shown, first of all, that a variety of sophisticated molecular mechanisms of hereditary change (including genetic recombination) evolved very early, in the primitive bacteria. Moreover, many different mechanisms for generating mutations have been identified, and some have large phenotypic effects—a property that is obviously useful for effecting macroevolution. But an even greater contribution to evolutionary biology is the remarkably direct new evidence for macroevolution. This evidence is of two kinds: divergences in DNA sequences parallel divergences in phenotype, from the simplest to the most complex organisms; and small blocks of DNA occasionally move between species, reinforcing their genetic continuity.

To be sure, since evolution involves an inordinate number of variables, with complex interactions, we cannot predict its future course in detail. But this limitation does not undermine the scientific rigor of the theory, any more than the inability of meteorologists to predict the weather shakes our confidence in the underlying physical principles. Accordingly, the term "theory of evolution," with its overtone of tentativeness, is outdated: evolutionary theory (quite a different concept) has a position in biology comparable to that of atomic theory in chemistry.

L'Envoi: Life and Information

In closing, we should note that molecular genetics has brought great unity to biology, through the development of concepts and techniques that now permeate many fields. In addition, it has provided a rigorous, marvelously simple answer to one of the deepest questions in biology: what fundamentally distinguishes living from nonliving matter? Earlier vitalists vainly sought novel forces, and more recently some physicists sought novel physical laws. However, we now see that the uniqueness of life does not derive from special physical forces or laws. It lies, instead, in the organization of its materials in a way that generates a unique property: *the storage of information within molecules.*

The concept of molecular information arose from studies on nucleic acids, which store such information in *sequences.* However, the concept applies as well to allosteric proteins, which store information in their *conformation*. They sense concentrations of a metabolite or a hormone, by changing their conformation when they bind it—and that information then regulates the activity of specific genes, enzymes, or cell-membrane components. This molecular mechanism, which developed early in bacteria, later evolved in animals into the nervous system, where the transduction of sensory stimuli, conduction along nerve fibers, and synaptic transmission all depend on allosteric proteins.

The interactions between allosteric proteins and genes also have epistemological implications, for they demonstrate concretely a continuity, both evolutionary and functional, between two kinds of knowledge: inherited knowledge, programmed in our genes and expressed through development (including that of the brain), and learned knowledge, acquired by experience and programmed largely in our brain. This insight tells us that Kant was right, and the naive empiricists were wrong. We start our lives with a great deal of inborn information, which is a priori for individuals but a posteriori as a product of evolution, and we then add a second class of learned information.

Evolution ties together the extraordinarily coherent, esthetically satisfying, and practically valuable body of knowledge that we have acquired about the living world, ranging from bacteria to man. And these insights do not destroy our sense of awe, as is often charged; rather, they shift the focus. For the miracle is not simply the wondrous complexity and beauty of the firmament, the living world, and man; rather, it is the ability of evolution to produce a creature that can learn to understand so much about these matters. Yet we are surrounded by a sea of unbelievers—including many highly educated people.

The problem of teaching evolution effectively is exacerbated by the long separation of two main streams in biology: the naturalists, asking how the diverse organisms arose and are distributed, and the experimental physiologists, asking how they function. Molecular genetics has brought these streams together in principle, but so far, the occupants eye each other warily. The evolutionists warn of excessive reductionism, and the molecular geneticists are turned off by the metaphysical verbiage at the fringes of the evolutionary literature. Perhaps it will take a new generation of graduate students, exposed to both disciplines, to consummate the intermarriage. Meanwhile we will surely improve our teaching of evolution, even at an elementary level, by building on the molecular evidence.

REFERENCES

1. BETHELL, T. Darwin's mistake. *Harper's*, February 1976.
2. POPPER, K. R. *The Logic of Scientific Discovery*. London: Hutchinson, 1959.

3. POPPER, K. R. *Dialectica* 32:344, 1978.
4. POPPER, K. R. Letter. *New Sci.* 87:611, 1980.
5. RUSE, M. Karl Popper's philosophy of biology. *Phil. Sci.* 44:638, 1977.
6. *New York Times,* August 29, 1982.
7. *Science and Creationism.* Washington, D.C.: National Academy of Science Press, 1984.
8. MAYR, E. *The Growth of Biological Thought.* Cambridge, Mass.: Harvard Univ. Press, 1982.
9. PROVINE, W. B. *The Origins of Theoretical Population Genetics.* Chicago: Univ. Chicago Press, 1971.
10. MAYR, E. *Animal Species and Evolution.* Cambridge, Mass.: Harvard Univ. Press, 1963.
11. DOBZHANSKY, T. *Genetics of the Evolutionary Process.* New York, Columbia Univ. Press, 1970.
12. HUXLEY, J. S. *Evolution, the Modern Synthesis.* London: Allen & Unwin, 1942.
13. MAYR, E., and PROVINE, W. B. (eds.). *The Evolutionary Synthesis.* Cambridge, Mass.: Harvard Univ. Press, 1980.
14. DOVER, G. A., and FLAVELL, R. B. *Genome Evolution.* New York: Academic Press, 1982.
15. BENDALL, D. S. (ed.). *Evolution from Molecules to Man.* Cambridge: Cambridge Univ. Press, 1983.
16. REANNEY, D. Genetic noise in evolution? *Nature* 307:318, 1984.
17. ANDERSON, N. G. Evolutionary significance of virus infection. *Nature* 227:1346, 1970.
18. REANNEY, D. Extrachromosomal elements as possible agents of adaptation and development. *Bacteriol. Rev.* 40:552, 1976.
19. GOULD, S. J., and ELDREDGE, N. Punctuated equilibria: the tempo and mode of evolution reconsidered. *Paleobiology* 3:115–151, 1977.
20. STANLEY, S. M. *The New Evolutionary Timetable.* New York: Basic, 1981.
21. GOULD, S. J. Irrelevance, submission and partnership: the changing role of paleontology in Darwin's three centennials, and a modest proposal for macroevolution. In *Evolution from Molecules to Men,* edited by D. S. BENDALL. Cambridge: Cambridge Univ. Press, 1983.
22. GOULD, S. J., The meaning of punctuated equilibrium and its role in validating a hierarchical approach to macroevolution. In *Perspectives in Evolution,* edited by R. MILKMAN. Sunderland, Mass.: Sinauer, 1982.
23. EIGEN, M., and SCHUSTER, P. The hypercycle: a principle of natural self-organization. *Naturwissenschaften* 64:541–565, 1977; 65:7–41, 341–369, 1978.
24. BASS, B., and CECH, T. *Nature* 308:820–823, 1984.
25. GUERRIER-TAKADA, C. et al. *Cell* 35:849–857, 1984.
26. WILSON, A. C.; CARLSON, S. S.; and WHITE, T. J. Biochemical evolution. *Annu. Rev. Biochem.* 46:573–639, 1977.
27. KING, M. C., and WILSON, A. C. Evolution at two levels in humans and chimpanzees. *Science* 188:107–116, 1975.
28. DIAMOND, J. M. Taxonomy by nucleotides. *Nature* 305:17–18, 1983.
29. JACOB, F. Evolution and tinkering. *Science* 196:1161–1166, 1977.
30. DOOLITTLE, W. F., and SAPIENZA, C. Selfish genes, the phenotype paradigm and genome evolution. *Nature* 284:601–603, 1980.
31. ORGEL, L. E., and CRICK, F. H. C. Selfish DNA, the ultimate parasite. *Nature* 284:604–607, 1980.
32. ZAMBRYSKI, P.; HOLSTERS, M.; KRUGER, K.; et al. DNA structure in plant

cells transformed by *Agrobacterium tumefaciens*. *Science* 209:1385–1391, 1980.

33. MARTIN, J. P. JR., and FRIDOVICH, I. Evidence for a natural gene transfer from the ponyfish to its bioluminescent bacterial symbiont *Photobacter leiognathi*. *J. Biol. Chem.* 256:6080–6089, 1981.
34. BUSSLINGER, M.; ROSIONI, S.; and BIRNSTIEL, M. L. An unusual evolutionary behavior of a sea urchin historic gene cluster. *EMBO J.* 1:27–33, 1982.

HUMAN EVOLUTION AND MOLECULAR BIOLOGY

JEROLD M. LOWENSTEIN* and ADRIENNE L. ZIHLMAN†

The molecular biology revolution is having a profound impact on medical and biological progress and has also changed the rules of the game in the study of human evolution. Under the old rules, the primate evolutionary tree was deduced from anatomical similarities between living and fossil primates, including man, and genetic relationships were likewise deduced from these anatomical resemblances. Thus, from the time of Darwin it has been recognized that among all living primates modern man is anatomically most like the African apes, the chimpanzee and gorilla.

During the past 2 decades the prevailing interpretation of the fossil record, believed to contain "manlike" apes 14 million years old, indicated that the human lineage had separated from that of the apes at least 20 million years ago. During these same 2 decades, mounting evidence from immunological and molecular biological studies of living primates strongly contradicted this "early-divergence" hypothesis and suggested instead that man and the African apes diverged from a common ancestor a mere 5 million years ago.

Evolution and Molecular Biology

The most direct evidence for the genetic similarity between two or more species derives from comparing their genetic material, the DNA. Additional evidence comes from comparisons of RNA, of protein amino acid sequences, of protein immunology, and of karyotypes and chromosomal banding patterns. All these kinds of comparisons have been made, and all have shown remarkable and equal genetic similarity among man, chimpanzee, and gorilla. The DNAs of the three species, from DNA hybridization experiments, are 99 percent identical [1].

*Clinical professor of medicine, University of California, San Francisco, California 94143.
†Professor of anthropology, University of California, Santa Cruz, California 95064.

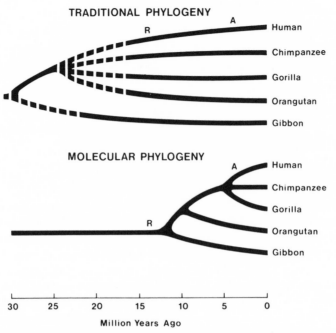

TRADITIONAL PHYLOGENY

Human
Chimpanzee
Gorilla
Orangutan
Gibbon

MOLECULAR PHYLOGENY

Human
Chimpanzee
Gorilla
Orangutan
Gibbon

| 30 | 25 | 20 | 15 | 10 | 5 | 0 |

Million Years Ago

Fig. 1.—Comparison of the traditional phylogeny of hominoid evolution with that derived from molecular data. The traditional view was strongly influenced by the conviction that *Ramapithecus* (*R*), a fossil genus dated as old as 14 million years ago, was on the human lineage. This opinion was based on teeth that resembled those of humans in some but not all respects. Recent discoveries of *Ramapithecus* face and limb bones have been interpreted as showing closer relations to the orangutan. *Australopithecus* (*A*), with fossils going back almost 4 million years, is the oldest undisputed hominid. In contrast with the traditional phylogeny, the molecular one places *Ramapithecus* somewhere on the lineage ancestral or parallel to that of all living hominoids and has *Australopithecus* near the divergence of the human line from that of the African apes.

When these data first emerged, they were surprising in two respects: first, in the degree of genetic similarity between humans and African apes; second, in that chimpanzee and gorilla are genetically no more similar to each other than either is to humans—anatomically, of course, the two apes are much more like each other than they are like us. These findings carry two logical implications: that morphological and genetic changes have occurred at different rates in these three lineages, and that the man-ape evolutionary divergence was much more recent than the 15–20 million years previously thought (fig. 1).

If it is assumed that primate DNA has evolved at more or less the same rate as that of other animals whose divergence times can be inferred from the fossil record, then humans, chimpanzees, and gorillas had a common ancestor, presumably in Africa, about 5 million years ago. The "molecular clock" of DNA base-pair and amino acid sequence changes in animals descended from a common ancestor provides a badly needed quantitative control on estimates formerly based on morphological differences. For example, two different paleontologists, using anatomical characteristics, estimated the chimpanzee-gorilla divergence times at 3 million years [2] and 15 million years [3] ago.

MOLECULAR CLOCK

Historically, immunological techniques led the way in demonstrating evolutionary relationships based on molecular rather than gross anatomical similarities. Around 1900 George H. F. Nuttall [4] injected the sera of thousands of different species of animals into rabbits and used homologous and heterologous precipitin reactions as an indicator of evolutionary distance. About 1960 Morris Goodman refined this technique by using Ouchterlony immunodiffusion plates. Sera from man, chimpanzee, and gorilla reacted so nearly identically that Goodman suggested this triad be classified in the same family, the Hominidae [5]. In 1967 Vincent Sarich and Allan Wilson further quantitated the immunological similarities by using purified serum albumin in a microcomplement fixation technique. Their results suggested that albumin sequence differences increased linearly with the time of divergence from a common ancestor and that man and the African apes shared a common ancestor only 5 million years ago [6].

During the 1960s a number of anthropologists believed that the fossil record required a human lineage distinct from that of the apes for at least 15 million years. Since they could not dispute the data showing 99 percent identity of the proteins and DNA of man, chimpanzee, and gorilla, they attacked the conclusion that molecules evolve at a steady rate [7]. The early-divergence hypothesis could be rescued by assuming that primate DNA and proteins evolved at only about one-third the average rate demonstrated for other mammalian lineages [8] such as bats, horses, cows, whales, and carnivores, and even such nonmammalian groups as frogs [9].

The conflict between the traditionalists and the new breed of molecular evolutionists has given rise to an enormous literature on the molecular clock controversy [1, 9–13]. The traditionalists who study teeth and bones have largely continued to maintain that molecular clocks do not keep good time, and therefore they have felt free to ignore the results of molecular evolutionary studies. The biochemists have continued to pile up volumes of data showing (*a*) that different proteins evolve at different rates, like hour, minute, and second hands on a clock—for instance, cytochrome c is slow, hemoglobin moderate, and fibrinopeptides fast— but the results of different protein comparisons are concordant with each other [14, 15]; and (b) that, contrary to what most of us have been taught, a considerable number of mutations in the genetic material (DNA) are neither deleterious nor favorable but are selectively "neutral" [16]. Two examples of these neutral mutations are those that occur in the long stretches of noncoding DNA known as introns and in the "silent codon" mutations that change a base without changing the resultant amino acid, owing to the redundancy of the genetic code [13, 15]. Also, it appears that many amino acid substitutions do not significantly alter protein function: the more rapidly evolving proteins, such as fibrinopeptides, can tolerate more such substitutions [15].

Our picture of evolutionary processes in the eucaryotic genome has changed rapidly in the past few years. Whereas previously it was thought that DNA base-pair mutations were relatively rare and that most such mutations would be deleterious or lethal and thus eliminated by natural selection, we now see a veritable beehive of mutational activity, much of it taking place in the long stretches of "selfish" or "parasitic" DNA which have little or no effect on the phenotype. It might be expected that such silent mutations, not subject to natural selection, would be seen more frequently than those which result in protein amino acid substitutions, and it has been demonstrated that intron and silent codon mutations are four or five times as common [13].

Furthermore, each eucaryotic cell has two kinds of DNA—nuclear and mitochondrial. Mitochondria may have evolved from bacteria-like procaryotic invaders of early eucaryotic cells hundreds of millions of years ago [17]. Residing in the cytoplasm, mitochondria are derived solely from the mother. Mitochondrial DNA has been found to be evolving at about 10 times the average rate of nuclear DNA and thus has provided molecular anthropologists with another "second hand" for clocking the later stages of primate evolution [18]. From these kinds of data it appears that the modern human races diverged between 200,000 and 500,000 years ago [19].

A good scientific hypothesis makes predictions which can be verified or falsified by subsequent observation. The molecular clock, as applied to human evolution, predicted that the oldest hominid fossils would be found in Africa, would be no more than 5 million years old, and would resemble chimpanzees or gorillas. In fact, the oldest undisputed hominid fossils to date, *Australopithecus*, have been found in East Africa, dated by the potassium-argon method as between 3 and 3.5 million years old, and have pelvises and lower limbs indicating upright bipedal locomotion but a brain the size of a chimpanzee's [20].

Hominid Evolution

Why, then, has the recent-divergence hypothesis of 5 million years been so strongly resisted by much of the anthropological establishment worldwide? The principal objection has been the existence of a fossil genus *Ramapithecus* in Africa, Asia, and Europe, dated between 8 and 15 million years ago. Until recently it was widely believed to be hominid, that is, on the human lineage as distinct from that of the apes [21]. For several decades in this century *Ramapithecus* fossils consisted of only a few jaws and teeth. There were no skulls to evaluate brain size or cranial and facial features, no pelvic or limb bones to confirm the bipedal stance it was invariably depicted to have in artistic reconstructions [22]. It was argued that the teeth, especially the small canines, were like those of early hominids, not like those of apes. On that basis, molecular studies pointing to a more recent divergence of apes and humans were rejected.

If the molecular evidence is correct in dating the ape-human divergence at about 5 million years ago, then *Ramapithecus* at 15 million years ago obviously could not be hominid. In the past few years new fossil finds in China, Pakistan, and Europe have brought forth *Ramapithecus* skulls which resemble those of the orangutan and limbs which suggest that the creature was a tree-living quadruped rather than a ground-dwelling biped. A number of anthropologists who formerly considered *Ramapithecus* as hominid now believe it was an ape ancestral to living orangutans [23, 24]. Thus the implications of the molecular clock data have in this instance been vindicated. With *Ramapithecus* removed from the human family tree, we can now focus on the oldest undisputed hominids, *Australopithecus* of Africa, going back in time almost 4 million years.

AUSTRALOPITHECUS

When the first *Australopithecus* ("southern ape") was recognized and named by Raymond Dart in 1925, combined with Robert Broom's subsequent discoveries in the 1930s and 1940s, a "missing link" [25] emerged that was not expected by the scientific community—a creature with small canines, large molar and premolar teeth, and a human-like pelvis, but with a small, ape-sized brain. Later discoveries of *Australopithecus* by Louis and Mary Leakey at Olduvai Gorge in Tanzania confirmed these findings from South Africa and also added significant, new pieces to the human evolutionary puzzle—a radiometric date of almost 2 million years was at first greeted by disbelief in the anthropological community as being much too old [25].

The most recent fossil discoveries from Tanzania at Laetoli by Mary

Leakey now push back the age of the human family to somewhat over 3.5 million years [26]. The Hadar fossils in Ethiopia, well known because of one relatively complete individual, "Lucy" [27], are dated to about 3 million years ago. Lucy is nearly identical morphologically with some South African australopithecines which are also estimated to date to about 3 million years ago, although the dating is less certain than that of the East African discoveries. Like the South African finds of 40 years ago, the Ethiopian fossils confirm bipedalism as a defining human characteristic which emerged early; and Mary Leakey's discovery of 3.5-million-year-old "human" footprints at Laetoli in Tanzania provides graphic support for this view.

The fossil data so far are entirely consistent with the recent-divergence hypothesis, the inference from molecular studies that the human lineage diverged from that of the apes about 5 million years ago.

During the 1960s, while the molecular comparisons between apes and humans were being compiled, another line of evidence was revealing unexpected similarities between humans and chimpanzees in the behavioral realm. Jane Goodall and several Japanese primatologists were carrying out research in East Africa on the life of chimpanzees (*Pan troglodytes*) under natural conditions. To the world's amazement, it was learned that chimpanzees used tools, shared food, ate meat, had close mother-infant ties and enduring social bonds, and communicated with a complex of gestures and vocalizations [28–30]. All these forms of behavior had been thought unique to humans.

When we consider the mere 1 percent genetic difference between chimpanzees and humans, it is perhaps surprising that the behavior and the anatomy are not even more alike. What is implied is a long period of common ancestry prior to the split about 5 million years ago and a common developmental, neurological, and morphological substrate. Hominids of 3–3.5 million years ago must have been considerably more apelike than we are.

The earliest hominids had a brain size about 400 cm^3, within the range of those of living chimpanzees and about one-third the size of those of modern *Homo sapiens* [31]. The hand bones are in some respects like those of chimpanzees with modifications toward those of tool-using hominids [32]. The pelvis and lower extremities are of particular interest because of the striking changes required to achieve bipedalism [33]. When *Ramapithecus* was widely regarded as an early hominid at 14 million years ago, some anthropologists considered bipedalism at 3 million years ago to be essentially "complete," like that of modern man [34]. But if the ape-human divergence took place about 5 million years ago, as the molecular evidence suggests, bipedalism must have emerged rather rapidly in evolutionary time, and we now recognize many apelike features in the australopithecine pelvis and bones of the lower extremities [33, 35].

If humans, chimpanzees, and gorillas diverged from a common ancestor some 5 million years ago, it is natural to ask what that common ancestor must have been like. We must speculate and reconstruct, for there are no eligible candidates in the fossil record: in fact, there are virtually no primate fossils for the critical period from 4 to 8 million years ago, and obviously one good fossil would be better than many volumes of theories. There are, however, a variety of clues—molecular, anatomical, behavioral, and paleontological—which may be plausibly woven together to provide a glimpse of that African progenitor.

First, we might ask what kind of creature would have been most likely to give rise to three lines of descent as diverse as the human, chimpanzee, and gorilla. Which of the three, or what mixture of the three, would provide the best starting point? Is it probable that one has changed less in morphology than the other two in the past 5 million years?

Anthropologists have generally considered the chimpanzee the best model for the common ancestor because it is the least specialized and seems to have the potential for evolving toward the other two. It is at home in the trees and on the ground, is intelligent and social, uses tools, and has an omnivorous diet. Early australopithecines are very chimplike in many respects. Gorillas and humans, however, have many specialized, "derived" traits. Gorillas are specialized vegetarians and are quite large—much larger than the earliest hominids—and modern humans are bipedal, large-brained, and specialized tool users.

Two species of chimpanzees live in Africa today: the common species, *Pan troglodytes,* and a rarer species confined to a small enclave in Zaire, the pygmy chimpanzee, *P. paniscus.* First described in 1933 [36], little was known of the pygmy chimpanzee's anatomy and behavior until the past decade. A case has been made that *P. paniscus,* not *P. troglodytes,* represents the best living prototype of the common ancestor [37]. The molecular clock indicates that the two chimpanzee species diverged from each other about 3 million years ago, but pygmy chimpanzees appear to retain a number of anatomical traits of the common ancestor and thus may provide us with a "living link" to that ancestor. Pygmy chimpanzees show a number of similarities with the early australopithecines [38]. Comparison of a female pygmy chimpanzee with the 3-million-year-old Lucy shows the striking resemblances between the two, in overall size and body build, and in the ankle and foot bones (fig. 2). Differences occur in features of the pelvis, quadrupedal in the ape and bipedal in the hominid, and the premolar and molar teeth are much larger in the hominid, probably due to a diet of tough, fibrous plant foods in contrast with the largely fruit diet of chimpanzees.

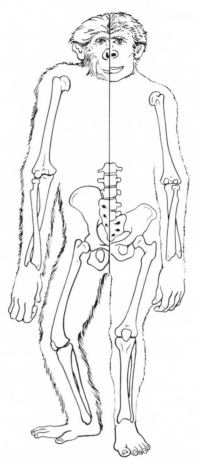

Fig. 2.—Comparison of the *Australopithecus* "Lucy," a 3-million-year-old fossil, with a female pygmy chimpanzee (*Pan paniscus*). On the right is Lucy, with soft anatomy reconstructed; on the left is a pygmy chimpanzee, both drawn to the same scale. The major anatomical difference between the two is seen in the pelvis and reflects the quadrupedal locomotion of the chimpanzee and the bipedalism of the early hominid. There is no significant difference in brain size. (Modified from [39].)

Punctuation versus Gradualism

A hotly debated topic in evolution over the past decade has been the issue of punctuation versus gradualism. Paleontologists have noted for a long time that the fossil record does not seem to be continuous: there may be long periods of stasis during which species remain morphologically unchanged, then suddenly there is another stage of evolution without graded, intervening forms. Generally this phenomenon has been attributed to the incompleteness of the fossil record, which is rather like a few pages torn at random from a long biography. Niles Eldredge and Steven Jay Gould, however, have suggested that the abrupt transitions reflect the way evolution works, by a process of "punctuated equilibria"—stasis followed by relatively short outbursts of rapid morphological change [40]. Several books and numerous articles have been written on this topic, in defense of the punctuationist or gradualist thesis, but only rarely has the underlying molecular basis of evolutionary change been discussed in this context, which is rather like writing extensively about architecture without ever mentioning building materials.

Aside from all the silent DNA mutations discussed above, it has been estimated by Kimura that only approximately 5 percent of the mutations resulting in amino acid substitutions cause any significant functional or morphological change [16]. Hence the concept that most mutations are "neutral." For instance, all amphibians, reptiles, birds, and mammals have homologous proteins such as cytochrome c, hemoglobin, albumin, and fibrinopeptides. Frogs have changed relatively little in appearance during the past 100 million years, yet the albumins of two species of frogs may be as different from each other as are those of a bat from those of a whale [9]. All the evidence suggests that frog DNA and proteins have evolved at the same rate as mammalian DNA and proteins. Because frog morphology is just about optimal for doing what frogs do, natural selection has kept the morphotype relatively constant. By contrast, birds have changed morphologically so much that they are classified in many suborders, but their DNAs are quite similar [41]. Mammals, too, during the past 100 million years have moved into new niches on land, in the air, and in the water, and their morphologies have undergone appropriate adaptive modifications.

From the molecular view, the punctuation versus gradualism debate has a Scholastic ring. The known rate of mutations in nuclear DNA provides plenty of grist for the mill of natural selection, so that organisms may remain morphologically static for eons, as frogs have, or undergo rapid change and diversification, as birds and mammals have, when new environmental opportunities open up.

Human evolution provides an example both of rapid evolutionary change, from ape to man in 5 million years, and of phyletic gradualism, for we see in the hominid fossil record the continual increase in brain size from 400 cm^3 in *Australopithecus*, to 600 cm^3 in *Homo habilis*, to 850–1,050 cm^3 in *H. erectus*, to an average of 1,350 cm^3 in *H. sapiens* [42].

The origin of bipedalism can also be viewed either as a punctuational event—for it is difficult to imagine a stage intermediate between walking on four legs and walking on two—or as merely one of many adaptational shifts between a forest-dwelling ape which obtains most of its food from the trees and a savanna-dwelling hominid which walks long distances to exploit scattered food resources. In terms of molecular biological mechanisms, it is obvious that relatively tiny genomic changes could account for both the increase in brain size and the origin of bipedalism, when one considers that the *total* difference between modern human and African ape DNA is only 1 percent.

As we learn more about the molecular basis of evolutionary change, such popular controversies as punctuation versus gradualism appear to be arguments over definition that cannot be settled from the fossil record. Punctuation and gradualism, and particularly stasis, apply to morphology *only*, because there is continual change at the DNA level. Depending on the environmental context and genetic background of the organism, morphological change—which is what the paleontologists are arguing about—can take place rapidly or slowly.

Major breakthroughs in science often bring a theoretical unity to disparate disciplines, as quantum mechanics provided the common framework for physics and chemistry. The advances in molecular biology of the past 2 decades have brought a new and common basis for understanding cellular processes, disease, and the evolution of all life, including human evolution.

REFERENCES

1. GOODMAN, M., and TASHIAN, R. (eds.). *Molecular Anthropology*. New York: Plenum, 1976.
2. WALKER, A. Splitting times among hominoids deduced from the fossil record. In [1, pp. 63–77].
3. SIMONS, E. The fossil record of primate phylogeny. In [1, pp. 35–62].
4. NUTTALL, G. H. F. Blood immunity and blood relationships. Cambridge: Cambridge Univ. Press, 1904.
5. GOODMAN, M. Man's place in the phylogeny of the primates as reflected in serum proteins. In *Classification and Human Evolution*, edited by S. L. WASHBURN. Chicago: Aldine, 1963.
6. SARICH, V. M., and WILSON, A. C. Immunological time scale for hominid evolution. *Science* 158:1200–1203, 1967.
7. UZZELL, T., and PILBEAM, D. R. Phyletic divergence dates of hominoid primates: a comparison of fossil and molecular data. *Evolution* 25:615–635, 1971.
8. GOODMAN, M. Toward a genealogical description of the primates. In [1, pp. 321–353].
9. WILSON, A. C.; CARLSON, S. S.; and WHITE, T. J. Biochemical evolution. *Ann. Rev. Biochem.* 46:573–639, 1977.
10. CRONIN, J. E., and SARICH, V. M. South American molecular systematics, evolutionary clocks, and continental drift. In *Origin of the New World Monkeys and Continental Drift*, edited by A. B. CHIARELLI and R. CIOCHON. New York: Academic Press, 1981.
11. CORRUCCINI, R. S.; BABA, M.; GOODMAN, M.; et al. Non-linear macromolecular evolution and the molecular clock. *Evolution* 34:1216–1219, 1980.
12. JACOBS, L. L., and PILBEAM, D. R. Of mice and men: fossil-based divergence dates and molecular "clocks." *J. Hum. Evol.* 9:551–555, 1980.

13. Jukes, T. H. Silent nucleotide substitutions and the molecular evolutionary clock. *Science* 210:973–978, 1980.
14. Dickerson, R. E. The structure and history of an ancient protein. *Sci. Am.* 226(4):58–72, 1972.
15. Doolittle, R. F. Protein evolution. In *The Proteins*, vol. 4. New York: Academic Press, 1979.
16. Kimura, M. The neutral theory of molecular evolution. *Sci. Am.* 241(5):98–126, 1979.
17. Grivell, L. A. Mitochondrial DNA. *Sci. Am.* 248(3):78–89, 1983.
18. Brown, W. M.; Prager, E. M.; Wang, A.; and Wilson, A. C. Mitochondrial DNA sequences of primates: tempo and mode of evolution. *J. Mol. Evol.* 18:225–239, 1982.
19. Denaro, M.; Blanc, H.; Johnson, M. J.; et al. Ethnic variation in *Hpa* endonuclease cleavage patterns of human mitochondrial DNA. *Proc. Natl. Acad. Sci. USA* 78:5768–5772, 1981.
20. Washburn, S. L. The evolution of man. *Sci. Am.* 239(3):194–208, 1978.
21. Simons, E. L. Ramapithecus. *Sci. Am.* 236(5):28–35, 1977.
22. Zihlman, A. L., and Lowenstein, J. M. False start of the human parade. *Nat. Hist.* 88:86–91, 1979.
23. Andrews, P., and Cronin, J. E. The relationships of *Sivapithecus* and *Ramapithecus*. *Nature* 295:185, 1982.
24. Pilbeam, D. R. New hominoid skull material from the Miocene of Pakistan. *Nature* 295:232, 1982.
25. Reader, J. Missing links: the hunt for earliest man. Boston: Little, Brown, 1981.
26. Curtis, G. H. Establishing a relevant time scale in anthropological and archeological research. *Philos. Trans. R. Soc. Lond.* B292:3–5, 1981.
27. Johanson, D., and Edey, M. *Lucy: The Beginnings of Humankind.* New York: Simon & Schuster, 1981.
28. Goodall, J. The behaviour of free living chimpanzees in the Gombe Stream Reserve. *Anim. Behav. Monogr.* 1:161–311, 1968.
29. Itani, J., and Suzuki, A. The social unit of chimpanzees. *Primates* 8:355–381, 1967.
30. Nishida, T. The social group of wild chimpanzees in the Mahali mountains. *Primates* 9:167–224, 1968.
31. Kimbel, W. H. Characteristics of fossil hominid cranial remains from Hadar, Ethiopia. *Am. J. Phys. Anthropol.* 50:454, 1979.
32. Marzke, M. W. Joint functions and grips of the *Australopithecus afarensis* hand, with special reference to the region of the capitate. *J. Hum. Evol.* 12:197–211, 1983.
33. Zihlman, A. L., and Brunker, L. Hominid bipedalism: then and now. *Yearbook Phys. Anthropol.* 22:132–162, 1979.
34. Lovejoy, C. O.; Heiple, K. G.; and Burstein, A. H. The gait of *Australopithecus*. *Am. J. Phys. Anthropol.* 38:757–780, 1973.
35. Stern, J. T., and Susman, R. L. The locomotor anatomy of *Australopithecus afarensis*. *Am. J. Phys. Anthropol.* 60:279–317, 1983.
36. Coolidge, H. J. *Pan paniscus*, pygmy chimpanzee from south of the Congo River. *Am. J. Phys. Anthropol.* 18:1–59, 1933.
37. Zihlman, A. L.; Cronin, J. E.; Cramer, D. L.; and Sarich, V. M. Pygmy chimpanzees as a possible prototype for the common ancestor of humans, chimpanzees and gorillas. *Nature* 275:744–746, 1978.
38. Zihlman, A. L. Pygmy chimpanzees and early hominids. *S. Afr. J. Sci.* 75(4):165–168, 1979.

39. ZIHLMAN, A. L. *The Human Evolution Coloring Book.* New York: Harper & Row, 1982.
40. ELDREDGE, N., and GOULD, S. J. Punctuated equilibria: an alternative to phyletic gradualism. In *Models in Paleobiology,* edited by T. J. M. SCHOPF. San Francisco: W. H. Freeman, 1972.
41. SIBLEY, C. G., and AHLQUIST, J. E. The relationships of the yellow-breasted chat (*Icteria virens*) and the alleged slowdown in the rate of macromolecular evolution in birds. *Postilla* 187:1–19, 1982.
42. CRONIN, J. E.; BOAZ, N. T.; STRINGER, C. B.; and RAK, Y. Tempo and mode in hominid evolution. *Nature* 292:113–122, 1981.

HUMAN GENETICS: INTRODUCTION

The foundations of human genetics are firmly rooted in concepts from discoveries made with peas, fruit flies, bread mold, and bacteria. The transfer of ideas from transmission and population genetics, cytogenetics and the one gene–one enzyme relationship to human and medical genetics validates the worth of basic research for human affairs. Stern reviews the transfer and discusses the difficult problems faced by geneticists in dealing with simple genetic models for mental disease and intelligence.

Kalckar emphasizes the significance of the discovery of a molecular basis for sickle cell anemia in relating biochemical lesions to the genetic diseases of man. His analysis of different types of mutations in the bread mold *Neurospora crassa* leading to a biochemical block as models for understanding "inborn errors" of human metabolism illustrates the fundamental unity of biochemical processes in different life forms. On the other hand, Kalckar predicts complications can be expected in the complex systems of interacting pathways of development leading to pleiotropic manifestations of single mutations in mammals.

Brewer uses sickle cell anemia, the first molecular disease, to illustrate the relation between basic research and its application to the practical problems of medical treatment. While the molecular basis of a genetic disease and the pleiotropic effects of the mutant polypeptide may be known and appreciated, the management or treatment of the disease can baffle the best efforts of medical scientists. This original article provides an antidote to promoters of easy solutions for difficult medical problems.

Neel discusses the role of genetic theory in advising parents at risk in having children with a genetic disease. The progress in amniocentesis technology and the detection of missing or variant enzymes in cultured cells provide an escape either by the therapeutic abortion of the mutant fetus or by the delivery of a normal child. For all of the scientific input, the social and political problems of any type of abortion still require solution.

McKusick identifies pleiotropism, the multiple effects of a single mutant gene on development, and genetic heterogeneity, two or more fundamentally distinct genetic entities with essentially the same phenotype, as the two principles of the nosology (classification) of genetic diseases. A detailed discussion of the clinical, genetic, and biochemical approaches to recognizing genetic heterogeneity illustrates the interdisciplinary nature of modern medicine. Even with the current high technology of molecular genetics, McKusick points to practical difficulties in recognizing distinct genetic entities.

Neel and Schull undertake an analysis of some trends in understanding the genetics of man in an original article published in 1968. Their emphasis on the population genetics of man may reflect the then relatively recent development of electrophoresis to detect variant enzymes and proteins. The information explosion is controlled by the application of advanced computers to analyze the data. Who could have anticipated amniocentesis, recombinant DNA, and nucleotide sequencing for the human genome?

Ajl and Mori present an eloquent case for supporting basic research as the wellspring of ideas and concepts to cope with the problems of birth defects. Their views of basic research in the biomedical sciences include an assault on the "known unknowns" of molecular biology to provide the wedges into the future. The range of approaches and the choice of experimental species are left to the competent imaginative investigator in the optimistic expectation that advances usually come from unexpected quarters.

Neel discusses human evolution in the bright light of molecular biology. Man had been a poor candidate as an experimental species for geneticists in the pre-DNA age. With the flowering of molecular genetics, man has become a prime target for geneticists and medical scientists. The new medical geneticists have opened novel and promising approaches for the diagnosis and treatment of the genetic diseases of man.

While biomedical scientists continue to provide new and sensitive tests to detect genetic defects in infants, the questions of voluntary or compulsory screening continue to be debated by social scientists, political scientists, and legislators. Kopelman cogently discusses the ramifications of screening specific groups or populations from the ethical, social, and political aspects. Human genetics and the social problems arising from genetic diseases are intertwined and not likely to be easily dissociated.

GENES AND PEOPLE

CURT STERN*

Genes are the units of inheritance—thus postulated the Mendelians. Genes form a linear array in the chromosomes—thus discovered the Morganists. Genes are composed of nucleic acids—thus proved Avery. Nucleic acids are macromolecules composed of four kinds of sub-units assembled in sets of a few hundred or a few thousand—thus it was shown by Wilkins, Watson, and Crick.

All this we have learned from peas and mice, corn and flies, bacteria and viruses. The narrow hereditary bridge between parents and offspring which egg and sperm represent is now understood in terms of two micro-packages of molecular information whose code has been revealed to the inquiring mind.

Human genetics had its part in these triumphs. At the very beginning of genetics, Garrod's book *Inborn Errors of Metabolism* directed attention to the molecular basis of inherited diseases. Again the later recognition in normal and abnormal hemoglobins of amino acid substitutions was an important step in the "gene to polypeptide" theory. Beyond such discoveries at the physical-chemical level, human genetics par excellence brings back into recognition the fact that genes do not act in isolation but are imbedded in an interrelated harmonious whole, the organism. It is a commonplace to say that even a bacterium is so much more than the sum of its thousands of constituents and reactions, but this commonplace is all-too-easily forgotten. This is the more so since the fundamental discoveries relating to feedback and other regulatory systems may deceptively suggest that the commonplace statement is no longer true. The existence of an organism asks for the comprehension of its parts, of its evolutionary

* Departments of Zoology and Genetics, University of California, Berkeley. This paper is a revised version of the Public Evening Lecture which was presented under the auspices of the National Foundation at the Third International Congress of Human Genetics in Chicago, 1966. The original version is being included in the published *Proceedings of the Congress*.

history which put them together, and it asks even for the metaphysical realization that the universe is miraculously endowed with the potentiality of bringing forth organisms. If all this is valid when one contemplates a bacterium, how much more is an organismic attitude an essential aspect of viewing Man. The money value of a man was once calculated from his composition of iron, calcium, phosphorus, and other elements. He turned out to be worth less than a dollar. His money value as an economic social entity is much higher but still infinitely inadequate for characterizing his value as an organism. A million times the expenditure of the billions of dollars which go into sending a space capsule to the moon or beyond and a million of the brightest minds could not create the astronauts who man the vehicle. It is amusing to say, as J. B. S. Haldane once did, that all love poetry owes its existence to a microscopic bit of matter, the sex-chromosome—but have you ever seen a naked Y chromosome compose a sonnet?

There is no limit to the number of stories which one could tell about genes and people. Genes are involved in every baby's being well born even though most happy parents forget to think of their and the child's good fortune. Genes are involved in all-too-many tragic events when malformations, inborn diseases, or mental deficiencies happen. Sometimes, the involvement of genes brings a smile to one's face. I remember a young lady who wanted genetic counseling calling from out of town. What was the danger that her future children would be afflicted with a serious neurological condition present in several of her close relatives? I requested her to furnish medical diagnoses of the illness and in addition to collect exhaustive pedigree data on her family. When the day of her appointment had arrived, she telephoned that she would not come to see me. "What a marvelous thing I found out," she told me instead. "You asked me for information on my family, and I went to my mother to get it. Upon questioning, she broke down and told me that I was an adopted child. I am so happy since there is no more need for me to worry."

I do not intend to tell you other personal stories on genes and people. Instead I shall select a few items of general significance. They will fall into two different groups. The first will deal with certain types of birth defects and spontaneous abortions whose nature has recently been clarified to the benefit of many individuals. The second group of topics will encompass discussions on genes and mental illness; genes and differences in mental endowment between people, between subpopulations, and between racial groups. In all these latter topics one recurrent theme will be the two-sidedness of our being as a product of genetic *and* non-genetic influences, of heredity *and* environment, of nature *and* nurture. Another theme will be a plea for openmindedness concerning the conduct of research in areas where further knowledge is needed, even if the results of such research are liable not to fit the expectations of some people or be misinterpreted by others.

Let us begin with the effect of a surplus of genes on mental development. The prime example is Down's syndrome (also called mongolism). It is the most frequent single type of severe congenital affliction, occurring as it does in more than one in a thousand births. Down's syndrome has long been known to have a peculiar relation to age of mothers. It appears at low frequency among the children of young mothers but at increasingly higher frequencies among those of mothers thirty-five years old and beyond. The age of a parent should not change appreciably his or her genetic constitution. Thus, it seemed reasonable to suppose that not genic changes but some kind of physiological alterations take place during the later years of a woman's reproductive period, alterations which during pregnancy may affect unfavorably the development of the embryo.

Changes in the balanced machinery of the endocrine systems, known to occur with increasing age, seemed prime candidates for being agents which might decrease the quality of the intrauterine environment. From this hypothesis, another assumption easily suggested itself. The endocrine balance can be influenced strongly in a psychosomatic fashion. Psychological attitudes may keep it in equilibrium or may push it into disharmony. Was it not reasonable to believe that the birth of a Down's child often is the result of an emotional upset in the expectant mother? If so, should not the mother have kept her emotions under better control? Was the mother, and by interaction between the parents the father also, morally responsible for the tragic event? The stage was set for life-long feelings of guilt.

In reality, the case for abnormal intrauterine environment as the sole agent causing Down's syndrome was never convincing. It did not fit several facts concerning the abnormality, particularly those derived from studies on twins. Many twin pairs in which Down's syndrome occurs are concordant, that is, both twins are affected. Some of the pure environmentalists cited these twin pairs as proof of an ill-effect of the pregnant mother on development. Both embryos, they emphasized, had been subjected to the same intrauterine influences and thus both came to suffer. Even more twin pairs, however, are discordant, that is, only one of the twins has the syndrome. Why did the same uterine environment lead to such different results in these twins?

Genetics had a ready answer. Whenever the diagnosis was conclusive, the twins of concordant pairs proved to be identical in genetic constitution, having arisen from a single fertilized egg, while the twins of the discordant pairs were not-identical genetically, having come from two separately fertilized eggs. This meant that a genotype making one twin develop with Down's syndrome would force a twin partner with the same genotype to develop likewise, but permit a twin partner with a different genotype to develop well. A purely genetic interpretation, however, could not be the whole story. It had to be fitted to the environmental variable "age of mother." This was done by postulating that a special genotype predisposing to Down's syndrome has to interact with a special intrauterine environment to result in the disease.

Such a synthesis of hereditary and environmental views was not the only interpretation of the origin of afflicted children. Many years ago it was suggested that the ovaries of relatively older women may tend to produce "bad" eggs whose development would lead to Down's syndrome. A single such egg if giving rise to twins would be likely to result in ill-development of both partners. In contrast, of two eggs fertilized at the same time, usually only one would be expected to be of the poor variety, and only one twin would suffer. The nature of the specific badness of an egg remained unknown until rather recently when some simple improvements in the technique of cell studies were introduced. By their use, in 1956 a study of tissue cultures of legally aborted Swedish embryos established the chromosome number of man as twice 23, equal to 46, instead of the long-accepted number of twice 24, equal to 48. This was not only a welcome correction to a mistaken count but it laid the basis for the recognition of abnormal chromosome numbers (and other chromosome aberrations) as causes of congenital diseases. Only a few years after the Swedish study it was demonstrated that individuals affected with Down's syndrome have 47 chromosomes, the forty-seventh member being an extra specimen of one of the smallest chromosomes known as number 21.

Normal persons have two chromosomes number 21, one obtained from each parent. Down's persons have three such chromosomes.

From where does the extra chromosome come? It was natural in view of the maternal age effect to think of the mother as its source, rather than the father. Moreover, a specific parallel in Drosophila had long been known as a classical example of abnormal chromosome constitution and its origin. In the fly, mature eggs ready for fertilization occasionally retain both chromosomes of a pair instead of assigning, as is normal, only one chromosome to the egg nucleus. Such a process of "non-disjunction" involving chromosome pair 21 would explain the origin of the "21-trisomy" responsible for Down's syndrome. Whether this is indeed the true origin is unclear at present. There is evidence from mice that non-disjunction may occur not only prior to fertilization but also soon after fertilization of a chromosomally normal egg. If the same is true in man, again a trisomic Down's embryo could develop.

Whatever the detail, the roles of heredity and environment are now specified in a fundamentally new fashion. It is the environment of a relatively older ovary that may lead to an egg in which non-disjunction takes place, and it is the genetic fact of having three chromosomes 21 in the cells of the embryo which leads to its abnormal development.

Within close limits, the fate of a child in regard to Down's syndrome is decided ultimately at the time of conception. We can now assure the parents of such children that any feelings of personal guilt about events which happened after conception, that is, during pregnancy, are without foundation. This guilt-relieving discovery alone justifies the great effort which has gone into the study of the causes of congenital defects among which Down's syndrome is so important. More work, however, is needed. Age of the ovaries alone is not decisive; otherwise many more children from relatively older mothers would be affected. What then singles out an egg for non-disjunction? Chance, we used to answer, for lack of more precise information. Now, new leads are coming from epidemiology. There appear to be variations in time and geographic area in the incidence of Down's syndrome among the newborn. What is responsible for these variations? It is too early to be certain. It is suggestive, however, that in one population at least the variation in incidence of the syndrome paralleled the variation in incidence of a virus infection. Thus, there seem to exist specific external agents which in relatively older women tend to disturb chromosome distribution in maturing or fertilized eggs. Once the nature of these preconceptional factors has been established with certainty, it will be possible to advise older women on how to plan their pregnancies to avoid children from ill-predisposed eggs.

Whatever the initiating agents, the immediate cause of Down's syndrome trisomy is a misfunction of nuclear division. Clearly, there is a need for new attention to research on the familiar processes of mitosis and meiosis. Much of our study has centered on the normal features of these basic cellular events. Yet, their occasional breakdown is responsible each year in our country alone for six thousand new tragedies of Down's syndrome and for many more if one includes abnormalities in chromosomes other than number 21.

It would, of course, be highly desirable to discover a treatment for Down's syndrome. The knowledge, however, that a whole extra chromosome is involved in the abnormality implies great difficulties in the way of such treatment. Defective development in trisomy is not due to a defective gene. On the contrary, the extra chromosome 21 is no worse genetically than any other chromosome of its kind. What is wrong is the dosage of the genes located in the chromosome. Normal development is dependent on the thousands of biochemical reactions controlled by two of each kind of genes. When three of each of the genes in a chromosome are involved while the genes in other chromosomes are acting in their normal double dose, the relations between the activities of all genes are disturbed. Treatment would have to counteract the imbalance of not just one gene but of very many. If one is optimistic, one may hope that relatively few genes among all those in a chromosome are mainly responsible for the ill-effects of the trisomy. In this case, treatment needs to encompass relatively few remedies. If one is pessimistic, one may fear that very many genes participate in the ill-effects of genic redundancy. If this is true, a whole battery of simultaneous remedies might be required. In either case, treatment will probably have to begin long before birth since, in Down's syndrome at least, the defects seem to be initiated during the first trimester of pregnancy.

Understanding the chromosomal basis of Down's syndrome permits well-founded genetic counseling. In general, the birth of a 21-trisomy child is a sporadic event. Leaving aside the general age effect, no other members of the family group and not even the parents of the child need to fear a recurrence. There are, however, certain "built-in" chromosome aberrations which lead to the appearance of Down's syndrome in some of the brothers and sisters of affected individuals and to its recurrence in successive generations. These are the instances in which one of the chromosomes 21 has become permanently attached to some other chromosome. A normal appearing individual will have one such attached and one free chromosome. In a man who carries such a translocation, most sperm will contain only one chromosome 21, the free or the attached, and his offspring will be normal. In a woman, chromosome distribution—for reasons unknown—frequently leads to eggs in which an attached and a free chromosome remain together in the mature nucleus, the sperm bringing in a third chromosome 21. The resulting "translocation trisomy" accounts for about 3–4 per cent of all newborn affected with Down's syndrome. This type of the syndrome occurs with equal frequency in mothers of different ages.

Chromosome studies in Down's syndrome of blood cells of normal and affected family members can show whether normal or translocation trisomy is involved. If the latter is found, parents may have to be told that there is a high chance that subsequent children will be affected, and certain normal relatives may have to be told the same. Other relatives, however, may be given the good news that they will have no more to fear than most other people, even though they themselves may have affected siblings and other affected close relatives.

Imbalance of genic action results not only from excess but also from losses of chromosomal material. Loss of a whole chromosome, other than that of a sex chromosome, is not compatible with completion of development. Loss of part of chromosome number 5 is compatible with survival beyond birth but results in an infant severely defective physically and mentally. The peculiar sounds made by such a child have given its phenotype the designation "cat's cry syndrome."

In addition to the role of chromosomes in Down's and other syndromes, we now know that abnormal chromosome numbers affect the lives of many people who are normal themselves and have only normal children and other normal relatives. How many healthy married couples experience the disturbing fate that the warmly welcomed conception of a child ends in a spontaneous abortion? Recent chromosomal studies throw a completely new light on spontaneous abortion. A large fraction of early aborted fetuses, perhaps some 20 per cent or even more, has abnormal chromosome constitutions. If trisomy for the small chromosome 21 is compatible with development, albeit defective up to term and beyond, trisomy for most of the larger chromosomes with their larger genic content nearly always leads to a complete breakdown of the embryo. Similar breakdowns occur in embryos in which 3 times 23 or 4 times 23 chromosomes are present in each cell, or in embryos which are mosaically composed of cells with normal and abnormal chromosome numbers. With these embryos, early abortion, regrettable as it is as a personal event, must be welcomed as the selective agent which eliminates the most severe malformations before they come to term. The parents, moreover, may be assured that they were hardly responsible for the maldistribution of chromosomes which led to the loss of the fetus. Nature does not treat humans any more kindly than she treats all other organisms which also bring to term only a fraction of their progeny.

The topic of genes and chromosomes and their effects on individual persons forms only one aspect of the relation of genes to people. Another aspect concerns genes and their effects on populations. People vary in most manifold ways, and populations can be characterized by the distributions of the variations, their averages, their spread of variants, and other statistical measures. Comparisons of populations must pay attention to many different traits. For ages the question has been asked, "Why do people vary?" and for ages dogmatism has given answers. Some assigned all observable differences to inborn genetic variation, others would see in the observable differences nothing but the effect of external variation impinging on equally endowed developing individuals. Still others, aware of the influences of both heredity and environment, would try to answer the question, "Which is the more important of the two?" The controversies arising from these primitive approaches were by no means sterile. They gradually led to a refinement in asking questions and a refinement in answering them. It became clear that the heredity-environment problem has to be studied separately for each trait, for each genetic constitution, and for each environment. Variations in many traits turned out to be strictly determined by genes. Blood groups and eye colors are examples of this, and so are the variations in proteins which are essential parts of our enzymes, hemoglobins, and the structural and metabolic bases of our body tissues.

The mechanism by means of which our DNA genes have their coded information transcribed and translated into appropriate polypeptide chains suggests a strictly genetic determinism of the primary products of genic activity. From then on, however, the bonds of heredity loosen. In the various regions of the embryo, different genes are called into action, and the products of their activities are subject to enhancing or diminishing influences. These influences vary in a way which may express itself in an unpredictable, indeterministic mode. Take the pattern of ridges forming the whorls, loops, and arches of our fingerprints. The cells of each finger of a given embryo contain the same genes that predispose the limb buds toward specific patterns fixed long before birth. Nevertheless, the prints of none of our fingers are alike, not even those of the corresponding digits on the two hands. Subtle differences within the developing embryo enter into the determination of the finger patterns for which genes provide limits and preferences of variations.

The pattern of finger ridges is fixed early. When the child is born, heredity and environment have long ceased to exert any influence on the pattern. Other traits are malleable throughout life. Thus, body weight in man can vary greatly not only between individuals but within limited time in the same person. Nurture in its simplest sense conditions our weight, but nature too plays its role as everyone is aware from common experience: some of us remain thin despite high food intake, and others gain weight despite restricted diets. Do such observations, however, prove that variation in body weight is the product of interactions between nature and nurture? Could it not be true that early non-genetic influences, perhaps going back to fetal life, have fixed the type of our adult reactions to variable nutrition? How do we actually know that differences in form and function are co-controlled by genes?

There are various approaches to this problem. An answer is easily obtained when the appearance of a trait in pedigrees follows simple Mendelian rules. This was the way in which the all-nature causation of blood-group properties, color-blindness, and many other similarly clear-cut traits was proven. Only slightly less easy is the recognition of genetic determiners if special circumstances in the environment are required to bring out differences among people. Favism, the often severe type of anemia caused by inhalation of pollen or the eating of the raw seeds of the broad bean, is clearly caused by external agents. Only certain persons, however, are susceptible to the disease. They are now known to be genetically different from non-susceptible ones, this difference being based on a specific gene in the X chromosome.

Many human traits do not fall into two or a few alternative classes as those just mentioned. Or, even if they do, pedigrees fail to yield the proportions expected from simple genetic situations. Other traits vary in a continuous fashion from low to high like body weight, from short to tall, from one end of a scale to the other. In plants and animals, the geneticist can analyze the variance by growing the organisms under controlled conditions and by studying the variation in separate strains and in the populations derived from crosses between them. In man, where experiments of this kind are not feasible, a first approach to nature-nurture analysis is made by studies of twins. What does it signify if two identical twins are alike in a specific trait such as body weight or size? Not very much! Their being alike may mean exclusive determination by their identical genes, or exclusive determination by the like environments which they encountered during their common childhood. What does it signify if two identical twins are unlike in a specific trait? Something very important! It proves that non-genetic influences are instrumental in causing differences in the trait. To use a specific example, it has been found in some studies that, among one hundred identical twin pairs each of which includes a schizophrenic patient, about sixty pairs are concordant. If we concentrate our attention on these concordant pairs, we can draw only scant conclusions since we cannot distinguish between the effects of like genes or like life experiences. But the forty discordant pairs show beyond doubt that *if* there are genetic influences in schizophrenia they are not all powerful, since one partner of each of these pairs of twins escaped the fate of the other. "*If* there are genetic influences in schizophrenia," we said. Are there? Help in answering this question comes from a consideration of non-identical twin pairs, each of which includes a schizophrenic patient. When a hundred pairs of these are studied, it may be found that only ten are concordant. Why is this frequency so much lower than in identical twins? A ready explanation is based on the genetic makeup of identical twins as contrasted with non-identicals. If there is a genetic predisposition to the mental illness in one of a pair of identical twins, the same genetic predisposition is present in the other. If the predisposition exists in one member of a pair of non-identical twins, it will not be present in the other in a large number of cases.

This genetic interpretation of twin data is not immune to criticism. It presumes that the environment, the life experiences of non-identical twins are no more different than those of identical ones. In support of this presumption, it used to be emphasized that the two partners of both kinds of pairs grow up at the same time in the same family and social setting. This argument is not airtight. The same family setting leaves room for a great variety of specific niches. Two identical twins who have so much in common might be much more inclined to select and experience similar niches than two non-identical twins who are so much more different. If then a given niche has disease-causing attributes, both identical partners often will experience them, but both non-identical partners usually will not. To many students this reasoning appears constrained, but in itself it cannot be shown to be wrong. It loses conviction if other facts are taken into account. It is striking, for instance, that concordance for schizophrenia in identical pairs is the rule even if the two partners were separated very early in life. It is impressive that the frequencies of morbid risk among parents, sibs, children, and other relatives are greatly increased over those in the general population. It is noteworthy that the birth order of schizophrenic sibs in families with two affected among three or more children is a random one, instead of being clustered as might be expected if unfavorable home conditions played a primary role. Nevertheless, these and other aspects, all of which would be predicted from a genetic predisposition to the illness, are to some degree susceptible to alternative non-genetic interpretations. These alternative interpretations focus on the very small number of schizophrenic identical twins reared apart, on the lack of unanimity involved in diagnosing the psychosis, and on the difficulties of devising a specific genetic theory in terms of a dominant or recessive, a single or a multiple gene system. Significantly, any bias the hereditarian may have in favor of a purely genetic interpretation is counteracted by his knowledge of discordant identical twins which proves the existence of non-genetic components in schizophrenia. The environmentalist on the other side encounters no intrinsic check to a purely non-genetic interpretation so that he may remain unchallenged in the premise that lack of final proof for genetic causation is equivalent to its exclusion.

The language in which some psychiatrists describe the symptoms of their patients is very unlike that used by geneticists. This difference, however, should not be divisive. Just as introspective psychology implies no contradiction to neurophysiology, so the analysis of personality changes in mental illness would not be in denial of a material biological basis which as such might be subject to genic influence.

Much of the controversy could have been laid to rest long ago if data on adopted children had been accessible to study, but unfortunately confidentiality usually made this impossible. It was obvious that in large populations there must be many children who have had a schizophrenic biological parent but who were raised by sound adoptive parents. Is the incidence of the illness in these children as high as if they had never been separated from their mother, or is it as low as if they were the own children of the adoptive parents, or is it somewhere between the two values? Also, in large populations there must be a good number of children from healthy biological parents who were adopted into homes in which a parent later developed schizophrenia. What would be the incidence of the illness in these children?

It is gratifying that we do not need to wait any longer for at least some answers to these questions. During the last year a report was published by Leonard L. Heston in which the fate of children born to institutionalized schizophrenic mothers is described. These children were born in Oregon between 1915 and 1945 and thus were adults at the time of the study. They had been separated from their mothers within the first two weeks after birth—usually within the first few days—and subsequently had had no contact with their natural mother nor lived with maternal relatives. Most of the children had been placed in foundling homes initially but eventually were placed in adoptive or foster families. The "experimental" subjects were carefully matched with control subjects for sex, type of eventual placement, birth dates within a few days of the experimental subject, and other characteristics. The "controls" were selected from the records of the same foundling homes that received some of the experimental subjects. After eliminating certain individuals for various reasons, there remained forty-seven persons in the experimental and fifty in the control group. Their psychosocial adjustment was determined by reviewing the records of schools, police agencies, Veterans Administration, hospitals, and other sources. With most subjects, a personal interview was held. Furthermore, scores on the Minnesota Multiphasic Personality Inventory test (MMPI), on an IQ test, and on social classification were obtained. Final evaluations of psychosocial ability as well as psychiatric diagnosis where indicated were made independently by two psychiatrists who were given dossiers on each subject from which genetic and institutional information had been excluded. A third evaluation was made by the author himself.

What was the outcome of this study? There was a multiplicity of answers. As to schizophrenia itself, none of the fifty children from non-schizophrenic mothers, but no less than five of the forty-seven children from affected mothers, were schizophrenic. Three of the five were chronic patients who had been hospitalized for several years; the other two had been hospitalized and were taking anti-psychotic drugs. The probability of all schizophrenic offspring coming by chance from schizophrenic mothers only is less than 2.5 in a hundred. Among other differences between experimental and control groups, the number of rejections or discharges from the armed forces for psychiatric or behavioral reasons may be cited: only one out of seventeen among the controls, but eight out of twenty-one among the experimentals, excluding three schizophrenics, which would have raised the rejected or discharged group to eleven out of twenty-four. There were still other types of psychosocial disability for which there was an excess among the persons born to schizophrenic mothers. Altogether Heston's paper marks a breakthrough in providing information for a heretofore inaccessible area. One can only hope that what has been possible to do in Oregon will also be possible in other states so that additional data will become available. Moreover, it must be emphasized that Heston's study refers to only one of the two questions raised initially. It provides data on the children of schizophrenic women, who had been separated from their mothers. We still need to know the fate of children from healthy women, who grew up in adoptive homes in which a parent developed schizophrenia.[1]

A decisive advance in the study of schizophrenia would also be made if the existence of one or more specific predisposing biochemical errors could be established. No such errors have been demonstrated unequivocal-

[1] It may be wondered whether the prenatal environment of the children of schizophrenic mothers acts to the later detriment of these children. This seems unlikely in view of other data which indicate that there is no difference in the frequency of schizophrenic children from affected mothers as compared to affected fathers.

ly in schizophrenia, but this situation is not unique even among clear-cut genetic diseases. This may be illustrated by Huntington's chorea. The severe degeneration of the nervous system which characterizes it is beyond doubt caused by the action of a single dominant gene. The biochemical error by means of which this gene expresses itself is still unknown, and the presence of the gene cannot yet be determined during the decades of normal life of a later affected person.

A skeptical attitude in regard to the possible involvement of genes in mental illness is largely based on a still-lingering fear that proof of genic causation may interfere with attempts at non-genic prevention or healing. A similar unjustified fear also distorts discussions of the variations in intellectual performance among individuals in populations at large and of subpopulations. It is, of course, a fact that people vary greatly in the scores which they achieve in the so-called intelligence tests, be it a general IQ determination or a battery of tests for presumably independent components of intelligent performance. It is equally a fact that the mean scores obtained by children from different socioeconomic layers are not the same but are positively correlated with the level of their group. It is also true that the IQ of an individual is not a fixed quantity but can decrease or increase parallel with environmental deprivation or enrichment. It follows that the better environment which many homes of the upper socioeconomic groups provide for the intellectual stimulation and motivation of their children must lead to better average test performance as compared with that of children from lower layers. It does not follow, however, that the differences observed among individuals and among subpopulations are due exclusively to non-genetic circumstances. On the contrary, there has long been evidence from a variety of approaches for genetic components in the variability of intelligence. This is particularly clear in some specific instances of low mental performance. Take, for instance, the rare but widely known disease "PKU," phenylketonuria. This mental deficiency syndrome is the result of a single recessive gene. Its biochemical mode of action is known. Persons homozygous for it are unable to synthesize a specific liver enzyme which in normal individuals prevents accumulation of a specific amino acid. Lack of the enzyme results in a high level of the substance in the blood, and this in turn causes damage to brain function.

PKU is present in approximately 1 per cent of the patients in institutions for severe mental defectives. What accounts for the 99 per cent majority? May we expect many other single genes to be responsible for mental defect, each one in its different way disrupting the normal biochemical machinery so as to result in damage to the brain? We have only partial answers to these questions. Some mental defect in addition to PKU is indeed caused by one or another of a whole array of rare single genes. Other severe mental defects are, as in Down's syndrome, the consequences of chromosomal imbalance. Still other individuals with mental defect, particularly those with relatively small impediments, almost certainly do not differ from normal persons in a single major gene. Rather, normal brain function depends on the collaboration of many genes each of which may exist in a variety of types. Some will act toward high degrees of functioning, others will tend toward interference with normal function, and still others may be relatively neutral in their effect. The average person may have many such neutral genes or a more or less equal number of genes with positive and negative effects on mental function. If the shuffling of the genes in the egg and sperm which enter into the conception of a child happens to assign to it a majority of genes with negative effects, subnormality may result.

There are thus single- and multiple-gene defectives. Still other defectives owe their status to external injury suffered before birth, during birth, or later. We do not know precisely what fractions of the mentally deficient belong to the different classes of causation, but it seems that the sum of the various genetic types far outweighs the non-genetic ones. This statement applies particularly to the medium and low grades of feeblemindedness. Undoubtedly the upper grades include a considerable number of people whose innately normal mind has been stunted by excessive cultural deprivation.

It has sometimes been suggested that there is a fundamental difference in the causation of truly low-grade defect and the variability of mental performance in the range from somewhat below average to very high. The genetic nature of severe defect is granted, but a genetic predisposition to more or less normal mental performance is denied. This judgment is based on the higher quality of direct evidence for a genetic basis of severe defect as compared to that for normal variability. The difference in the quality of the evidence, however, depends on the ease with which it can be obtained rather than on a basic difference in causation. It is intrinsically easier to recognize single-gene substitutions, as in PKU, which block normal biochemical pathways so effectively as to lead to disaster than to unravel a whole constellation of genes whose joint effects result in a performance potential somewhat below or above normal.

Research with experimental organisms demonstrates that all genes exist in a variety of forms which can be ordered in their effectiveness concerning a specific chemical reaction. It would be strange indeed if the genetic control of brain function in man were fundamentally different from the genetic control of innumerable functions in other organisms, including brain function in non-human mammals.

It is not necessary, however, to argue from analogy only. We are now acquainted with a number of phenomena which provide unequivocal evidence for genetic determination of mental performance within the more or less normal range. For one, deviations from the normal sex-chromosome number result in some lowering of mental function. Many persons with such unbalanced chromosomal constitutions are normal, and only statistical comparisons show that on the average they are somewhat inferior in mental ability. Among these kinds of individuals are women with only one X chromosome and women with three X chromosomes as well as men with XXY constitution. It would have been difficult to prove the genetic basis of some of these slight impairments of mental performance had they not occurred as by-products of readily recognizable chromosome aberrations. The discovery of the genetic basis under these circumstances justifies the conviction that other genotypes more difficult to analyze are involved in similar slight variations in mental function.

It is worthwhile to go into some details concerning one of the types with unusual chromosomal constitution, that with only one X and no Y chromosome. Samples of these XO women whose complex of specific physical features is known as Turner's syndrome have been given a variety of psychological tests which include over-all measures of IQ performance as well as specific measures for intellectual and cognitive function. The striking result of these studies which were carried out by Dr. Money and his associates at The Johns Hopkins University is that as a group the Turner individuals are specifically superior—even if only slightly so—in certain aspects and specifically inferior—and strikingly so—in others. Areas of strength are verbal IQ and verbal comprehension. Areas of weakness are space-form perception and ability to deal with numbers.

These weaknesses are not only apparent in tests which involve design copying from memory or sight and human figure drawing but also show up in everyday life where a poor sense of direction and related difficulties may lead to special personal problems. Some of the performance patterns characteristic for individuals with Turner's syndrome are well known in certain types of brain damage. It seems then that such damage in Turner persons is in some fashion produced by the chromosomal anomaly.

We know of no chromosome aberration responsible for increased overall mental performance, but very recently a suggestive relation has been reported between a genetically conditioned, abnormally high production of a hormone and higher than average IQ test intelligence. A rare disease, the adrenogenital syndrome, results from a recessive gene. This induces a metabolic error inhibiting the production of cortisone. As a result, the adrenals produce excessive amounts of male sex hormone. In a sample of seventy patients affected with the syndrome, a variety of intelligence tests yielded a mean 10 points above the average. Moreover, 60 per cent of the individuals scored above 110 instead of the expected 25 per cent. It seems that the high mean performance is causally connected with the special hormonal situation, but, it must be stressed, further evidence is required.

The existence of genetic factors influencing test intelligence is also proven by the performance of children from consanguinous marriages as compared to children from unrelated parents. Both in Sweden and in Japan, where relevant studies have been made, formal mental tests showed a small but significant depression in the average performance by the inbred children. Where different degrees of inbreeding could be distinguished, increased inbreeding led to increasingly poorer ratings. Inbred children also performed slightly worse at school. These inbreeding effects were the residue of greater differences after socioeconomic factors had been taken into account. The data demonstrate the existence of recessive genes which influence mental performance in the normal range and which have been made homozygous as a consequence of consanguinity.

Test intelligence has long been the object of family studies and of investigations of twins both reared together and apart. Again the results fit genetic interpretations after giving due weight to the presence of clearly apparent, powerful environmental components. In these types of studies, however, proof of the presence of genetic components is less unassailable than in those reported above.

It has often been stressed that test intelligence is strongly influenced by personality traits such as drive and motivation and that such traits are dependent on the cultural milieu which varies among families and among various subpopulations such as ethnic groups and socioeconomic classes. Simultaneously, it is frequently implied that individual and mean differences in personality traits are solely the results of non-genetic factors. Recent studies, however, show this not to be true. On the average, genetically identical twin partners score much more similarly in batteries of personality tests than non-identical partners, and observations on behavior of twin infants agree with the test results. This points to genetic components in the variance of personality traits. As in intelligence scores, the genetic basis of personality variance consists presumably of a multitude of genes and is therefore not easily broken down into individual genetic units. There are, however, some special types of personality in which specific genetic influences are discernible. Thus, the extra chromosome in Down's syndrome is responsible not only for physical abnormalities and mental defect but typically provides its bearers with cheerful and friendly personalities. In contrast, the dominant gene for Huntington's chorea seems often to lead to personality traits of less desirable character. Perhaps some cases of the specific personality trait "aggressiveness" represent an unusually striking example of genetic co-determination. If preliminary findings should be confirmed, inmates in institutions for particularly aggressive mental defectives seem to include a significantly higher proportion of males with two Y chromosomes than found in the general population and in less aggressive mental defectives. In a specific institution, XYY males were convicted on the average at an earlier age than XY psychopaths of undetermined cause in the same institution, and their offenses were of different kind from those of the XY group. Moreover, the sibs of the XYY males were practically free from crime, whereas this was far from true of the sibs of the XY's. These observations suggest that the double Y condition predisposes to aggressiveness. It is tempting to extrapolate from here and to speculate that the normal male owes his moderate, socially approved aggressiveness to his single Y and the female sex its gentleness to the absence of a Y chromosome.

The nature-nurture discussion of mental attributes has been called futile. It is futile only if, in compassion for our underprivileged fellowmen, we close our minds to the facts of life. It seems to me that Galton's dictum of 1869 still stands: "I have no patience with the hypothesis occasionally expressed, and often implied, especially in tales written to teach children to be good, that babies are born pretty much alike, and that the whole agencies in creating differences between boy and boy, and man and man, are steady application and moral effort. It is in the most unqualified manner that I object to pretensions of natural equality."

A genetic interpretation of the fact that the mean test intelligence of different socioeconomic layers varies, is primarily based on studies of adopted and of orphanage children. On the average, children born in the upper social layers score considerably higher than those of the lower layers, but children adopted into the upper groups score only moderately higher than those adopted into the lower groups. Apparently the own children do not only experience the positive or negative influences of their homes on test intelligence but also share to some measure the positively or negatively acting genes of their parental classes. On the other hand, the adopted children are genetically not correlated with their adoptive parents and experience only environmental differences. Limitations of the power of non-genetic factors are also apparent from a study of IQ scores of children raised from infancy in an orphanage under relatively uniform conditions. In spite of the non-genetic uniformity, the children of parents from upper occupational groups on the average scored higher than those from lower groups. Methodologically, none of these studies on adopted and orphanage children are beyond criticism, but in the aggregate they leave little doubt as to the existence of genetic differences in the mean endowments of different socioeconomic subpopulations.

The inequality of the mean test performance of members of different socioeconomic layers is important in several ways. Its genetic aspects are related to the differential fertility of the different layers. During the last hundred years in many different countries, the average number of children per family was low in the top layers and increasingly higher the more one descended in the social scale. The consequence was that proportionally more of the people comprising a later generation were offspring from parents of the lower than the higher groups. This would be of no genetic consequence if the mean endowments of the various groups were alike. Since, however, they are not, the inverse relation between fertility and test intelligence of the different layers would, so it seemed to many, result in "gene erosion" of the population.

Gene erosion is one of the possible changes in genic content in successive generations of groups of mankind. Evolutionarily, there has been a striking rise in intelligence endowment between the times when there were only prehuman primate ancestors and the emergence of *Homo sapiens*. What has happened since the appearance of modern man, specifically in the last 50,000 years, we do not know. Has the rise been continued, though perhaps at a lesser pace or has it leveled off completely? Could the genetic endowment even have decreased? What may be the trends for the next hundred years or the next thousand? Few attempts have been made to get direct information on these important matters. The most extensive study was the Scottish survey of the test intelligence of two complete cohorts of Scottish eleven-year-old schoolchildren, one consisting of those born in 1921, the other of those born in 1936. The advocates of the gene erosion theory who predicted a fall in performance during the fifteen-year test interval were not sustained by the results. The mean scores of the second group, tested in 1947, were significantly higher than those of the first group, tested in 1932. Did, then, the Scottish national intelligence rise intrinsically? A very careful analysis of the data did not lead to a decision. We do not know whether the genetic endowment had improved or whether an unchanged endowment resulted in better performance due possibly to such non-genetic factors as improved educational background, test sophistication, accelerated physical and mental maturation of the children, or still other environmental forces. The results do not even exclude the possibility that the better performance in 1947 was due to a combination of decreased endowment and increased environmental quality.

During the last decades, the gaps between the fertilities of different socioeconomic layers have diminished, partly due to the spread of contraception to the lower layers which earlier had practiced it least, partly due to volitional raising of larger families in upper layers. Moreover, data on the fertility and mean test performance of members of different population layers give only a coarse and possibly misleading measure of the situation. Within each layer there is a wide and greatly overlapping range of endowments. It is very likely that fertility differentials exist between differently endowed persons within the same layer, and it is conceivable that in part the differentials are in opposite direction to those between layers. Some recent studies, limited in extent but nonetheless highly noteworthy, have been concerned with the fertility of individuals classified by intelligence test performance independently of their socioeconomic status. By using data on the completed fertility of women whose IQ scores were available from the time they had been at school, it has become apparent that not only the fertility of the poorest test performers is high but also that of the best performers. In other words, the relation is not inverse throughout but shows a bimodal distribution. Calculations indicate that in these sample populations the mean IQ range holds its own reproductively. Much more information is needed to judge the situation in general in order to know what happens to our genes in the course of generations. It is unlikely that a permanent trend exists. A continuous watch for fertility changes is necessary just as general census data have to be collected on a continuing basis. It may be true that at the present time no striking change in the pool of our intelligence genes is taking place, and this may remain so in the foreseeable future. Even then the question is bound to come up whether society should be acquiescent in the absence of gene erosion or should institute measures of improving its endowment.

I have not yet touched on the question whether different racial groups are differently endowed in intelligence. The answer is: we do not know. Our tests are worthless if used to compare genetic components of groups in which social and historical differences alone account for an overwhelming part of differences in test performance. This state of affairs is neither compatible with pronouncements alleging inferiority of some races nor with pronouncements postulating strict equality. Examples of the former are well known and all too frequent. A recent pronouncement of this type may be cited here. It starts with the indisputable fact that the mean scores in intelligence tests of American Negroes are significantly lower than those of Whites. And it continues with the statement: "I believe that at least half of this difference is related to genetic variables." On what is this belief based? As far as I know, on nothing but private opinion. And let us compare this opinion with one of opposite bias. It comes from an impressive study by the Office of Policy Planning and Research of a U.S. federal agency on "The Negro Family." The study demonstrates convincingly certain social non-genetic circumstances which act to the detriment of Negro youth in the United States in the performance of mental tests. And then it extrapolates as follows: "There is absolutely no question of any genetic differential. Intelligence potential is distributed among Negro infants in the same proportion and pattern as among Icelanders or Chinese or any other group." On what is this declaration based? As far as I know, on nothing but private opinion. Indeed it seems very improbable that the genetic endowments of groups which have been relatively isolated from one another for thousands of years are exactly alike. The prediction, however, that two racial groups will be found to be different does not include a specific guess concerning which of the groups is endowed more highly than the other with reference to a given quality. In the present context, a suggestion by Washburn is pertinent. He writes as follows. "If one looks at the degree of social discrimination against Negroes and their lack of education, and also takes into account the tremendous amount of overlapping between the observed IQ's of both, one can make an equally good case that, given a comparable chance to that of Whites, their IQ's would test out ahead." In the face of so much lack of knowledge on the one side and beliefs of such divergent kind on the other, it is surely desirable to search for methods to obtain the facts.

There are many who would discourage such a search in the present climate of emotionalism on race. But will continued lack of knowledge lead to less emotionalism? It seems better to me to continue the search for facts and simultaneously stress the social consequences of whatever findings will be made. The opinion has been expressed at times that the mean innate endowment of the great Oriental peoples may exceed that of Caucasians. Of course, again we lack solid facts. If they were available and if they confirmed the opinion concerning the excellence of the Orientals should we then encourage feelings of racial inferiority of the Whites? Clearly, the goal in all such comparisons should be the abandonment of judgments of groups and the concentration on the individual. We will then accept the simple fact that we are all unique and imperfect.

Genes and people! We are different in the genes we received at conception, different in health and disease, intelligence endowment, and personality traits. Like all other organisms, we belong to a polymorphic species. We must educate ourselves to accept the many-faceted inequalities of man. We must not forget also that an individual person who ranks high in some respects may rank average or below in others. The same multidimensional aspects would apply to subpopulations such as socioeconomic layers or racial groups. If there are somewhat more persons in group A endowed with one desirable quality, there may be more persons in group B endowed with a different desirable one.

There is little new in what I have told you about the genetic diversity of people. To paraphrase Goethe: Who can say something clever or something stupid that has not been said before? Why then did I choose to speak to you as I did? Because truth and the search for it can be suppressed not only by ill-will but also by good will. In our justified fear of mankind's misuse of its powers the cry for a moratorium on research has often been sounded. Those who defend such a moratorium do not make entries on both sides of the balance sheet. From the discovery of fire to that of nuclear power, the destructive consequences have been accompanied by beneficial ones. It is the moral tragedy of man that though the extinction of one life cannot be compensated by the preservation of even many lives he cannot avoid making choices. Moreover, it is impossible to wait with new explorations until man is a more moral being. It is not likely that he will soon attain the status of an angel, either by changing his basic nature or by being born into a perfect society. Rather, for a long time the inhuman nature in each of us will have to be overcome by slow individual effort. While this is the case, can we deny man the benefits of possible new discoveries on his genetic variability because there is also the possibility of harm arising from such discoveries?

We must learn to realize that changes in our inequalities are going on incessantly, often independent of our conscious actions and dependent on the social system under which we live. As facts become known, we or our children must become willing to use them in our planning for the future. There is no urgency about this. For a long time to come, the immense gene pools of human populations will include a reservoir of genes from which selection in any desired direction will be possible.

I have abstained from proposing a eugenic program. The need for more knowledge, the danger of rash political action, and the slowness with which genetic changes impinge on large populations justify a waiting attitude. Nevertheless, responsible thinking on problems of eugenics should be encouraged. Past errors of proponents of eugenic measures and the crimes committed under the pretense of eugenics should not stand in the way of new approaches. Culture and social organization are not the ultimate forces which form us. They themselves are made possible by our genes.

A FEW REFERENCES

Schizophrenia

Genetics of schizophrenia. (Annotations.) Lancet, **2**:692, 1966.

LEONARD L. HESTON. Psychiatric disorders in foster home reared children of schizophrenic mothers. Brit. J. Psychiat., **112**:819, 1966.

Turner's Syndrome

J. MONEY and DUANE ALEXANDER. Turner's syndrome: further demonstration of the presence of specific cognitional deficiencies. J. Med. Genet., **3**:47, 1966.

Adrenogenital Syndrome

J. MONEY and VIOLA LEWIS. IQ, genetics and accelerated growth: adrenogenital syndrome. Johns Hopkins Hosp., Bull., **118**(5): 365, 1966.

Extra Y Chromosome

The YY syndrome. (Editorial.) Lancet, **1**:583, 1966.

IQ and Fertility

SCOTTISH COUNCIL FOR RESEARCH IN EDUCATION. The trend of Scottish intelligence. London: Univ. London Press, 1949.

———. *Social implications of the 1947 Scottish mental survey*. London: Univ. London Press, 1953.

J. HIGGINS, E. REED, and S. REED. Intelligence and family size: a paradox resolved. Eugen. Quart., **9**:84, 1962.

C. BAJEMA. Estimation of the direction and intensity of natural selection in relation to human intelligence by means of the intrinsic rate of natural increase. Eugen. Quart., **10**:175, 1963.

———. Relation of fertility to educational attainment in a Kalamazoo public school population: a follow-up study. *Ibid.* **13**:306, 1966.

Race

S. L. WASHBURN. The study of race. Amer. Anthropol., **65**(3): 521, 1963.

SOME CONSIDERATIONS REGARDING BIOCHEMICAL GENETICS IN MAN

HERMAN M. KALCKAR, M.D., Ph.D.*

I. *Apology*

I was once lucky enough to attend a social gathering of the National Convention of Professors and Teachers of English. I overheard the remark that it was regrettable that a promising young humanist, Dr. X, was unsuitable for a position at a university since his field was eighteenth-century English poetry and the chair called for a specialist in seventeenth-century English poetry. Compartmentalization is probably not so bad in the natural sciences. Yet I would greatly hesitate to pose problems outside my working field in the laboratory, and especially those exposed in the present essay, were it not for the invitation to write in this new journal *Perspectives* about "inborn metabolic errors in man." Moreover, I take the liberty of broadening this topic considerably by dealing with borderline subjects between the fields of enzymology and genetics.

II. *Biophysics and Genetics in General*

The impact of physics, lately especially of crystallography, on biology and the application of physical chemistry and biochemistry to genetics have fused all disciplines of natural sciences together. This makes the scientific thinking of today more demanding, yet also more inspiring. As a biologist who has studied mainly metabolic enzymes, I want nevertheless to discuss some topics which happen at the present time to occupy my mind.

What have we learned and what can we learn about genes and the mode of action of genes from human genetics? The trend of our time is certainly to study the genetics of simple micro-organisms—especially the viruses,

* The author is chief of the Section on Metabolic Enzymes, National Institute of Arthritis and Metabolic Diseases, U.S. Public Health Service, Bethesda, Maryland.

since their relative simplicity, their short generation time, and the application of methods of selection make it possible to obtain an enormous number of results within a reasonable span of time. The reader is referred to the brilliant discussions by Benzer (1), Hartman (2), and Lederberg (3) summing up basic problems concerning the subdivision of genes and the integration of these subunits into functional units. Offhand, we cannot expect, from a study of the genetics of higher animals, to make contributions to this area, although we should never disregard the possibility of unexpected novel approaches.

III. *Effect of Heterozygotism on the Properties of Gene Products*

The study of human genetics and especially of inheritance of human diseases has, however, been able to contribute fundamental knowledge toward our understanding of the action of genes. The most important example is the successful introduction of the notion of "molecular diseases," which was first based on studies by Pauling, Itano, Singer, and Wells (4) of a human hereditary disease. Not only has the introduction of the name "molecular disease" been a great stimulus, but the study on which it was based represents one of the most elegant and refined pieces of work in the area of genetics in general, especially with regard to our understanding of the dependence or independence of heterozygous alleles. Sickle-cell hemoglobin might never have been discovered if it had not been for the fact that the altered globin in the form of a sickle-cell ferrous hemoglobin has a lowered solubility. This, in connection with the fact that the red blood cells have a very high concentration of hemoglobin (normal or sickled cells), brings about the crystallization of the sickle-cell hemoglobin at oxygen pressures present in normal venous blood. The crystallization gives rise to morphological manifestations, like tactoid formation and sickle-cell deformation followed by destruction of the cells. The increased rate of destruction of red blood cells results in a manifestation of disease. It is worth emphasizing that the function of the hemoprotein is unaffected; i.e., the oxygen dissociation curve is the same as that of normal hemoglobin.

Individuals with "sickle-cell trait"[1] are heterozygous with respect to the gene for sickling, whereas those with sickle-cell anemia are homo-

---

1. Defined here as a condition in heterozygotes similar to, but less severe than, the homozygous state.

zygous with respect to this gene (5). Although its function is unaltered, sickle-cell hemoglobin moves faster in an electric field than normal adult hemoglobin because of a few extra charges sufficiently large to make a detectable difference. In this way it was possible to show clearly that the erythrocyte in a heterozygous individual contains two molecular species: part of it as ordinary adult hemoglobin and the other part (usually smaller) as sickle-cell hemoglobin. Thus red blood cells from parents of individuals with the disease could be shown by means of electrophoresis to contain two molecular "populations" of hemoglobin molecules which migrated independently—normal adult and sickle-cell hemoglobin. Normal adult individuals contain only normal hemoglobin, and individuals afflicted with sickle-cell anemia contain sickle-cell hemoglobin in their erythrocytes. The last two cases correspond to the homozygous states (5). Perhaps Gregor Mendel and a couple of enlightened friends may have suspected that hybrids contain a mixture of two populations of gene products, whereas the homozygotes contain only one or the other gene product. However, it took a physiochemical study of human genetics to furnish the first complete demonstration of this idea.

We might now pose the question whether this picture is generally applicable. In metabolic disorders in man (inborn errors in metabolism) it has been found that relatives of individuals with diseases show decreased tolerance to the respective metabolites involved in the disorders. It has even been possible to demonstrate a change in sweat composition (water transport?) in parents of individuals afflicted with a hereditary disease called "mucoviscidosis" because the mucus secreted from all the glands is more viscous than normal (6). However, in neither of these cases can it be stated whether the traits produced two molecular species of gene products (enzymes, hormones, etc.) or a hybrid product with lowered activity.

In the field of human blood groups experimental studies have contributed to this problem. Studies on the molecular size of β-agglutinins (7) have been interpreted as an example of interaction of genes at a single locus (8). The molecular weights of the agglutinins of OO_1, A_1A_1, and A_1O individuals were found to be 170,000, 300,000, and 500,000, respectively. Another example of possible interaction of genes in the same locus has been reported by Morgan and Watkins (9) on AB blood groups. "In the heterozygous AB individuals the A and B genes give rise to a molecular species which differs from that produced by either gene in the homozygous state." The O gene does not seem to interfere, since some of the A or B products presumably stemmed from OA or OB individuals. If, for instance, anti-A serum was added to unfractionated or fractionated AB, stemming from mucus of AB secretors, both activities were precipitated out, leaving no antigen in the supernatant. If a corresponding experiment was performed on an artificial mixture of A and B antigens obtained from the corresponding individuals, resolution could be achieved; i.e., if anti-A was added, only A activity was precipitated out, and the B activity could be accounted for in the supernatant. The same lack of segregation of gene products could be demonstrated if anti-B was used. In other words, in AB heterozygotism, according to the authors Morgan and Watkins, no segregation of gene products of the type which we know from the sickle-cell trait could be detected, rather one molecular hybrid species endowed with both characters.

IV. *Synthesis of Metabolic Enzymes in Heterozygotes and in Heterokaryons*

How do we know which of the two types of heterozygotism is operating in metabolic diseases? A lowered tolerance might be the result of a reduction of the number of active enzyme molecules or might be due to the formation of hybrid gene products with lowered catalytic activity.

The problem of the possible existence of a recombination of gene products in bacteria transduced with provirus has been discussed elsewhere (10). Although sexual recombinants of bacteria may well be particularly suitable for the study of such a problem, there is another system which is in a way simpler and, so far, has yielded the most superior experiments in biochemical genetics.

Ever since the pioneer work of Beadle and Tatum, studies on *Neurospora crassa* have provided a wealth of valuable information about the interrelation of genes and enzyme synthesis. This relationship can be further pursued, thanks to the peculiar existence of heterokaryons in which haploid nuclei from two different strains of *Neurospora* share a common cytoplasm without recombination or other interaction.

The discovery of thermolabile genes (11) and the isolation of thermolabile and thermostable enzymes from the corresponding mutants and normal strains of bacteria (12) added an additional approach to biochemical genetics. Horowitz and Fling (13) in an important study of differences in thermostability of tyrosinase in *Neurospora* mutants have clarified the following points. Tyrosinase can exist in two forms which differ in thermostability. The difference is inherited in a simple Mendelian way. Since the stable and the labile forms are recognized by a qualitative, rather than a quantitative, effect on the enzyme, it was feasible to look into the question of allele interference versus non-interference in heterokaryons. In other words, does a heterokaryon composed of one strain with stable and one with labile enzymes produce a hybrid product, a dominant stable or labile form, or a clear-cut mixture of unaltered stable and labile enzyme molecules? Horowitz and Fling showed that the latter possibility, which corresponds to the situation found in sickle-cell trait, is the correct one. The heterokaryon contained about 52 per cent thermostable and 48 per cent thermolabile tyrosinase. A further demonstration of the Mendelian inheritance and the lack of cytoplasmatic interference or inheritance was performed through recovery of the original strains by isolation of the respective nuclei from terminal points (hyphal tips) of the mycelium. The resolved strains showed only one type of tyrosinase—either thermostable or thermolabile.

Does this mean that a parent of a child afflicted with phenylpyruvate oligophrenia possesses two populations of·proteins with respect to the hydroxylation of phenylalanine—one which is catalytically active and one which is inactive? In other words, is the lowered tolerance to phenylalanine a result of a reduction in the number of otherwise unaltered active enzyme molecules? We are not as yet in a position to answer this question. It seems that there are types of interaction between gene products made from the corresponding alleles which have not been described until recently. Once more, experiments on *Neurospora* have been able to carry an analysis very far, partly because, as mentioned, heterokaryons of this organism do not have any nuclear fusion.

Giles and co-workers (14) have described various mutants in *Neurospora* which are blocked in the enzyme adenylosuccinase. They are closely linked mutations, all affecting one and the same enzyme—adenylosuccinase. It was therefore surprising and unexpected that a heterokaryon between two of these mutants brought about synthesis of adenylosuccinase. Mixing of extracts of the two separate mutants did not perform any reactivation. On the other hand, nuclear interaction is out of the question, and cytoplasmatic inheritance was likewise disproved by the subsequent resolution of the blocked mutants from the heterokaryon. Giles therefore posed the question of interaction of gene products, or, to use his own words: "Thus the mechanism could be visualized as involving cytoplasmatic interaction between two types of imperfect templates resulting in the formation of perfect templates for enzyme formation. . . ."

Such types of interaction should also be considered when dealing with metabolic traits in man. It seems obvious that the burden of proof is bound to be heavier on the studies in which segregation of polymers cannot be demonstrated. Yet this should not make us prejudiced.

Regardless of whether we have heterozygotes of one kind or another or whether we deal with the homozygous mutant, we are faced with a host of challenging physicochemical problems. Is the "error" made in the area of catalysis, specificity, or immunospecificity, not to forget the center of induction? In sickle-cell hemoglobin the "error" is apparently located in an area which is relatively inert with respect to these functions. Ingram (15) recently discovered the "error": sickle-cell globin contains a valine where normal adult globin contains a glutamate. The "error" affects, as mentioned, mainly the migration in an electric field and, what is physiologically more critical, the solubility.

V. *Variations in Manifestations of Inborn Errors*

A number of inborn errors are the result of a combination of single recessives. Phenylpyruvate oligophrenia is a typical example of a disease transmitted as a single recessive. It has recently been found that parents show a distinctly lowered tolerance toward phenylalanine, as judged from blood levels, after administration of this amino acid (16).

Other inborn errors have a much more complex type of inheritance. This seems, for instance, to be the case with congenital galactosemia. Holzel and Komrower (17) have brought evidence for an inheritance pattern of galactosemia quite different from that of a single recessive. They found that in the great majority of cases investigated one, and only one, of the parents showed lowered galactose tolerance. Dr. Glückson-Waelsch, at Albert Einstein College, pointed out to me in a private discussion that the possibility of so-called incomplete penetrance—or, expressed more specifically, action of suppressor genes (18)—ought to be considered. According to such an interpretation, the subject of an afflicted family which shows decreased galactose tolerance without having the disease may well be homozygous with respect to the defect locus. Yet the presence of one or more other genes (suppressor genes, epistatic genes) prevents the development of a marked or complete enzyme defect. In another instance an individual might possess a different gene makeup, so composed as to allow full penetrance, i.e., development of disease. The theory is of interest because it implies that lowered galactose tolerance or, to be more specific, a moderate lowering of the specific enzyme involved (see next section) might occur fairly frequently among several members of families not afflicted with the disease. This could eventually be tested, although it would require better methods than are now available. Precise enzymic assays which furnish data for initial rates should be worked out.

VI. *Variability and Heterozygotism*

A study of enzyme kinetics—for instance, of the enzyme under inquiry in congenital galactosemia—would be complicated by the fact that most phenotypic expressions are subject to considerable quantitative variations; there is nothing called the "normal" individual. Garrod (19) realized fully twenty-four years ago that what he first termed "inborn errors" are merely extreme expressions of smaller but significant variations which are bound to occur in a heterogeneous population like the human race. Williams in his recent monograph *Biochemistry and Individuality* (20) compiled a very telling collection of variations of morphology, chemistry, and psychology in mammals. His own pioneering studies show strikingly that the tendency of rats to consume alcohol on a specified diet depends on the strain. One finds a certain order of alcohol consumption within one particular inbred strain of rats. A different strain may show a much higher or a much lower alcohol consumption under otherwise identical conditions.

It should not be forgotten that inbred strains of mammals are homozygous only to a limited number of genes. In order to insure, on an absolute basis, that extranuclear, especially extracellular, factors are solely at play in diploid organisms, complete homozygotism is required. This is Johannsen's thesis (21, 22). Johannsen also found the device for such an undertaking: the use of self-fertilizing plants like beans or peas. Variability in such a population can be said to be independent of what he called the "genotype," a name which has "stuck" ever since. Since self-fertilization cannot be used in animals, we shall never know for certain whether the individuals of a highly inbred strain are genotypically identical. One-egg twins constitute, of course, a special exception.

VII. *Phenotypic Expression of Genes in Gametes and the Possibility of Screening Gametes from Heterozygous Organisms*

The implication of the concept of molecular enzyme species and their coexistence in heterozygous individuals has, to my knowledge, not as yet been extended to one area, the gametes. The problem in particular is the possible influence of the genotype of the haploid mammalian cell on the viability or general manifestation of this cell. Let me try to explain what I mean by briefly referring to the biochemistry of spermatozoa. It is known that the spermatozoön has an intense carbohydrate metabolism (23). How would a simple metabolic error in carbohydrate metabolism influence the properties of the spermatozoön? It may not have any influence whatsoever, but suppose it has? Let us take an example which may or may not apply. Fructolysis is one of the main energy sources for the maintenance of viability of spermatozoa in man (23). Suppose the inborn error, fructosuria, were a heterozygous state in which the normal somatic cells contain two types of fructokinase: a catalytically active one and a catalytically inactive one, as in the case of sickle-cell trait. Such a mixture might well lower the tolerance to fructose more or less markedly, but it would not harm the individual as such, since fructose appears only as a natural, but not irreplaceable, nutrient in the fluids of the reproductive system. Fructose is the main component in the seminal fluid, but there are other nutrients present. Let us assume that enzyme synthesis can take place after meiosis, i.e., in the spermatocytes II and in the spermatides. This is not unreasonable at all. The conversion of histones to protamines in salmon spermatozoa takes place very late in meiosis (24). If fructokinase were synthesized after meiosis and the gene controlling this enzyme given phenotypic expression in the haploid cells, it should be possible to screen the spermatozoa of a heterozygous individual, again assuming that fructosuria is a heterozygous state in which the normal gene makes active, normal fructokinase and the abnormal gene makes a catalytically inactive kinase protein. After meiosis there would be two types of spermatozoa with respect to the locus for fructokinase. One type would get the normal F gene and the other the defective f gene. If the spermatozoa from such a heterozygous individual were washed and placed in a medium in which fructose is the only energy source (and under anaerobic conditions which do not permit utilization of intracellular lipids), the viability, motility, membrane potential, etc., ought to be vastly different for the F and f gametes—that is, provided that there is not too much phenotypic carrying-over from the diploid state and provided that fructokinase is being synthesized in the haploid state. A moderate carrying-over from the diploid state would probably not prevent screening procedures, but it is mandatory that at least the spermatocytes II and spermatides (or at least the former) perform a biosynthesis of fructokinase. An attempt at obtaining a visible segregation by means of autoradiography, using C^{14}-labeled fructose, might be the first approach. A tissue-culture approach would further facilitate such an approach. I mentioned that the conversion of histones to protamines in salmon sperm was found to take place mainly after the second division after meiosis. P. M. Bhargava from T. S. Work's institute described experiments which directly show the occurrence of protein synthesis and exchange in bull spermatozoa. The determinations were performed by means of C^{14}-labeled amino acid (25).

Dr. George W. Beadle kindly turned my attention to a very important example of polymer biosynthesis in gametes. This is the phenotypic segregation in pollen from maize. Since this phenomenon is so well described in Sturtevant and Beadle's textbook (26), I prefer to quote from their text:

There are a few instances in which segregating genes produce effects that may be identified in the haploid state of the life history. In such cases the segregation is very striking since it is not necessary to deduce what has happened by taking into account the contribution from the other parent. Such an example is that of waxy character in maize (Indian corn). The starch in waxy kernels gives, on treatment with iodine, a red color and not the usual blue color found in normal starchy strains. If normal and waxy are crossed, the F_1 is normal and in F_2 a ratio of 3 normal to 1 waxy is obtained. In other words, the waxy character is due to a recessive gene (wx). The peculiar red color on staining with iodine also occurs in the starch present in the pollen grains, i.e., the male haploids (or male gametophytes). If the pollen grains of an F_1 plant (from the crossing of normal and waxy strains) are stained, it is at once evident that half of them are blue, half are red. In a heterozygote +/wx, half of the pollen grains contain + the other half wx. The iodine test shows that the starch in each kind stains according to its genetic constitution. Here then it is possible to study the results of segregation without the intervention of fertilization between the occurrency of meiosis and the time the observations are made. The simple 1:1 ratio results directly.

The textbook of Srb and Owen (27) has an illustration of this type of experiment as carried out by Dr. Demerec at Cold Spring Harbor.

It is very possible that such a phenotypic segregation of gametes from heterozygotes may apply to the blood group antigens. Landsteiner and

Levine in 1926 described the occurrence of the ABO blood groups in spermatozoa (28). In view of the above-mentioned complexity with respect to the gene product of AB heterozygotes, studies on a possible screening of spermatozoa of AB individuals by means of immunological techniques would merit special consideration.

Perhaps one day our knowledge of biochemical genetics will be called upon not only for the maintenance or selection of inborn errors but, in case methods acceptable to a free human society can be found, to serve as a basis for selective screening. This might make it feasible to select against at least a few of the biochemical errors and by a principle which circumvents the conflict "nature against nurture."

VIII. *Are Hereditary Metabolic Diseases Simple Enzyme Blocks?*

A detailed knowledge of enzymology and intermediary metabolism is a prerequisite to the study of human hereditary diseases. The fundamental work of the Coris on the hereditary glycogen storage disease (29, 30) revealed that this group of hereditary diseases is most heterogeneous. Space here does not permit me to go into a description of the various forms. I shall confine myself to one particularly intricate and interesting form which the Coris resolved with great elegance by means of enzymatic methods. They showed that one type of glycogen storage disease gave rise to glycogen molecules with abnormally short end branches (30). This complex situation scarcely looked like a simple enzyme block. However, the Coris had studied the enzymology of glycogen branching thoroughly, described the chain-forming enzyme, the branching enzyme, and the debranching enzyme. They found that it was the last which was missing in the type of storage disease in which the terminal branches were short. Because of the lack of "debrancher," only the most peripheral tier of the glycogen molecule was available to the organism. The debrancher defect may be inherited as a single recessive.

Udenfriend and Cooper (31) showed that the defect in hereditary phenylpyruvate oligophrenia is located in the enzymatic hydroxylation of phenylalanine. This is an important finding which might have bearing on many other problems.

Congenital galactosemia in children has lately been characterized on a biochemical basis by Komrower and co-workers (32, 33) as a state in which the second metabolic product of galactose metabolism, galactose-1-

phosphate, accumulates, which has been inferred frequently as being due to a block in Leloir's galacto-waldenase. This seemed unlikely to us, for the simple reason that the afflicted children recover dramatically if they are put on a strict galactose-free diet. If the enzymatic transition of glucose or its metabolites to galactose were blocked, how would they make brain galactolipids without administration of galactose? During the first months after birth a very substantial deposition of this brain constituent takes place. The accumulation of galactose-1-phosphate might, of course, be the result of a complicated disturbance in hormonal regulation. We were, however, interested in the possibility of an enzyme block of the step which makes galactose-1-phosphate available to the organism. We had found such an enzyme—Gal-1-P transuridylase (PGal uridyl transferase)—in galactose-adapted yeast (34, 35). If this enzyme were defective, the infant would still be able to make galactolipids from glucose, whereas galactose would only be phosphorylated and pile up as galactose-1-phosphate. We found that the red blood cells contain the enzymes which convert galactose into glucose-6-phosphate. A study on red blood cells from galactosemic individuals revealed in a striking way that it was indeed the Gal-1-P transuridylase which was defective and that the other members of the chain were present (36, 37). The same relationships were later demonstrated in liver biopsies (38); hence congenital galactosemia represents another example of a metabolic block.

There may well be family diseases which may be more involved than a simple metabolic block. For instance, Stetten has found that patients with gout make uric acid from glycine much more readily than can be explained by the nucleotide pathway (39). He has therefore suggested that gout may represent a revival of the uricolytic pathway of birds. Perhaps gout is to be considered an atavism to the bird or reptile stage, an unhappy back-mutant to the "wild type" without the ability to live a wild life because the individual does not possess an avian kidney.

IX. *Accumulation Rather than Deprivation as a Basis for Metabolic Diseases*

Certain types of metabolic disturbances in which the lack of production of an essential substance cannot be compensated for by supplementation would presumably cause cessation of growth and development as early as the stage of the fertilized egg or the early embryo. Most of the metabolic errors bring about disease due to accumulation of products prior to the enzyme block. Because of the block, a particular pathway has become a blind alley and perhaps a trap for valuable biological groups, like phosphate or high-energy phosphate. This could well be the case in congenital galactosemia, in which galactose can be phosphorylated (32) but the galactose-1-phosphate formed cannot be incorporated in the Leloir type of nucleotides (36). Moreover, the accumulation product may inhibit directly. Galactose-1-phosphate, for instance, inhibits enzyme reactions of glucose metabolism (40, 41). It is still hard to understand, though, the biological basis for feeble-mindedness in metabolic diseases like phenylpyruvate oligophrenia and galactosemia. Bickis, Kennedy, and Quastel recently described inhibitions of tyrosine metabolism by 1-phenylalanine, and they raise the question once more whether interference of a biosynthesis of hormones stemming from tyrosine may be the main factor in the mental defect in phenylpyruvate oligophrenia (42). Concerning the inhibition of mammalian tyrosinase, it may also be worth citing the work of Miyamoto, Fitzpatrick, and Fellmann (43, 44). These authors believe that tyrosinase inhibition may account for the curious fact that phenylpyruvate oligophrenics are frequently blond and blue-eyed, even among Mediterranean people.

Our understanding of several mental disturbances will undoubtedly be greatly enriched by studies on metabolic enzymes, especially those operating in the brain. Yet it should be obvious that enzymatology alone will not do. A knowledge of endocrinology, pharmacology, and even of biophysics is needed.

X. *Concluding Remarks*

I am afraid that the professor of eighteenth-century English poetry would have more right to discuss seventeenth-century poetry than I have to discuss genetics. The words of a great biologist deserve to be remembered; Niels Steensen (Nicolaus Steno), in 1665, upon invitation by the royal librarian in Paris to lecture about the anatomy and physiology of the brain, initiated his discussion with the following words (45): "Au lieu de vous promettre de contenter votre curiosité, touchant l'Anatomie du Cerveau, je vous fais icy une confession sincère et publique, que je n'y connois rien."

I should like to express the hope that *Perspectives* will provide space for "Letters" for competent readers to add their criticism of articles such as this one.

REFERENCES

1. BENZER, S. In McELROY, W. D., and GLASS, B., The chemical basis of heredity, p. 70. Baltimore: Johns Hopkins University Press, 1957.
2. HARTMAN, P. E. In McELROY, W. D., and GLASS, B., *op. cit.,* p. 408.
3. MORSE, M. L., LEDERBERG, E. M., and LEDERBERG, J. Genetics, 41:758, 1956.
4. PAULING, L., ITANO, H. A., SINGER, S. J., and WELLS, I. C. Science, 110:543, 1949.
5. NEEL, J. V. Science, 110:64, 1949.
6. DI SANT AGNESE, P. A. Pediatric Research Conference on Fibrocystic Disease of the Pancreas, p. 45. Columbus, Ohio: Ross Laboratories, 1955.
7. FILLITI-WURMSER, S., AUBEL-LESURE, G., and WURMSER, R. J. chim. phys., 50:236, 1953.
8. HALDANE, J. B. S. Proc. Roy. Soc. London, s.B, 144:217, 1955.
9. MORGAN, W. T. J., and WATKINS, W. M. Nature, 177:521, 1956.
10. KALCKAR, H. M. In McELROY, W. D., and GLASS, B., *op. cit.,* p. 463.
11. MITCHELL, H. K., and HOULAHAN, M. B. Am. J. Bot., 33:31, 1946.
12. MAAS, W., and DAVIS, B. D. Proc. Nat. Acad. Sc., 36:785, 1952.
13. HOROWITZ, N. H., and FLING, M. Proc. Nat. Acad. Sc., 42:498, 1956.
14. GILES, N. H., PARTRIDGE, C. W. H., and NELSON, N. J. Proc. Nat. Acad. Sc., 43:305, 1957.
15. INGRAM, V. Nature, 178:792, 1956.
16. HSIA, D. Y. Y., DRISCOLL, K. W., TROLL, W., and KNOX, W. E. Nature, 178:1239, 1956.
17. HOLZEL, A., and KOMROWER, G. M. Arch. Dis. Childhood, 30:155, 1955.
18. HOULAHAN, M. B., and MITCHELL, H. K. Proc. Nat. Acad. Sc., 33:223, 1947.
19. GARROD, A. E. Inborn errors of metabolism. London: Oxford University Press, 1923.
20. WILLIAMS, R. J. Biochemical individuality. New York: John Wiley & Sons, 1956.
21. JOHANNSEN, W. Kgl. danske vidensk. selsk. medd., 3:235, 1903.
22. JOHANNSEN, W. Elemente der exakten Erblichkeitslehre. Jena: Fischer, 1926.
23. MANN, T. Biochemistry of semen. London: Methuen & Co., Ltd., 1954.
24. ALFERT, M. J. Biophys. & Biochem. Cytol., 2:109, 1956.
25. BHARGAVA, P. M. Nature, 179:1120, 1957.
26. STURTEVANT, A. H., and BEADLE, G. W. An introduction to genetics, p. 43. Philadelphia: W. B. Saunders Co., 1940.
27. SRB, A. M., and OWEN, R. D. General genetics, p. 134. San Francisco: W. H. Freeman Pub. Co., 1952.
28. LANDSTEINER, K., and LEVINE, P. J. Immunol., 12:415, 1926.
29. CORI, G. T., and CORI, C. F. J. Biol. Chem., 199:661, 1952.
30. ILLINGSWORTH, B., CORI, G. T., and CORI, C. F. J. Biol. Chem., 218:123, 1956.

31. UDENFRIEND, S., and COOPER, J. R.   J. Biol. Chem., 203:953, 1953.

32. SCHWARZ, V., GOLBERG, L., KOMROWER, G. M., and HOLZEL, A. Biochem. J., 62: 34, 1956.

33. KOMROWER, C. M., SCHWARZ, V., HOLZEL, A., and GOLBERG, L. Arch. Dis. Childhood, 31:254, 1956.

34. KALCKAR, H. M., BRAGANCA, B., and MUNCH-PETERSEN, A. Nature, 172:1039, 1953.

35. MUNCH-PETERSEN, A., KALCKAR, H. M., and SMITH, E. E. B. Kgl. danske vidensk. selsk. biol. medd., 22: No. 7, 3, 1955.

36. KALCKAR, H. M., ANDERSON, E. P., and ISSELBACHER, K. J. Biochim. & biophys. acta, 20:262, 1956.

37. ISSELBACHER, K. J., ANDERSON, E. P., KURAHASHI, K., and KALCKAR, H. M. Science, 123:635, 1956.

38. ANDERSON, E. P., KALCKAR, H. M., and ISSELBACHER, K. J. Science, 125:113, 1957.

39. STETTEN, D. W., JR. Bull. New York Acad. Med., 2d ser., 28:664, 1952.

40. SIDBURY, J. B., JR. Proc. Am. Chem. Soc., September, 1957, p. 27 C.

41. GINSBURG, V., and NEUFELD, E. F. Proc. Am. Chem. Soc., September, 1957, p. 27 C.

42. BICKIS, I. J., KENNEDY, J. P., and QUASTEL, J. H. Nature, 179:1124, 1957.

43. MIYAMOTO, M., and FITZPATRICK, T. B. Nature, 179:99, 1957.

44. FELLMANN, J. H. Proc. Soc. Exper. Biol. & Med., 93:403, 1956.

45. Discours de Monsieur Stenon, sur L'Anatomie du Cerveau, Robert de Ninville, au bout du Pont St. Michel Paris, MDCLXIX (1669) avec privilège du Roy, re-edited by Rask-Ørsted Foundation, Copenhagen (1950).

DETOURS ON THE ROAD TO SUCCESSFUL
TREATMENT OF SICKLE CELL ANEMIA

*GEORGE J. BREWER, M.D.**

Brief Review of the Disease

Sickle cell anemia (SCA) is one of medicine's most fascinating diseases. It ushered in the era of molecular medicine. This inherited disease is one of the cornerstones of human genetics. The mechanism of gene action was worked out not in bacteriophage, drosophila, maize, or mice, but in man himself, and SCA was the instrument of that important step. The sickle gene offers one of the best examples of the operation of natural selection in man. Discussions of SCA have, from time to time, been highly politicized, since the disease occurs primarily in blacks. In spite of all of this, we still do not have effective treatment for SCA.

The first case was described in 1910 by a Chicago physician named Herrick [1] in a paper entitled "Peculiar Elongated and Sickle-shaped Red Corpuscles in a Case of Severe Anemia." This patient had anemia, jaundice, dark urine, shortness of breath, palpitations, leg ulcers, "muscular rheumatism," and severe abdominal pain. Most striking was the presence of sickle-shaped red cells. Subsequently, it was shown that this type of anemia was a fairly common occurrence in certain populations derived from Africa, such as the American black. In the intervening years, it has been found that the sickle cells tend to obstruct blood vessels. This leads to pain episodes called pain crises, probably the explanation for the muscular rheumatism and abdominal pain of Herrick's case. Blood vessel obstruction also leads to organ damage and a variety of complications including leg ulcers and heart failure, the latter the probable basis for shortness of breath and palpitations. These complications

*Professor of Human Genetics and Internal Medicine, University of Michigan Medical School, Ann Arbor, Michigan 48104. Our work in sickle cell anemia has been supported in part by contract no. 1-H62-2918 and Center grant HL 16008 from the Sickle Cell Branch, National Heart, Lung, and Blood Institute, DHEW, and by the Herrick Foundation, the Sage Foundation, Meyer Laboratories, Inc., the Meyers Foundation, the Upjohn Center for Clinical Pharmacology, and the Clinical Research Unit of University Hospital, which is supported by USPHS grant SMO1-FR-42.

are so varied that SCA, like syphilis, has been called the "great imitator" of other diseases.

Further progress was slow for several decades, although Emmel [2] showed in 1917 that normal appearing red cells of anemic patients and of healthy blacks could be made to sickle reversibly in the laboratory. The terms "sickle cell anemia" and "sickle cell trait" were derived from these studies. Huck [3] seems to have been the first to assert, in 1923, that sickling was inherited.

After World War II, progress in understanding SCA became more rapid. The correct genetic mechanism of sickle cell had been explained by 1949 by two investigators independently, Neel [4] and Beet [5]. It was demonstrated that the individuals whose blood showed sickle red cells under appropriate test conditions (deprivation of oxygen) were of two types. Those who were nonanemic and asymptomatic had one dose of the gene; these individuals were said to carry the gene (heterozygous state) and to have sickle cell trait. Those who were anemic and symptomatic had two doses of the gene; such homozygous patients have SCA. The sickle gene is on one of the nonsex chromosomes (autosomes), and thus SCA is an autosomal recessive disease. Most cases of SCA occur in children resulting from the marriage of two people with sickle cell trait. On the average, one out of four children from such a mating will have SCA, two out of four will have sickle cell trait, and one out of four will be normal.

In 1949, a remarkable event occurred. This was the ushering in of the era of molecular medicine, with the demonstration by Pauling and his colleagues [6] that the abnormality in SCA involves the hemoglobin molecule. It was shown that the electrophoretic migration of the hemoglobin of sickle cell patients was different than that of normal individuals, thus indicating a molecular abnormality in hemoglobin as the cause of a disease in man. This was the first demonstration of a molecular disease. It was found that sickle cell trait individuals have both sickle hemoglobin (Hb S) and normal hemoglobin (Hb A), while sickle cell anemics have essentially only Hb S.

The mechanism of gene action was first shown in 1956, resulting from Ingram's [7] studies of sickle hemoglobin. Hemoglobin consists of four polypeptide chains, two of the alpha type and two of the beta type. Ingram was able to show that sickle hemoglobin had an amino acid substitution in the sixth position of the beta chain such that the normal glutamic acid in that position was replaced by valine. This showed that a mutant gene, the sickle gene in this case, caused a mistake such that the wrong amino acid occurred in a polypeptide. This was the first demonstration that genes determine which amino acids occur in particular positions in proteins. Thus, the sickle hemoglobin gene, in man himself, was the instrument by which the mechanism of gene action was first established. It is apparent, then, that since 1956 we have known the specific molecular defect in SCA. We have known the molecular basis for this disease longer than for any other, yet we still have no effective treatment.

The sickle cell gene is of considerable interest from other standpoints, particularly the factors which led to its prevalence in certain populations. In certain African populations, upward of 20 percent of the population are carriers of the gene. Current theory, backed by a reasonable amount of circumstantial evidence, indicates that the gene reached a high prevalence as a result of favorable selection in malarious environments [8]. The mechanism, we believe, is that individuals with sickle cell trait have a selective advantage in environments in which falciparum malaria is highly endemic, as it was in Africa. This favorable selection for the sickle gene in heterozygotes is balanced by an unfavorable effect and loss of sickle genes in the homozygote with SCA, who in the absence of modern medical care would not usually live long enough to reproduce. This circumstance is called a balanced genetic polymorphism. Outbreeding species show a remarkable number of genetic polymorphisms, and this is one of the few in which we have some concept of selective or balancing forces in operation.

The sequence of pathogenic effects leading to the abnormalities in SCA is only partly known. It is established that Hb S, when it is deoxygenated, polymerizes, forming long fibers, called tactoids, of a very precise structure [9]. The amino acid substitution in the Hb S is at one of the contact points for this polymerization. Apparently, the presence of glutamic acid in Hb A, rather than valine as in Hb S, at this potential contact point prevents polymerization. Oxygenated Hb S does not polymerize. It is well established that the conformation of oxy and deoxy hemoglobin molecules is different, and it appears that the conformation of oxygenated Hb S does not allow polymerization. Carrying this process from the molecular to the red cell level, when homozygous sickle red cells are deoxygenated, they begin to take on morphologically abnormal appearances, some of which are of a classical sickle shape, while others are holly leaf or have a variety of pointed configurations. It appears that the fibers of polymerizing hemoglobin simply push the membrane ahead of it into points. Such cells have lost much of the pliability and easy deformability which is so characteristic of normal red cells, and they are quite rigid. Because of the problems discussed above, sickle red cells do not survive in the circulation as long as normal cells [10]. The cells either break down in the circulation or get trapped because of their rigidity. This hemolytic anemia results in decreased hemoglobin levels, usually ranging between 5 and 11 g per 100 ml. The cause of the variability in the severity of the anemia of SCA is unknown, but it is patient specific, that is, patients with the lowest hemoglobin levels always have the lowest hemoglobin levels, and vice versa.

The rigidity of sickle cells as they sickle is thought to cause another major problem in this disease to which we have already alluded, the sickle cell pain crises. It is thought that if cells begin to sickle in small blood vessels they begin to logjam, slowing flow, decreasing oxygenation, and leading to more sickling of additional cells. This logjamming effect eventually causes obstruction of larger blood vessels. If the process is extensive enough, it causes pain. Severe episodes of pain, sickle cell pain crises, are a hallmark of SCA [11]. Pain crises may involve any part of the body, but typically involve the joints, back, chest, and abdomen. They may last only a few hours, or up to 2 or 3 weeks. They occur quite frequently in some patients, perhaps 12 or more times per year, and only rarely in other patients. Some patients with SCA never have pain crises. The cause of the variability in the frequency and severity of pain crises among patients with SCA is unknown.

Also largely unknown are the precipitating causes of pain crises. Why, for example, does a patient get along well for several months with no pain and suddenly have an excruciating pain crisis? We know that infections, dehydration, altitude exposure, and emotional stress can frequently cause pain crises, but with many pain crises there is no known antecedent precipitating cause.

A specific treatment does not exist for pain crisis. Treatment consists of providing symptomatic relief with the use of analgesia, such as narcotics, and the appropriate treatment of any identified precipitating factor such as infection or dehydration.

In addition to the anemia and the pain crises, another major problem is organ damage [11]. Organ damage is presumed to result from chronic small blood vessel obstruction, which may gradually lead to heart failure, renal failure, leg ulcers, or damage to almost any organ of the body. The SCA patients, particularly during their pediatric years, are very susceptible to infections, particularly with pneumococcus and salmonella organisms [12]. This is thought to be due to a combination of splenic autoinfarction and partial failure of other systems combating infection.

In summary, then, the disease SCA can be thought of as producing three major types of abnormalities. These are anemia, the sickle cell pain crises, and organ damage. It is within the context of these areas of malfunction that therapeutic modalities can be considered. At first glance this tripartite scheme may not seem to cover all of the possible problems in this disease, which include such diverse areas as increased susceptibility to infection, coagulation abnormalities, zinc deficiency, decreased growth, hypogonadism, etc., yet these all fall, one way or another, under the aegis of organ damage.

It is my plan for the rest of this essay to do as follows: First, I will briefly review certain features of the history and current status of development of treatments in SCA. Next, I plan to give my perspective on where our best hopes lie for developing effective therapeutic control over the disease in the not-too-distant future. Then, I would like to use SCA as an illustration of the problems in development of therapeutic agents when the disease is not common enough to excite commercial interest.

Treatment Development in SCA

I should make it clear at the outset that this is not a review of the management of SCA. Thus, we will not discuss treatment of acute episodes or consider the problems of management related to complications in this disease, such as infections, accumulated lesions of bone, lung, liver, kidney, heart, and skin, etc. It is also not my intention here to review all of the prophylactic treatments for SCA which have been proposed and/or attempted and found wanting (a listing of many will be found in [13]). Rather, I will be selective, to illustrate potentially useful approaches, and I will set the stage for presenting my view on where our best hopes for the near future lie.

INCREASING HEMOGLOBIN OXYGEN AFFINITY

The oxygen affinity of hemoglobin is one possible approach to treating SCA. The theoretical basis of this approach is the principle we have already discussed, that is, that only deoxy Hb S will polymerize, not oxy Hb S. Thus, increasing the affinity of the hemoglobin for oxygen will cause a higher proportion of a red cell's hemoglobin to be in the oxy form at any given oxygen tension, and this will cause the cell to be more resistant to sickling.

This effect on hemoglobin oxygen affinity is the basis for the old treatment with alkali. An increase in pH will increase hemoglobin oxygen affinity through what is known as the Bohr effect. In short, an increase in pH tends to drive the hemoglobin molecule toward the oxy conformation.

Using this rationale, physicians have for years treated crisis with intravenous administration of sodium bicarbonate or other alkaline salts. However, in 1973 and 1974 the National Institutes of Health (NIH) sponsored a double-blind clinical trial of, among other things, alkali treatment to shorten crisis. This study found alkali treatment to be ineffective for this purpose [14]. One possible reason for this failure is that it may be impossible to reverse sickle cell crisis once the pathogenic events have been set in motion. Beyond this, however, a theoretical problem with this approach is that an increase in pH stimulates red cell glycolysis, which will result in an increase in levels of 2,3-diphosphoglycerate (DPG), which binds to hemoglobin and decreases oxygen affinity. Thus, the increased synthesis of DPG resulting from an increased pH will, after several hours, more than compensate for the Bohr effect and result in a net decrease in oxygen affinity. The DPG response will possibly take about 12 hr, and beyond that time treatment with alkali would certainly be expected to be counterproductive.

Other approaches to increasing hemoglobin oxygen affinity include the use of agents which increase the level of methemoglobin or carboxyhemoglobin. Neither methemoglobin nor carboxyhemoglobin can transport oxygen, but both result in an increase in the oxygen affinity of the remaining hemoglobin. Carboxyhemoglobin results from exposure of the hemoglobin to carbon monoxide and methemoglobin is formed from exposure to a variety of oxidizing agents. Sporadic efforts to use these agents on a short-term basis in SCA have not met with significant success [15–17]. One of the drawbacks to this approach is that the conversion of a certain amount of hemoglobin to methemoglobin or carboxyhemoglobin takes away that amount of the hemoglobin from oxygen transport.

In spite of these negative results with the use of increasing oxygen affinity as a therapy for SCA, this approach remains viable. We will discuss shortly a certain amount of success with the drug cyanate which proves that this approach is potentially useful. First, we will briefly discuss the use of urea in SCA.

UREA

As a result of the hypothesis of Murayama [18, 19] that polymerization of the deoxy Hb S molecule occurs because of hydrophobic bonding between molecules, Nalbandian and colleagues [20–22] began work with urea as an agent which might interfere with this type of aggregation. It was soon demonstrated that in the laboratory urea inhibited and actually reversed morphologic sickling [23]. Nalbandian et al. [24] then reported, with confirmation from McCurdy and Mahmood [25], that urea administered intravenously in a solution of invert sugar would abort sickle cell pain crisis. That is, patients who were said to have prolonged crises normally were found to have shorter crises when treated with intravenous urea. Of course, these studies were uncontrolled. These reports, plus considerable publicity, led the NIH to fund a double-blind study of the use of intravenous urea to abort sickle cell crises. In this carefully controlled study, crises were not found to be shortened or symptoms alleviated in urea-treated versus control (placebo-treated) patients [14, 26]. Thus, at least in the hands of these investigators, urea was not found to have efficacy. The intravenous-urea approach also suffers from the defect of not being prophylactic against crisis. As pointed out earlier, once a crisis is underway, it may be too late to stop the sequence of pathogenic events. Trials of oral urea on a prophylactic basis against crisis have also been attempted [27], but efficacy has not been confirmed.

One of the problems with the use of urea may be that very high levels are required in order to obtain antisickling effects. Some workers [28] claim that levels of the order of 1.0 molar urea are required, which would be equivalent to a blood-urea nitrogen of 2800 mg per 100 ml. It is obvious that blood-urea nitrogen levels of this magnitude cannot be obtained in patients.

CYANATE

The work with urea stimulated a team of investigators at Rockefeller University to investigate the drug cyanate [29]. They reasoned that the clinical benefit of urea reported by Nalbandian and colleagues was not due to the urea, since they believed very high concentrations of urea were required, but, rather, due to a trace contaminant which forms in urea solutions, namely, cyanate. It is now thought that this reasoning was probably incorrect on two bases: first, that the urea actually had clinical efficacy, and second, that the freshly prepared urea solutions in use contained cyanate. Nonetheless, laboratory work with cyanate was carried out, and it was soon demonstrated that cyanate would bind to hemoglobin (called carbamylation) and inhibit sickling in vitro [29]. It now appears that the major mechanism of this effect is due to an increase in oxygen affinity. Cyanate binds to the N-terminal valines of the globin chains, and the DPG binding site includes the N-terminal valines of the beta globin chains. Thus the irreversible carbamylation of hemoglobin by cyanate partially blocks DPG binding and increases hemoglobin oxygen affinity. It was soon shown that labeled sickle cells carbamylated in vitro had improved survival when transfused back into the sickle cell patients [30, 31]. After a certain amount of animal toxicity testing, uncontrolled clinical trials were begun with oral cyanate [30]. Dosage levels were gradually increased up to about 30 mg of cyanate per kilogram of body weight. A minimal amount of hematologic improvement and some possible prophylaxis against pain crisis were noted [32, 33]. However, a double-blind clinical trial of the effect of this dose of cyanate on crisis prophylaxis was negative [34]. In the meantime, toxicity became evident in the form of neuropathy [35]. Oral cyanate therapy has now been completely discontinued because of this toxicity.

Subsequently, an evaluation of extracorporeal cyanate treatment has been carried out [36]. This approach involves withdrawing blood once weekly, treating it with cyanate, washing away unbound cyanate, and reinfusing the blood into the subject. This treatment has markedly improved the anemia and has been associated with marked reduction in crisis frequency, but the latter are uncontrolled observations. Extracorporeal cyanate treatment also reduces the count of irreversibly sickled cells (ISCs). Unfortunately the use of weekly extracorporeal treatment is difficult and tedious and will not be practical on any large scale unless there are new technological developments for administering drugs in this manner. However, the effectiveness of cyanate therapy, regardless of its route of administration, makes an important point. It demonstrates that the approach of increasing oxygen affinity is valid in treating SCA.

The experience with cyanate also illustrates another important principle [37]. The principle is that, while it will probably not be hard to find hemoglobin-binding agents which interfere with deoxy Hb S polymerization, it probably will be difficult to find such an agent which is also nontoxic enough to be administered directly to patients. With cyanate, up to 50 percent of the hemoglobin tetramers must have a cyanate bound to obtain effective antisickling results. Therefore, a large amount of the drug must be administered because of the very large amount of hemoglobin present in the blood. For instance, the in vitro concentration of the cyanate moiety, including that irreversibly bound to hemoglobin, is around 1.0 molar in the whole blood when 50 percent of the hemoglobin tetramers have a carbamyl group attached. Not many drugs in current use reach this concentration. To reduce the total amount of drug which must be administered to the patient, the specificity of the drug must be very high for binding to hemoglobin, or otherwise the binding of the agent to other proteins will be likely to cause toxicity due to interference with the function of one or more proteins. Actually, cyanate was quite specific for hemoglobin binding because cyanate's structure $(N=C=O)$ is analogous to that of CO_2 $(O=C=O)$, and for that reason cyanate has a reasonably high specificity for the N-terminal valine of the hemoglobin chains where CO_2 is normally transported as carbamino groups. Yet oral cyanate proved toxic. Presumably, enough of the administered cyanate bound to one or more proteins of nerve tissue to cause toxicity.

Summarizing, we should not be surprised that various protein-binding agents interfere with the polymerization of the deoxy Hb S because the polymerization is a very precise process involving multiple contact points, and there is a reasonably high probability that anything that binds to hemoglobin will interfere with this process. Thus, we should anticipate numerous reports of agents which interfere with deoxy Hb S polymerization in vitro and are said to have antisickling activity. On the other hand, it is my view that it may take considerable time to find an adequately nontoxic hemoglobin-binding agent that can be administered directly to patients in sufficient quantities for clinical antisickling effects.

DECREASING LEVELS OF DPG

One approach to increasing hemoglobin oxygen affinity that does not involve a hemoglobin-binding agent is through decreasing the levels of DPG in erythrocytes. As shown in figure 1, DPG is synthesized in a small shunt pathway which is part of the glycolytic pathway. If a practical and nontoxic way were found to decrease the synthesis of DPG in red cells, this could increase hemoglobin oxygen affinity. At the moment, no practical way exists. It is possible to slow overall red cell glycolysis, for instance, by lowering serum phosphorous levels, but this is probably not desirable since it would also decrease red cell ATP levels. As we will discuss, ATP levels are probably important in helping maintain proper function in the red cell. Further, inhibition of overall red cell glycolysis would also probably result in some inhibition of glycolysis in other cell types. However, another strategy may be possible in the red cell. It may be possible to decrease DPG by decreasing the activity of the DPGM enzyme, which results in synthesis of DPG from 1,3-DPG, or increasing the activity of the DPGP enzyme, which metabolizes DPG (fig. 1). This shunt pathway which produces DPG is a specialized feature of the red cell, since other tissues do not contain any appreciable activity of this type. However, no practical mechanism for affecting these enzyme activities is available at this time.

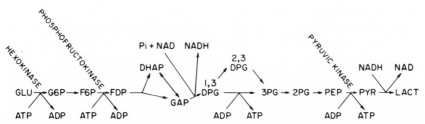

FIG. 1.—The glycolytic pathway in the human erythrocyte. I wish to call particular attention to the 2,3-DPG bypass. The 2,3-DPG, which I refer to in the text as DPG, is synthesized from 1,3-DPG by diphosphoglyceromutase (DPGM) and is catabolized to 3-PG by diphosphoglycero phosphatase (DPGP). The three regulatory enzymes of the overall pathway—hexokinase, phosphofructokinase, and pyruvic kinase—are indicated. Abbreviations: GLU = glucose; ATP and ADP = adenosine triphosphate and diphosphate, respectively; G6P = glucose-6-phosphate; F6P = fructose-6-phosphate; FDP = fructose diphosphate; DHAP = dihydroxyacetone phosphate; GAP = glyceraldehyde-3-phosphate; Pi = inorganic phosphate; NAD and NADH = oxidized and reduced nicotinamide adenine dinucleotide, respectively; 1,3-DPG = 1,3-diphosphoglycerate; 2,3-DPG = 2,3-diphosphoglycerate; 3-PG = 3-phosphoglycerate; 2-PG = 2 phosphoglycerate; PEP = phosphoenol pyruvate; PYR = pyruvate; LACT = lactate.

Recently, an exciting new avenue for possible therapy has become available. This involves the membrane lesion in the sickle cell. Increasingly, it is apparent that there is more to the damage and eventual destruction of the sickle cell and more to its occlusion of blood vessels than simple intracellular hemoglobin polymerization. It appears that damage to the cell is cumulative, such that repeated sickling episodes ultimately lead to so much damage that the cell is destroyed or, alternatively, takes part in vascular obstruction. But what is the nature of this cumulative damage? The evidence points to a lesion in the sickle cell membrane. In fact, there is a type of red cell circulating in the blood of all patients with SCA which appears to be membrane damaged. These are the so-called ISCs. These morphologically abnormal cells are irreversibly damaged in that they will not return to normal shape, even upon full oxygenation. Their shape is sometimes sickled, sometimes merely oblong, sometimes holly leaf, or a variety of other configurations. The percentage of these cells in the blood of a given sickle cell patient is fairly constant within narrow limits, but varies widely among patients from counts as high as 50 percent to as low as 10 percent [38, 39]. Following oxygenation of these cells, the internal hemoglobin appears to be solubilized (not polymerized), yet the cell remains permanently distorted. In fact, the cell can be hemolyzed, most of the hemoglobin allowed to escape, and the remaining cell membrane will remain in the shape of the original ISC. Recent evidence suggests that two important components of the membrane, spectrin and actin, are involved in this process. For example, most of the lipoproteins of the membrane can be removed with Triton-X, but the remaining cytoskeleton, composed largely of spectrin and actin, retains the abnormal shape [40].

Current work suggests that the damage to the cell membranes may be due to calcium accumulation [41]. Normally, the ratio of calcium in the plasma compared to the red cell is about 100:1, and the red cell has a very active ATPase pump to pump out excess calcium. It has been shown by other workers that excessive intracellular accumulation of calcium is harmful to the red cell membrane in terms of decreased deformability [42]. The ISCs contain considerably more calcium than the normal appearing red cells in the circulation of a sickle cell patient, and it has been shown that sickling induced by deoxygenation leads to an influx of calcium into the sickle red cell [41], as well as to a loss of potassium from the cell. Recent work indicates that it is possible to form permanently damaged membranes from normal (Hb A) red cells by using a calcium ionophore which allows calcium to leak into and accumulate inside the cell [43]. Other work indicates that potassium loss from the cell interior is important in membrane damage. If red cells are suspended in a potassium medium, such that any leak of potassium out will be compensated by leakage in, the membrane changes do not occur in spite of increased calcium [43].

Thus, one hypothesis for the pathogenic events in vivo for the formation of ISCs is as follows: First, deoxygenation or partial deoxygenation in the tissues leads to Hb S polymerization and a sickled red cell shape. This causes a momentary calcium leak into and a potassium leak out of the cell. Assuming that deoxygenation is temporary, as the cell oxygenates, cation levels are returned toward normal by the action of the cation ATPase pumps. It may be that cellular cation levels gradually become abnormal as a result of repeated sickle-unsickle cycles. Alternatively, if a cell is trapped for a time in partially deoxygenated condition, accumulation of calcium and leakage of potassium may become more severe. In any case, the effects of calcium gain and potassium loss eventually become severe enough to cause irreversible membrane damage, and the cell is then an ISC. It is not completely clear why patients vary so widely in their ISC counts. Perhaps the tendency to trap the red cell peripherally is a patient-specific characteristic. The red cells of the patients may have differing intrinsic sickleability. One factor of some importance is the level of fetal hemoglobin. Apparently the presence of fetal hemoglobin is partially protective against ISC formation, because fetal hemoglobin levels and ISC counts are inversely correlated in SCA patients [44]. Fetal hemoglobin is recognized as ameliorating Hb S polymerization because the dilution of the Hb S tetramers with fetal hemoglobin makes it necessary to have a higher deoxy hemoglobin concentration to obtain polymerization.

The number of ISCs in the circulation of patients is correlated with hemolytic rate, those patients with the highest ISC counts having the greatest hemolytic rate and tending to have the most severe anemia [44]. Additionally, the ISC count is positively correlated with conjunctival vascular damage [45] and with splenic autoinfarction [44]. Thus, ISCs can possibly be placed in a pathogenic sequence for both the severity of hemolysis and for organ damage. However, this is not the case for crisis. The number of ISCs in the circulation is not correlated with crisis frequency or severity.

Summarizing, we have traced a possible sequence of pathogenic events beginning with cellular deoxygenation, Hb S polymerization, cellular sickling, cellular calcium accumulation and potassium loss, membrane damage, and a patient-specific level of ISC accumulation. The circulating number of ISCs is inversely correlated with fetal hemoglobin levels and is directly correlated with hemolytic rate, vascular obstruction, vascular damage, and organ damage.

It may be possible to interrupt this pathogenic sequence of events in SCA not only at the primary lesion in the hemoglobin molecule but at any of the secondary or later steps at which symptoms or disease are produced. The involvement of the membrane in the pathogenic sequence may offer an early site for interrupting most of the pathology. This may have some advantage because the amount of a membrane-active drug required may be considerably less than the amount of drug required for direct action on hemoglobin, due to the lower concentration of the membrane components and the various reactants. This concept opens up the field of antisickling agents to those which affect membranes and, if the pathogenic involvement of calcium is correct, to those which counter the effect of calcium in the membrane.

ZINC THERAPY

The treatment of SCA with zinc may represent the first example of use of a class of therapeutic agents which attempt to interrupt the pathogenesis of SCA at the membrane level, possibly by an anticalcium mechanism. Historically, however, zinc was not used with that strategy in mind. One of my collaborators, Anada Prasad, at Wayne State University had hypothesized that SCA patients might be zinc deficient. This thought was largely stimulated by the hypogonadism and growth retardation which occurs in male SCA patients. We investigated and found that zinc deficiency does play a role in this aspect of SCA [46]. This interest in zinc and SCA resulted in initiation by my colleagues and me of a sequence of studies designed to elucidate any effects of zinc on sickle red blood cells. We first established that zinc binds to hemoglobin and increases hemoglobin oxygen affinity [47]. This seemed quite interesting in view of the potential therapeutic effect of cyanate, which, as I have discussed, has efficacy through an increase in hemoglobin oxygen affinity. However, the amount of zinc required to significantly increase hemoglobin oxygen affinity in vivo is too great for any practical realization of a therapeutic effect. The binding of zinc to hemoglobin remains an object of considerable theoretical interest [48]. The exact sites of binding of zinc to hemoglobin are not yet known.

In the meantime, investigations in our laboratory established that zinc at fairly low concentrations improved the ability of sickle red cells to pass through Nucleopore filters with small pore size (3μ–5μ) at intermediate oxygen pressures [49]. The improvement in filterability of sickle cells at low concentrations of zinc caused us to turn our attention to a possible red cell membrane effect of zinc. Since a role for calcium in the membrane damage of sickling had recently been proposed, as discussed earlier, we initiated studies to determine if zinc might have an anticalcium effect. The results of five types of studies indicate that this may in fact be so. In the first, the relative effects of calcium, zinc, and calcium plus zinc on the binding of hemoglobin to red cell membranes were studied. Calcium promotes hemoglobin binding, whereas zinc was found to decrease hemoglobin binding to membranes and to antagonize the effect of calcium [50]. Second, the direct binding of calcium (as ^{45}Ca) to red cell membranes was found to be inhibited by zinc [49]. Third, the relative effects of calcium and zinc on echinocyte production were studied. Echinocytes are red cells or red cell membranes with multiple roundish projections. A variety of agents, including calcium, can promote the development of echinocytes. We found that zinc inhibits the formation of echinocytes and, in particular, antagonizes the action of calcium in producing echinocytes [51]. Work recently reported by Eaton et al. [43] has shown a fourth calcium-antagonistic effect of zinc in red cells. Calcium, when its concentration is increased inside red cells, causes the cells to shrink, probably by some type of effect on the membrane proteins. Zinc inhibits or antagonizes this effect of calcium and prevents the cell from shrinking. Fifth, we [52] have recently observed that under some circumstances zinc can protect against probable calcium-induced ATP loss. That is, accumulation of calcium inside the red cell tends to cause rapid depletion of ATP. Zinc is capable of protecting against this effect. We believe this ATP-preserving effect is also mediated through some type of membrane effect of zinc.

The reciprocal, or antagonistic, relationships of zinc and calcium described above may have implications well beyond the erythrocyte membrane. Influx of calcium is a basic signaling mechanism to cells and cell membranes causing a variety of types of activation and cellular responses. Zinc may be a normal physiological inhibitor of some of these reactions. In fact, zinc has already been shown to inhibit many actions in a variety of cell types, including spermatozoa, platelets, granulocytes, and macrophages [53]. Thus, calcium and zinc may act in reciprocal fashion in many cell types. If this is so, it may open the way to the use of zinc as a modulator of various types of cellular reactions in a number of diseases. For example, it has recently been reported that oral zinc is effective in reducing joint inflammation in rheumatoid arthritis [54, 55]. If so, this may be due to zinc's inhibition of inflammatory cells, which in turn may be due to an anticalcium action. If the effect of zinc on rheumatoid arthritis is real, zinc may have general antiinflammatory properties.

As a result of the encouraging effects of zinc on sickle cell filterability, and as a result of the positive studies on anticalcium properties, we decided to try zinc as an antisickling agent. Our first problem was to develop an appropriate dosage schedule. Zinc had previously been administered in a wide variety of clinical studies as zinc sulfate, 220 mg three times a day, which is equivalent to 50 mg of elemental zinc three times a day. We soon established that oral doses of zinc of 50, 25, or 12.5 mg all caused a plasma increase in zinc lasting only about 5 hr [56]. We therefore reasoned that we should give zinc more often than three times a day in order to maintain a plasma elevation at all times. We have used six times a day administration and reduced each dose to 25 mg of elemental zinc to stay in the same daily dose range with which there has been previous broad clinical experience.

We also established early in the program that most foods will block zinc absorption [57]. Considerable work has been done previously on the effects of certain foods in blocking absorption of dietary zinc, but little has been reported on the effects of various foods on pharmacologically administered zinc. We found that hamburger buns and other breads, a variety of vegetables and fruits, milk, egg, coffee, tea, and several other foods all blocked zinc absorption very effectively when administered along with zinc. We have therefore instructed our patients to avoid food 1 hr before and 1 hr after each dose.

Gastric irritation has been a frequent problem with zinc administration, and the sulfate component has been suspected of causing part of the difficulty. We studied the relative absorption of zinc salts other than zinc sulfate and found that zinc acetate absorption was comparable to that of zinc sulfate, while zinc carbonate absorption was inferior [57]. We have used zinc acetate in our studies and believe, on the basis of uncontrolled observations, that it is tolerated better than zinc sulfate.

In summary, our present regimen is a zinc acetate tablet or capsule containing 25 mg of elemental zinc administered every 4 hr, with the patient avoiding food 1 hr before and 1 hr after each dose. This regimen is, of course, tedious, and compliance is not perfect. Nonetheless, we do not at present have any way to avoid this regimen. Enteric-coated tablets or capsules, for instance, are not absorbed at all, and it is currently believed that zinc absorption takes place in a relatively short window of the upper small intestine. This being the case, administration of the usual timed-release type preparations will probably not be of any value.

With this regimen we have administered zinc to approximately 40 patients for antisickling purposes. We will briefly summarize our results here in the context of the three disease areas, namely, hemolytic anemia, sickle cell pain crisis, and organ damage. With respect to hemolytic anemia the effect of zinc is slight and somewhat variable. There is an average, approximately 5–10 percent, improvement in red cell survival, red cell mass, and hemoglobin levels. With respect to pain crisis, zinc therapy has reduced the frequency in uncontrolled studies [56, 58]. However, pain crises are a subjective type of symptomatology, and there may be strong placebo effects. We currently have a double-blind clinical trial underway to evaluate the effect of our oral zinc therapeutic regimen on pain crisis. This will take additional time to complete.

One of the most exciting developments with zinc is the recent observation that zinc therapy suppresses the count of ISCs when administered to patients with SCA [39]. Since ISC slides are coded for counting, the ISC suppression is an objective result which demonstrates that zinc therapy has a definite effect on the sickling process in vivo. The most important implication from the suppression of ISCs is in regard to the prevention of organ damage. As we have previously discussed, the levels of ISCs are positively correlated with vascular damage in the conjuctiva and with splenic autoinfarction [44, 45]. It now remains to be tested whether suppression of ISC levels with zinc will slow vascular obstruction and organ damage in SCA.

Zinc is a safe medication, its only two complications being gastric irritation in an occasional patient and copper deficiency [58]. The latter is probably due to the competition for gastrointestinal binding sites by zinc and copper for absorption. This complication is easily circumvented by administering small doses of copper along with zinc. I might note in passing that this complication of zinc therapy suggests zinc as a new treatment for Wilson's disease, in which copper accumulation is a major cause of the pathology.

Summarizing the antisickling effects of zinc, we have laboratory evidence for certain properties which may be responsible for zinc's antisickling effects. The most promising lead seems to be an anticalcium effect at the red cell membrane level. Clinically, we have evidence of a small improvement in hematologic status with zinc therapy, and in uncontrolled studies we have seen a moderate reduction in crisis with zinc. However, subjective symptomatology such as crisis must be evaluated in a controlled manner, and double-blind studies are now under way. Most important, we have unequivocal objective evidence of a reduction in the number of circulating ISCs in the blood of sickle cell patients treated with oral zinc. This finding has implications for the prevention or at least the slowing of organ damage in SCA.

There are definite additional needs for the further exploitation of zinc in SCA. First, the regimen is excessively difficult for any widespread use. A more convenient regimen is urgently needed. Second, the blood levels of zinc currently obtained are probably not optimal in terms of producing maximum therapeutic effect from this agent. Thus, we need a more convenient way of giving this drug and a way of obtaining higher blood levels. Methods for accomplishing these ends may very well be useful not only in SCA but in other conditions in which pharmacological levels of zinc may prove to be desirable.

Best Hopes for Treatment of SCA in the Near Future

I wish to be clear in differentiating areas of research which may ultimately offer good therapeutic potential for control of SCA, but which are long range, from therapeutic approaches which have a relatively short-range potential. Regarding the former, there are a number of approaches which may ultimately give complete or almost complete control over this disease. One such would be development of the capability of switching the beta chain of hemoglobin synthesis off and the gamma chain on, which would allow synthesis of fetal hemoglobin in the red cells of sickle cell patients. Another approach would be the ability to use bone marrow transplantation effectively. A third would be the development of a nontoxic agent which binds to hemoglobin and interferes with deoxy Hb S polymerization.

These approaches are all theoretically excellent, but for a variety of reasons I suspect that none of them will be available in the relatively near future. Rather, I think it more likely that we will obtain some control by incremental steps, each of which relieves some aspects of the disease. Membrane-active drugs offer a good initial approach. Zinc is an example of a membrane drug which is nontoxic and has antisickling properties. However, it is apparent that, at least as currently used, it will not completely control SCA. The anemia is not normalized. Crises occur during zinc therapy and organ damage, while it may be slowed, will not likely be completely eliminated. Addition of other membrane agents to a zinc regimen may have some additive effect and be a desirable end to pursue. If calcium damage is proven to be important in the membrane damage of sickle cells, the membrane approach can specifically be directed toward anticalcium drugs. It seems quite likely that, if organ damage can be slowed by membrane-active drugs such as zinc, some of the other types of complications, such as susceptibility to infection, may also be decreased.

Another approach which may have short- or intermediate-range potential is manipulation of red cell metabolism. It may be feasible to develop an agent which causes a decrease in the red cell DPG levels. The resulting increase in oxygen affinity would be beneficial in SCA. The specialized nature of the DPG shunt, that is, its uniqueness in the red cell, offers the possibility of affecting this metabolic pathway without damage to other tissues.

Anticoagulant or antiplatelet therapy may offer partial control over certain aspects of the disease. Recent work has suggested a link between intravascular coagulation and crisis [59]. Data at this time, while not definitive, suggest that there is increased platelet turnover [60] and an increase in high-molecular-weight fibrinogen complexes during crisis [61]. Between crisis periods, these parameters are normal. One reservation with the data collected to date is that the fibrinogen and platelet studies have been done in different laboratories and the studies have not been done on the same patients between and during crises. What is needed is a simultaneous study of fibrinogen and platelet turnover in the same patients during and between crisis. Most recently, Leichtman and Brewer [62] have found that fibrinopeptide A (FPA) levels are elevated in the plasma of SCA patients during crisis but are normal between crises. The FPA is the small peptide which is split off of fibrinogen when coagulation is initiated. An elevation of FPA during crisis indicates that intravascular coagulation has been initiated. If an association between intravascular coagulation and sickle cell pain crisis exists, the next important question is whether or not clotting is a causal factor in the pathogenesis of crisis or is merely a secondary manifestation. We can not answer this question at present, but we must keep in mind that something special happens to trigger a pain crisis. Thus, we must search for associations and then examine each to see if it identifies a causal factor. The triggering of the coagulation system is a reasonable possibility for initiating crisis, since it would immediately and perhaps drastically increase viscosity of the blood in the microvasculature.

In a scenario where coagulation plays a role in triggering crisis, one might hypothesize that the sickling process in a small arteriole or arterioles causes more than usual damage to the endothelial surfaces, releasing tissue thromboplastins and activating platelets and the fibrinogen system. The development of fibrin polymers along with platelet aggregation would slow flow; the increased local hypoxia would enhance sickling further. The end result might involve occlusion of vessels which are somewhat larger than the microvessels which are more or less continuously undergoing occlusion in SCA. The sudden increase in hypoxic tissue would cause pain. The spread of humeral coagulation factors to distant sites would cause this process to spread to other parts of the body. In such a scenario, sickled cells continue to play a role. As blood flow slows in small vessels, hypoxia increases, sickling increases, and sickled red cells add to the sludging and eventually to the obstruction of the blood vessels. Patient variability in crisis frequency might then stem from variability in the coagulation system, as well as from variability in red cells and blood vessels.

A major implication of a sequence such as that just described would be the possibility of intervening therapeutically to prevent or stop crisis by antiplatelet or anticoagulant therapy. While there are scattered reports in the literature of trials of anticoagulants in SCA going back several decades, I think it is fair to say that no carefully controlled study of an anticoagulant or antiplatelet approach to prophylaxis of sickle cell crisis has ever been reported.

Figure 2 summarizes the pathogenic sequences we have been discussing and illustrates graphically where therapeutic intervention may be attempted. Thus hemoglobin binding agents, membrane active agents, and anticoagulant agents can be used to block at the sites indicated as 1, 2, and 3, respectively, in figure 2.

Of course, it is not possible to know at this point whether anticoagulant or antiplatelet therapy will be helpful. However, I am optimistic that if

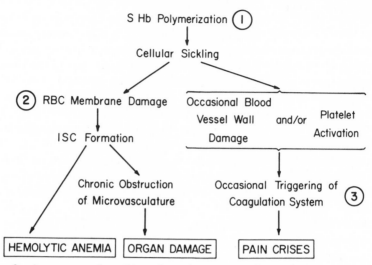

FIG. 2.—A summary of possible pathogenic sequences discussed in the text and sites where therapeutic intervention may be useful. At *1*, agents acting directly on hemoglobin or on hemoglobin oxygen affinity would be applicable. At *2*, agents acting on the membrane, or possibly as calcium antagonists, would be useful. At *3*, anticoagulant or anti-platelet agents would come into play. The efficacy of such therapeutic approaches, of course, depends on the validity of the postulated pathogenic sequence.

leads such as this are vigorously pursued we will make progress in controlling SCA. For example, it may be possible to combine a membrane or anticalcium agent such as zinc with an antiplatelet program such as aspirin and dipyrimadole and make the lives of some of these patients more normal. Whether or not the two types of therapy suggested above prove to be effective, the main point is that it is important to pursue all new possibilities persistently, vigorously, and exhaustively. I think a multi-pronged attack on this disease has the best initial chance of controlling it adequately so that patients have a more normal life and a more normal life expectancy. However, in the next section I wish to review the increasing difficulties facing clinical investigators trying to develop treatments and treatment approaches for relatively uncommon diseases such as SCA.

Increasing Difficulties with Development of Investigator-initiated Therapies

I wish to comment briefly on the types of problems which are emerging in attempts to develop therapeutic agents for disorders in which there is little commercial interest. Drug companies, of course, are not interested in developing agents unless the product will be profitable. In order for a product to be profitable, the disease to be treated must be common enough such that treatment will more than repay the investment in research and development. Most genetic diseases are sufficiently uncommon so that they do not fall in this category. Even SCA, with an estimated 30,000-plus cases in the United States, is not common enough to be of commercial interest to drug companies. This leaves the problem of drug development for these diseases to the physician-investigator. However, most such investigators are inexperienced and unskilled at drug development unless they happen also to be trained in pharmacology. The physician-investigator must rediscover the wheel, as it were, and learn about the FDA, the IND,[1] requirements for toxicity testing, etc. In addition to the paperwork, some of the other problems are formidable. For example, even if an investigator assumes that NIH would fund his grant application to develop his new treatment, it is unlikely that a relatively large sum of money would also be provided for routine animal toxicity testing through any of the usual NIH channels. The investigator must ask himself, How is this phase to be paid for? Because of these and other hurdles, most investigators tend not to get involved in drug development.

Another problem is organizational. The time has passed when it is easy for an individual or an individual laboratory to pioneer new therapeutic approaches at the patient level. Modern biomedicine is exceedingly complex, tending to require many types of expertise. At the same time, this expertise has become more organizationally compartmentalized in medical schools. One physician group sees a patient who has a therapy need which should stimulate new ideas. However, the expertise for thinking of the ideas for therapy may reside in the biochemistry, physiology, pharmacology, genetics, nutrition, engineering, etc., laboratories. The expertise for laboratory work to test the ideas in vitro, or on animals, may also reside in any of these laboratories, but not necessarily the same laboratory. The expertise for product development may reside in the pharmacy-pharmacology groups, but more likely it is not present at all at the medical school level. Finally, the expertise for clinical trials must involve a physician who can take care of the patients and the laboratory backup necessary to monitor and assess effects and need for changes. Left out of the above is the expertise needed to interact with the FDA, with NIH or other granting agencies, and with patenting procedures if such are involved. Actually, organizations of the type I have just described do exist—in drug companies. In short, it is difficult for the individual investigator to incorporate all the expertise of a drug company.

Another factor which has potentially ominous portent for developing

---

[1] Food and Drug Administration (FDA) and Investigational New Drug (IND) applications.

new treatments is the trend away from clinical investigation by younger physicians. Many individual clinical investigators have noted this trend anecdotally, but quantitative data also exist. First-time NIH grant awardees have fallen from 44 percent M.D.'s in 1966 to 22 percent in 1975 [63].

The solution to the problem of therapy development for relatively uncommon disorders is far from clear, yet the solution is important. While each of the disorders under consideration is relatively uncommon, there are literally hundreds of genetic disorders, as well as acquired disorders, that fall into this class, and cumulatively they account for much of human suffering. Who is to be concerned about the ultimate development of treatments for these patients? Nothing I see on the horizon indicates any improvement. Indeed, in our efforts to develop zinc as a treatment for SCA, we have encountered problems in many of the areas discussed above. These problems are of a sufficient magnitude to be quite discouraging to all but the most determined investigative team.

The solution to the problem of therapy development does not appear to reside in setting up new branches at NIH to deal with specific diseases. Whatever the accomplishments and merits of the Sickle Cell Branch of NIH, and there are many, they do not include aggressive pursuit of new therapeutic approaches. Actually, the Blood Resources Program of NIH, which was phased out when the Sickle Cell Branch was established, was considerably more aggressive in pursuing specific therapeutic approaches to disease, in spite of its more general mandate.

I suppose I should not stop before offering my own suggestions on what steps should be taken to counter the problems of therapy development just discussed. Here is my list:

1. Increase the "Medical Scientist" program. This is a program for training which results in both the M.D. and Ph.D. degrees. The graduates have a strong tendency to stay involved in clinical investigation. In addition, develop other incentives which encourage young M.D.'s to become involved in clinical investigations.

2. Develop a federal assistance program to help in the development of investigator-initiated therapies. This program should include availability of consultation on all phases of product development. It should include facilities and funding for toxicity testing once efficacy appears likely. It should also include funding, on a competitive basis, for investigators who propose research and development of new therapies. These should always be nonprofit enterprises, and the primary diseases involved should be those whose frequency is too low to be of commercial interest. These funds should be a pool separate from that of the usual research grant. The applicants should compete with one another, not with applicants for basic research grants. I believe that part of the reason a "bench

to bed" gap exists is lack of funding for research and development of new treatments. It appears that therapy-related research does not compete well with more basic research in the NIH study sections which deal with most of the applications concerning rarer diseases. Further, it appears inappropriate that these two types of applications be considered together.

3. Programs should be developed in medical schools which facilitate new therapy research and development. This involves regular communication between certain groups and commitments from these groups. The pharmacy and pharmacology programs can serve as core groups, but the other groups and individuals in the medical schools whose expertise is required must also be involved. Most important, clinicians with ideas on new treatments must have ready access to the program for consultation and advice, and, it is hoped, pilot assistance.

In short, a new climate needs to be created for research and development of new therapies for uncommon diseases.

REFERENCES

1. J. B. HERRICK. Arch. Intern. Med., 6:517, 1910.
2. V. E. EMMEL. Arch. Intern. Med., 20:586, 1917.
3. J. G. HUCK. The Johns Hopkins Hosp. Bull., 34:335, 1923.
4. J. V. NEEL. Science, 110:64, 1949.
5. E. A. BEET. Ann. Eugenics, 14:229, 1949.
6. L. PAULING, H. A. ITANO, S. J. SINGER, and I. C. WELLS. Science, 110:543, 1949.
7. V. M. INGRAM. Nature, 178:792, 1956.
8. A. C. ALLISON. Cold Spring Harbor Symp. Quant. Biol., 29:137, 1964.
9. J. F. BERTLES. Arch. Intern. Med., 133:538, 1974.
10. T. A. BENSINGER. Arch. Intern. Med., 133:624, 1974.
11. F. I. D. KONOTEY-AHULU. Arch. Intern. Med., 133:611, 1974.
12. P. A. BROMBERG. Arch. Intern. Med., 133:652, 1974.
13. S. CHARACHE. Arch. Intern. Med., 133:698, 1974.
14. COOPERATIVE UREA TRIALS GROUP. J.A.M.A., 228:1129, 1974.
15. J. A. SIRS. Lancet, 1:971, 1963.
16. E. BEUTLER. J. Clin. Invest., 40:1856, 1961.
17. H. R. PERUGGANAN. Lancet, 1:79, 1964.
18. M. MURAYAMA. Science, 153:145, 1966.
19. M. MURAYAMA. Clin. Chem., 14:578, 1967.
20. R. M. NALBANDIAN, R. L. TENRY, M. I. BERNHART, and F. R. CAMP, JR. U.S. Army Med. Res. Lab. Rep. no. 895 (DDC AD NO. 715250), 1970.
21. R. M. NALBANDIAN and T. N. EVANS. Mich. Med., 70:411, 1971.
22. R. M. NALBANDIAN. Lab. Med., 2:12, 1971.
23. R. M. NALBANDIAN, R. L. HENRY, M. I. BARNHART, B. M. NICHOLS, F. R. CAMP, JR., and P. L. WOLF. Am. J. Pathol., 64:405, 1971.
24. R. M. NALBANDIAN, G. SCHULTZ, J. M. LUSHER, J. W. ANDERSON, and R. L. HENRY. Am. J. Med. Sci., 261:309, 1971.
25. P. R. McCURDY and L. MAHMOOD. N. Engl. J. Med., 285:992, 1971.
26. COOPERATIVE UREA TRIALS GROUP. J.A.M.A., 228:1125, 1974.

27. J. M. Lusher and M. I. Barnhart. *In:* G. J. Brewer (ed.). Hemoglobin and red cell structure and function, p. 303. New York: Plenum, 1972.
28. J. M. Manning, A. Cerami, P. N. Gillette, F. G. deFuria, and D. R. Miller. *In:* G. J. Brewer (ed.). Hemoglobin and red cell structure and function, p. 253. New York: Plenum, 1972.
29. A. Cerami and J. M. Manning. Proc. Natl. Acad. Sci. U.S.A., **68**:1180, 1971.
30. P. N. Gillette, C. M. Peterson, J. M. Manning, and A. Cerami. *In:* G. J. Brewer (ed.). Hemoglobin structure and function, p. 261. New York: Plenum, 1972.
31. A. Cerami. N. Engl. J. Med., **287**:807, 1972.
32. P. N. Gillette, C. M. Peterson, Y. S. Lu, and A. Cerami. N. Engl. J. Med., **290**:654, 1974.
33. C. M. Peterson, A. C. deCiutiis, and A. Cerami. *In:* J. I. Hercules, A. N. Schechter, W. A. Eaton, and R. E. Jackson (eds.). Proc. 1st Natl. Symp. Sickle Cell Dis., p. 37. DHEW Publication no. (NIH) 75-723, 1974.
34. D. Harkness and S. Roth. *In:* E. Brown (ed.). Progress in hematology, p. 157. New York: Grune & Stratton, 1975.
35. C. M. Peterson, P. Tsairis, A. Ohnishi, Y. S. Lu, R. W. Grady, A. Cerami, and P. J. Dyck. Ann. Intern. Med., **81**:152, 1974.
36. D. Diedrich, P. Gill, P. Trueworthy, and W. Larsen. *In:* G. J. Brewer (ed.). Erythrocyte structure and function, p. 379. New York: Liss, 1975.
37. G. J. Brewer. Am. J. Hematol., **1**:121, 1976.
38. G. R. Serjeant, B. E. Serjeant, and P. R. Milner. Br. J. Haematol., **17**:527, 1969.
39. G. J. Brewer, L. F. Brewer, and A. S. Prasad. J. Lab. Clin. Med., **90**:549, 1977.
40. S. E. Lux, K. M. John, and M. J. Karnovsky. J. Clin. Invest., **58**:955, 1976.
41. J. W. Eaton, T. D. Skelton, H. S. Swofford, C. E. Kolpiu, and H. S. Jacob. Nature, **246**:105, 1973.
42. R. I. Weed and P. L. Lacelle. *In:* G. A. Jamieson and T. J. Greenwalt (eds.). Red cell membrane structure and function, p. 318. Philadelphia: Lippincott, 1969.
43. J. W. Eaton, E. Berger, J. G. White, and H. S. Jacob. *In:* G. J. Brewer and A. S. Prasad (eds.). Zinc metabolism: current aspects in health and disease, p. 275. New York: Liss, 1977.
44. G. R. Serjeant. Br. J. Haematol., **19**:635, 1970.
45. G. R. Serjeant, B. E. Serjeant, and P. I. Condon. J.A.M.A., **219**:1428, 1972.
46. A. S. Prasad, J. Ortega, G. J. Brewer, D. Oberleas, and E. B. Schoomaker. J.A.M.A., **235**:2396, 1976.
47. F. J. Oelshlegel, Jr., G. J. Brewer, C. Knutsen, A. S. Prasad, and E. B. Schoomaker. Arch. Biochem. Biophys., **163**:742, 1974.
48. J. G. Gilman, F. J. Oelshlegel, Jr., and G. J. Brewer. *In:* G. J. Brewer (ed.). Erythrocyte structure and function, p. 85. New York: Liss, 1975.
49. G. J. Brewer and F. J. Oelshlegel, Jr. Biochem. Biophys. Res. Commun., **58**:854, 1974.
50. S. Dash, G. J. Brewer, and F. J. Oelshlegel, Jr. Nature, **250**:251, 1974.
51. W. C. Kruckeberg, F. J. Oelshlegel, Jr., S. H. Shore, P. E. Smouse, and G. J. Brewer. Res. Exp. Med. (Berl.), **170**:149, 1977.
52. W. C. Kruckeberg, C. A. Knutsen, and G. J. Brewer. *In:* G. J. Brewer and A. S. Prasad (eds.). Zinc metabolism: current aspects in health and disease, p. 259. New York: Liss, 1977.

53. M. Chvapil, L. Stankova, P. Weldy, D. Bernhard, J. Campbell, E. C. Carlson, T. Cox, J. Peacock, Z. Bartos, and C. Zukaski. *In:* G. J. Brewer and A. S. Prasad (eds.). Zinc metabolism: current aspects in health and disease, p. 103. New York: Liss, 1977.
54. P. A. Simkin. Lancet, **2:**539, 1976.
55. P. A. Simkin. *In:* G. J. Brewer and A. S. Prasad (eds.). Zinc metabolism: current aspects in health and disease, p. 343. New York: Liss, 1977.
56. G. J. Brewer, F. J. Oelshlegel, Jr., and A. S. Prasad. *In:* G. J. Brewer (ed.). Erythrocyte structure and function, p. 417. New York: Liss, 1975.
57. F. J. Oelshlegel, Jr., and G. J. Brewer. *In:* G. J. Brewer and A. S. Prasad (eds.). Zinc metabolism: current aspects in health and disease, p. 299. New York: Liss, 1977.
58. G. J. Brewer, E. B. Schoomaker, D. A. Leichtman, W. C. Kruckeberg, L. B. Brewer, and N. Meyers. *In:* G. J. Brewer and A. S. Prasad (eds.). Zinc metabolism: current aspects in health and disease, p. 241. New York: Liss, 1977.
59. F. R. Rickles and D. S. O'Leary. Arch. Intern. Med., **133:**635, 1974.
60. M. J. Haut, D. H. Cowan, and J. W. Harris. J. Lab. Clin. Med., **82:**44, 1973.
61. N. Alkjaersig, A. Fletcher, H. Joist, and H. Chaplin, Jr. J. Lab. Clin. Med., **88:**440, 1976.
62. D. A. Leichtman and G. J. Brewer. Blood, **50**(suppl. 1): 111, 1977.
63. D. R. Challoner. Clin. Res., **24:**237, 1976.

THE WONDER OF OUR PRESENCE HERE: A COMMENTARY ON THE EVOLUTION AND MAINTENANCE OF HUMAN DIVERSITY

JAMES V. NEEL*

Introduction

The title of this address is deliberately ambiguous. In fact, there are not one, but many wonders to our presence here. The first is the wonder that none of us can escape as we walk the streets of Jerusalem, aware that this region is one of the cradles of civilization, with documented, probably continuous, human habitation for the past 12,000 years, and that the city in which we are meeting is seminal to Judaism, Christianity, and Islam. From no other comparably small area have concepts with greater worldwide impact originated. There is, next, a much more mundane connotation in which I use the word "wonder"; I marvel that in these difficult times so many of you have accommodated to the regional uncertainties and also that we have finally generated the financial support necessary to a Congress of this magnitude, and that it has been possible to generate a program of this scope. May I be permitted to express on behalf of all of us our deep appreciation to the many who have worked to bring this about.

The wonder I wish to address is neither of these but, rather, that sense of astonishment and amazement thrust upon all geneticists by the recent revelations concerning the complexity of our biological being. I am struck with wonder, as we gather here to open this Congress, at the sequence of events which, beginning some 4 billion years ago, must have transpired to create the genetic material which enables us to undertake

A shorter version of this paper was presented on September 14, 1981, as the Presidential Address at the Sixth International Congress of Human Genetics in Jerusalem, Israel. This expanded version permits a more rigorous documentation of my thesis than was then possible.

*Department of Human Genetics, University of Michigan Medical School, Ann Arbor, Michigan 48109.

such a meeting as this. In what follows, let us first refresh our memories as to the nature of that complexity and the many potentialities for its undoing. Next, we will examine our limited comprehension of how this complexity is not only maintained in good working order but is able to adapt and evolve in an ever-changing world. Finally, we will consider some of the hints now emerging for new directions in our thinking.

The Complexity of Our Genetic Organization

There is sufficient DNA in a human gamete to accommodate approximately 2×10^9 nucleotide pairs. The remarkable advances in recent years in assigning specific functions to specific regions of the chromosome in which this material is packaged will be summarized elsewhere at this Congress (see also [1]). Shortly after the nature of the genetic material became known, a number of geneticists made the obvious calculation that, if the average protein was a polypeptide composed of 300 amino acids, the requisite information could be carried by less than 1,000 nucleotide pairs, so that we humans have the coding potential for some 2,000,000 proteins. Since no one seems to believe humans have need for more than some 50,000 different proteins, then even with a seemingly generous allowance for the necessary nuclear "control" apparatus, the arithmetic suggested we might have DNA far in excess of our needs, and there was a rash of speculation about "junk DNA," an unnecessary and unloved relic of our evolutionary past. As in the past 10 years the true complexity of genetic organization has begun to emerge, it is clear no generalization was ever more premature, a tribute to our incurable intellectual arrogance in the face of the unknown.

Let us consider very briefly one example of the complexity of organization and function of the DNA, the still unfolding story of the γ-δ-β and α hemoglobin loci (reviewed in [2–8]). Figures 1 and 2 depict what genetic fine structure analysis has revealed concerning the organization of the DNA in the regions in which the γ-δ-β and α hemoglobin structural genes are found. This and the α-gene clusters code for the polypeptides for the three types of fetal hemoglobin which appear before the twelfth

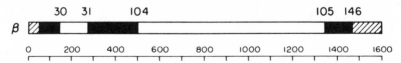

FIG. 1.—Fine structure of the gene coding for the β-polypeptide of human hemoglobin. Although the total length of the genetic code involved is, as shown on the scale, some 1,600 nucleotide pairs, only the portion shown in solid black finds expression in the hemoglobin molecule. The diagonally hatched sequences shown at the beginning and end of the gene are transcribed, from DNA to RNA (and that to the left apparently plays a role in initiating transcription), but do not find the functional hemoglobin molecule. (Reproduced by permission of Dr. T. Maniatis and Annual Reviews [7].)

week of gestation ($\zeta_2\epsilon_2$, $\alpha_2\epsilon_2$, $\zeta_2\gamma_2$), for the better-known fetal hemo-globins of late gestation ($\alpha_2^A\gamma_2$ and $\alpha_2^G\gamma_2$), and the two types of adult hemoglobin ($\alpha_2\beta_2$, $\alpha_2\delta_2$), all the hemoglobins being tetramers, composed of two identical chains coded by one member of the α gene cluster and two other identical chains coded by one member of the γ-δ-β gene cluster. Because this subject will be covered in extenso later in this Congress, I will be almost telegraphic in enumerating 12 salient aspects of our present knowledge, points relevant to the argument I will be developing. If the detail seems excessive, it is necessary to introducing some rigor into the arguments that follow.

1. There is a clustering of genes with like function. Clusters of this type provide evidence for evolution by gene duplication and subsequent divergence, as well as for evolutionary conservatism.

2. The "cascade" order of expression of these genes in ontogeny—from left (5′) to right (3′) in figure 2—is most readily explained if the essential control mechanisms are intrinsic to these regions (but note that the δ gene is "turned on" after the β).

3. All of the genes which find expression have two intervening se-quences, usually termed introns (see fig. 2)—stretches of DNA which fail to find expression in the polypeptide; the position of these introns has been conserved within the human line of descent for both the β-globin-like and α-globin-like genes since the time of duplication, some 450–500 million years ago, and, indeed, the nucleotide sequence at the intron-exon junctions and, for some introns, the intron length have also been conserved. In fact, two of the three introns found in the leghemoglobin of soybeans are in positions homologous to those in vertebrate hemoglo-bin [9].

4. These introns create the necessity for splicing signals and a type of precise editing of the primary RNA transcript, a burden on cell dynamics which presumably must be justified by some important func-tion. In the mouse, and presumably in man, at least two cleavage-ligation reactions are necessary to the excision of the larger intervening sequence from the primary β-globin mRNA precursor [10].

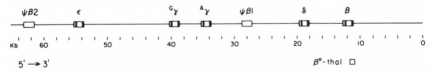

FIG. 2.—Fine structure of the region of chromosome immediately adjacent to the β-globin gene pictured in fig. 1. Within a sequence of some 60,000 nucleotides (60 kilobases) are found five genes expressed in hemoglobin at various stages in human devel-opment, plus two pseudogenes ($\Psi\beta1$, $\Psi\beta2$). These latter are sequences of DNA with unmistakable homologies to the DNA encoding for the β-polypeptide, but they do not find expression at any stage in development. All five of the genes that do find expression have the complexity of the gene shown in fig. 1. (Reproduced by permission of Dr. T. Maniatis and Annual Reviews [7].)

5. The introns appear to have evolved significantly more rapidly than the exons, as judged by interspecific differences in nucleotide sequence.

6. The translated products of the two α genes are identical; the translated products of the two γ genes differ by a single amino acid—but the α duplication is thought to have occurred at least 30,000,000 years ago, and the γ at about the same time. In contrast to such conservation of sequence within the species, the α chains of some primates differ from those of others by as many as 20 amino acids, that is, there has been considerable interspecific divergence.

7. At their 5′ ends, all the globin genes are characterized by a sequence of some 50–53 nucleotides, which sequence is transcribed but not translated. There is a similar sequence of some 87–135 nucleotides at the 3′ end. The lengths of these regions are, in general, evolutionarily conservative.

8. Outside the globin genes, on the 5′ side, there are two short repetitive homologies, about 31 and 77 base pairs from the beginning of the transcribed material; these same repetitive sequences are widespread throughout the human genome. The precise relationship of these sites to the promotion of RNA polymerase activity (i.e., of transcription) is being actively explored.

9. The thalassemias are a group of diseases due to deficient production, of varying degree, of either α or β hemoglobin polypeptides. There is evidence that some of these result from mutations that alter RNA processing. Because of the genetic linkage of thalassemias involving the β-polypeptide with mutants characterized by amino acid substitutions in the β chain, the locus controlling this abnormal processing must be placed within the hemoglobin complex. In one instance of β-thalassemia in which only small amounts of the β-polypeptide are made, the defect appears to be a single base change in the smaller of the two introns, which change creates a sequence much like that of the 3′ splice site of this intron and results in an inappropriate splice of the mRNA [11]. Similarly, a small deletion within the first intron of the α gene, just at the splicing site for the exons, results in a type of α-thalassemia in which no α-chains are made [12].

10. Both of these gene clusters contain "pseudogenes" (DNA segments that exhibit nucleotide sequences similar to the functional globin genes but do not appear to produce globin polypeptides); those which have been sequenced exhibit 75–80 percent homology when compared with the corresponding expressed gene. The loss of the ability to encode a functional globin polypeptide appears due to defective initiation sites or to the presence of small deletions and insertions or nucleotide substitutions that result in alterations of the translational reading frame, in stop codons, or in nonsense codons.

11. At still another level of organization, the recent statistical analysis

by Modiano, Battistazi, and Motulsky [13] suggests that, among the co-dons which find expression in the exons of the α- and β-globins, there is significant deficiency of those capable of mutating to chain termination codons, a situation which it would at present be very difficult to perceive as anything other than exceptionally fine tuning by natural selection (see also [14] on nonrandom coding strategies).

12. Finally, now that the details of the complex association of the α and β locus polypeptides in functional hemoglobin have been eluci-dated, we see that events at either of the corresponding loci will have implications for the other; major evolutionary changes for a dimer must proceed in a coordinated fashion, which introduces a new complexity to evolution.

I will return from time to time in this talk to some of the implications of this complexity, which, as it has emerged during the past 30 years, has completely shattered our earlier simplistic notions of genetic organiza-tion. Two other situations, emerging as equally complex, will also be dealt with in the course of this Congress: one is the genetic basis for antibody diversity, where increasingly it appears that at the pre-transcriptional level, as a result of mechanisms still poorly understood, there is a programmed juxtaposition of diverse DNA elements from multiple sets into an impressive array of structural genes, with further variety in the expression of this array introduced by posttranscriptional alternatives in RNA splicings (reviewed in [6, 15]). The number of dif-ferent "elements" available for this combinatorial process, in the C, V, and J regions, is still under investigation, but obviously sufficient to account for the tremendous diversity of antibodies. The phenomenon of allelic exclusion with respect to immunoglobin phenotypes (in any given cell only one of the two immunoglobin "genes" finds expression) re-quires the postulate of a mechanism so precise that one of the two alleles is not involved in these events. We have no meaningful insights at pres-ent into the nature of such a control mechanism.

The second example is provided by the human histocompatibility (HLA) gene cluster. Again, in view of other presentations to come of this Congress (see also [16]), it would be redundant to attempt a review. Let me mention just one point. In the mouse, the H-2 gene products (homologous to the HLA-A, HLA-B, and HLA-C products of the HLA gene cluster) are glycosylated proteins with an approximate molecular weight of 45,000, associated specifically but noncovalently with a non-glycosylated polypeptide chain of approximately 12,000 molecular weight. The differences between the products of the various H-2 alleles known to occur at each of the loci involved in the gene cluster reside in the polypeptide chain, comprised of some 350 amino acids. In the vari-ous inbred mouse strains, the gene products of the H-2 alleles associated with any specific gene of the gene cluster appear to differ by as many as 30 amino acids [17–19]. On the other hand, newly detected H-2 mutants, as might be predicted, differ from the parental gene product by only one or two amino acids (reviewed in [20]). One must assume that in a population of wild mice there is a storehouse of alleles whose products differ from one another by between one and 30 amino acids, with any particular inbred strain retaining only one of these alleles. Although Klein and co-workers [21] have established a minimum of 200 alleles at either the K or D locus in the mouse, the relatively crude serological approaches by which different allelic products are recognized presumably detect only a small fraction of these differences. The potential array of genetic diversity at this one complex locus is truly staggering. (The problem is only slightly less complex if each of the H-2 [and HLA] gene products reflects [by mechanisms not yet understood] the expression of a multilocus family of genes; cf. esp. [22].) Similar findings are emerging with respect to the human HLA loci, and there is no reason to believe humans will exhibit any less diversity at these loci.

A fourth example of this complexity, not to be treated at this Congress, is provided by the bovine POMC "locus," which we may assume has its counterpart in humans (reviewed in [23, 24]). This is the locus responsible for the synthesis of β-endorphin, melanocyte-stimulating hormone, adrenocorticotrophic hormone, and several other biologically active polypeptides. The primary gene product consists of a polypeptide with a molecular weight of approximately 28,500 daltons. This includes a "signal peptide" at the 5' end, of some 2,500 daltons, comparable to that found in many primary translation products, and quickly edited out. However, the remaining polypeptide is then edited into three pieces, a piece of approximately 16,000 daltons whose function is still obscure and two shorter polypeptides, adrenocorticotrophic hormone and β-lipotrophin. Depending on the precise tissue in which this translation and editing has occurred, there is then further editing, (a) the 16,000 dalton piece to yield γ-melanocyte stimulating hormone, (b) adrenocorticotrophin to yield α-melanocyte-stimulating hormone and a corticotrophin-like peptide, and (c) β-lipotrophin to yield γ-lipotrophin and β-endorphin. The γ-lipotrophin may even be edited further, to yield β-melanocyte-stimulating hormone. It would appear that a family of (genetically controlled) enzymes must be involved in this editing, the precise number yet to be determined. These facts reveal a sophistication of cell function in which the apparent economy of one polypeptide serving as the substrate for many functional proteins is offset by the need to maintain the several enzymes necessary to this editing. How did such a system evolve, and how many more such systems await recognition?

These four examples of genetic organization—one emphasizing a complexity of genetic fine structure, one a finely programmed sequence of DNA rearrangements, one a complexity of allele structure, and one an intricate editing of a primary gene product—may or may not be typical of the genome in general. It seems unlikely that investigators came upon the most complex of examples of genetic organization so early in the mapping of the DNA. On the other hand, in part these situations have been pursued with great vigor because of a perceived complexity, clearly a source of bias, and less complex situations have already been identified. Returning to our principal example, some 60,000 nucleotide pairs of DNA are involved in the hemoglobin γ-δ-β gene cluster. Given the many aspects of the total organization which have been maintained through extended evolutionary time—despite the disruptive forces to which I will be speaking shortly—it would be presumptuous in the extreme to refer to any large portion of this as "junk" DNA—and yet only a few percent of the total finds phenotypic expression at any time during development.

There is a very different line of evidence for a minority of human DNA falling into the junk category. Abrahamson and his colleagues [25] have suggested there is a straight-line relationship between "forward" mutation rate per locus per rad radiation and DNA content per haploid genome, over two orders of magnitude with reference to the amount of DNA per haploid genome. Although details of this treatment have been criticized [26, 27], the trend seems indisputable. One implication of this finding is that the amount of DNA associated with what in classical terminology we would call a locus varies widely between species, and there are with respect to radiation effects few silent areas of junk DNA, that is, the genetic targets have expanded in size in the evolutionary sequence, commensurate with the total amount of DNA.

The Accidents to Which Our DNA Is Subject

Even when the (diploid) human nucleus is in the so-called "resting state," the 4×10^{-9} nucleotide pairs are in dynamic equilibrium with their chemical milieu. They are subject to a variety of biochemical mishaps, conveniently cataloged by [28], including the "spontaneous" loss of the purine or pyrimidine moieties which impart the specificity to each nucleotide. Misrepair of this latter event is a possible mutation. The potentiality for mutation through misrepair emerges when we realize that, for example, the probability of spontaneous depurination of any specific site is 3×10^{-11}/sec in a mammalian cell. This low probability suggests a remarkable stability for any specific site, but because there are so many nucleotides, a mammalian cell should lose by spontaneous hydrolysis about 10,000 purine (and 500 pyrimidines) from its DNA during a 20-hour generation period [29].

As another example of this type of calculation, consider that in bacteriophage T4, the average "heat-induced" mutation rate (possibly due to cytosine deamination) is about 4×10^{-8} per G.C. (Guanidine-

Cytosine) base-pair per day. Baltz, Bingham, and Drake [30] have pointed out that at human body temperatures, since the haploid human genome contains about 1.4×10^9 gigacycle base-pairs, the endogenous heat-induced mutation rate would be about 100 per diploid cell per day in the absence of appropriate repair pathways (this calculation assumes the rates of deamination are similar for cytosine and 5-hydroxymethyl cytosine). Similar calculations can be made for the other types of spontaneous chemical mishaps to which DNA is prone.

In addition to this spontaneous instability, our DNA is constantly exposed to potentially disruptive extraneous influences. During the time span from birth to the middle of the reproductive period, background radiation results in a human gonadal exposure of approximately 3 roentgens. A roentgen produces about 2×10^{12} ions in a gram of gas. In water, which is essentially what we are, most of these ions recombine with another ion of the appropriate sign almost immediately and so do not travel very far, but if generated in close proximity to DNA, occasionally far enough to do some damage. Perhaps 5 percent of these ions escape immediate recombination and function briefly as free ions. That translates into approximately 3×10^{11} free ions released per gram of gonadal tissue because of background radiation. (I thank Professor H. R. Crane for this calculation.) Furthermore, in the "natural" environment, which I will equate with the nonindustrialized environment, there are compounds such as the aflatoxins, pyrrolizidine alkaloids, flavanoids, and the products of protein pyrolysis, which, to the extent they reach our germinal tissue and form adducts with DNA, could act as natural mutagens (reviewed in [31, 32]). Moreover, we are exposed from birth to the end of our reproductive period to a wide variety of viral infections, some, such as mumps, clearly prospering in reproductive tissues. In cell culture, many viral infections are associated with gross chromosomal damage (reviewed in [33]), and agents which break chromosomes, such as X-rays, are usually mutagenic.

Although it is over 20 years since the demonstration by Taylor, Woods, and Hughes [34] that homologous sister chromatids may undergo apparently reciprocal exchanges in somatic cells, termed sister chromatid exchanges (SCE's), the molecular basis of the phenomenon remains unclear. An especially attractive hypothesis by Shafer [35] postulates that SCE formation involves the bypass during replication of accidental or induced DNA cross-links. Since SCE's may reflect only about 1 percent of DNA adducts, and since the kinds of visible chromosome breaks that have long been our index of genetic damage are about 1 percent as frequent as SCE's, then the chromosome breaks previously scored in studies of spontaneous and induced cytogenetic damage may detect only one ten-thousandth of the total potential DNA damage of this type in a cell. In a normal cell in tissue culture, treated just as kindly as we know how, there will be six or seven SCE's. The points of exchange between the chromatids involved in an SCE, like the points of exchange between homologous chromosomes at meiosis, must be precise—otherwise the inevitable result is a change in the DNA sequence; here is a mechanism for spontaneous mutation. It may also be a misrepair of the process that leads to SCE's that occasionally results in the spontaneous exchanges between nonhomologous chromosomes that we term reciprocal translocations; and basically the same process may be involved in the exchanges between homologous chromosomes at meiosis which we term crossing-over. It would indeed be ironic if crossing-over, which geneticists have long seen as a superb adaptation to the need for reshuffling the genetic material each generation, was in part the inevitable result of chromosomal instability plus the juxtaposition of homologous chromosomes at meiosis.

As if all these threats to the integrity of the DNA were not enough, periodically the DNA must replicate itself—all 4×10^9 nucleotide pairs, and the duplicated set of chromosomes must undergo a very precise redistribution at mitosis. Furthermore, during meiosis, there are realized in the average human nucleus some 50 crossover events; if there is not to be a mutation, each of these must be accomplished with such precision (we believe) that not a single nucleotide is lost or duplicated, and then the chromosomes must execute their intricate meiotic quadrille perfectly—if one gets misplaced, the result is chromosomal non-disjunction.

Finally, during most of its existence the DNA is packaged in a way that would seem to be an invitation to total disaster. The length of the DNA strand which must be fitted into an interphase nucleus is about 2 m, over 200,000 times the diameter of the average cell nucleus. However this feat is accomplished, it must be a work of great precision, permitting on cue not only the replication of the strand, but the transcription of such stretches of the DNA as are essential to each cell's functioning. While some of this packaging may be inherent in the primary structure of the DNA, histone moieties are also intimately involved, the whole presenting a higher-order structure as complicated as that of the DNA itself.

Mutation

A germinal mutation is a transmitted change in the genetic code, be it large or small. It may result from any one of the accidents just mentioned, but our operational definition must include the result of such normal events as intracistronic recombination involving different alleles, as apparently exemplified by certain PGM_1 variants [36]. The frequency and phenotypic impact of mutation is, of course, a basic biological parameter. Until recently, estimates of mutation rates in higher eukaryotes have been based on gross phenotypic departures from normality. For man, for a variety of reasons, only the estimates based on dominantly inherited phenotypes, usually medical syndromes, are deemed reliable. The average of 18 such estimates, as summarized by Vogel and Rathenberg [37], is 2×10^{-5}/generation. Such estimates are usually given on a per locus basis, but for all the human traits used in such estimates, it is possible the consequences of mutation at several different loci are being measured, that is, that the trait is etiologically heterogeneous. A prime example of this is achondroplasia, repeatedly subdivided in recent years, to the extent that Morch's estimate in 1941 of a mutation rate of 4×10^5/generation has, for true achondroplasia, been reduced to half of the original estimate [38, 39]. Furthermore, some of the medical syndromes (aniridia, retinoblastoma) employed in these estimates have recently been shown to be sometimes associated with chromosomal deletions, so that the results of more than classical "point mutations" may be entering into these frequencies.

The appropriateness of these estimates as a general baseline has been repeatedly debated. That there is in these studies a strong element of selection for phenotypes known to arise through mutation is undoubted, and some have argued for an average frequency of mutation per locus at least one order of magnitude lower [40–42]. On the other hand, it could be that the loci involved are such that a high proportion of the changes manifest themselves in a dominantly inherited phenotype so that the estimate is a useful guide, although based on an atypical set of loci [43]. Such extreme views are possible because for none of these 18 traits is the biochemical basis known. A possible molecular model for dominant inheritance is supplied by Hgb Indianapolis, an unstable β-globin chain mutant that is synthesized normally but then very rapidly precipitated intracellularly, this and the resulting excess of α chains together causing membrane damage and hemolysis, the result being a "dominantly" inherited anemia [44]. But since mutants with apparently dominantly inherited phenotypes are a distinct minority among the hemoglobin variants, one must be cautious in postulating from this one example the existence of loci where mutation very frequently has this type of effect.

As the study of the hemoglobin variants began to reveal the many ways that mutation could alter a molecule, studies at the level of the gross phenotype, although important to defining the impact of mutation in a medical sense, began to lose their intellectual appeal because, since the protein basis of most of the syndromes was unknown, we could not visualize what was happening at the more basic DNA level. Furthermore, studies on species differences in mutation rates utilizing indicators for which the basis of the phenotype was unknown are of dubious value. It was to circumvent both these drawbacks that some 10 years ago our group embarked on a major effort to carry the study of human mutation to the biochemical level, initially searching for mutations altering the electrophoretic mobility of some 30 different proteins. This enables us to study any protein for which an electrophoretic technique exists, and so minimizes the bias mentioned for the medical syndromes. More recently, techniques for searching a subset of these proteins for "null" and "thermostability" mutations have been developed [45, 46]. ("Nulls" are mutations characterized by loss of enzyme activity and/or protein gene product; thermostability mutants are enzyme variants with significantly altered thermostabilities.) The battery of protein indicators employed include some with respect to which abnormalities are known to be associated with disease; to that extent the results of this approach may be related to the approach employing medical syndromes. Collectively these techniques—admittedly laborious—have the potential of revealing most of the mutational events affecting exons. Fortunately, the same technologies that have resulted in our new knowledge of genetic fine structure should within 10 years provide the framework for studies of mutation at the ultimate, the DNA level.

Thus far only the electrophoretic approach has yielded a significant body of data. Estimates of mutation rates can be generated from electrophoretic data by either a direct or an indirect method. The first method involves few assumptions: One examines a series of children, preferably consecutive newborns, for rare electrophoretic variants and, when one is encountered, conducts family studies. Absence of the variant in both parents implies mutation, the more so if the various genetic markers one can employ reveal no discrepancy between stated and biological parentage. Under favorable circumstances, exact probabilities can be developed for the proportion of putative mutations actually due to nonpaternity [47]. The second, the indirect, method defines the number of different rare variants per locus for a series of loci in isolated, highly endogamous populations and from a number of different simple relationships now available [48–50], calculates the mutation rate necessary to maintain these variants. Some key assumptions must be made, principally that the variants are essentially neutral selectively, that the populations in question have been numerically stable for many generations, and that no variants have been introduced to the populations under study by in-migration. Although these assumptions are not strictly met by any tribal population today, at a time when we are still attempting to fix an order of magnitude for mutation rates, the method appears sufficiently robust to yield useful approximations even in the face of small departures from these assumptions. The average number of generations a neutral mutant can be expected to survive also enters into the calculation; we have obtained that value by successive introductions of new mutations into a computer model of a population with the demographic characteristics of an Amerindian tribe [51].

Table 1 summarizes the results of the direct approach as practiced among civilized populations. A total of 815,803 locus tests has yielded a single putative electrophoretic mutation. While one hesitates to calculate a "rate" on the basis of a single finding, the data do yield an order-of-magnitude, of 0.12×10^{-5}/locus/generation. One must bear in mind, however, that electrophoresis detects only approximately one in three amino acid substitutions. Furthermore, it does not reliably detect nulls. In *Drosophila*, the only higher eukaryote for which there is an estimate of the ratio of electrophoretic:null mutations in germinal tissue, this ratio is approximately 1:5 [56]. If the same ratio obtains for humans, then the observed rate corresponds to a total rate of 1×10^{-5}, but because of the extrapolation, the rate carries an error equal to the estimate! We anticipate that the present laboriousness of this approach will be substantially alleviated by the advent of two-dimensional polyacrylamide gel electrophoresis and the development of suitable algorithms for the automated analysis of such gels [57].

The results of the indirect approach thus far are shown in table 2. There is considerable variation among the three estimates, the causes of which we cannot stop to explore; the average of the three estimates is 0.9×10^{-5}/locus/generation. This rate, based for the most part on tropical dwelling/nonindustrialized/tribal populations, is significantly higher than the maximum rate excluded by the direct studies on largely temperate dwelling/industrialized/civilized populations. Because of the many assumptions which enter into the indirect estimates, we propose to be cautious in our interpretation of the apparent finding. On the other hand, the frequency of the kinds of rare variants on which these estimates are based is some fivefold greater in the tribal populations than in civilized populations. We feel this higher frequency is not an artifact of the contagious sampling characteristic of fieldwork among tribal

TABLE 1

THE RESULTS OF FOUR EFFORTS TO DETECT BY THE DIRECT METHOD
ELECTROPHORETIC VARIANTS DUE TO SPONTANEOUS MUTATION IN
CIVILIZED/INDUSTRIALIZED POPULATIONS

Population	Equivalent Locus Tests	Mutations	Author
United Kingdom (London)	113,478	0	Harris, Hopkinson, and Robson [52]
Japan (Hiroshima/Nagasaki)	365,738	0	Satoh et al. [53]
United States	111,587	0	Neel, Mohrenweiser, and Meisler [54] and unpublished
West Germany	~225,000	1	Atland [55]
Total	815,803	1	...

populations living in widely dispersed villages [61]. This finding strongly reinforces the validity of the postulated difference in the two estimates of mutation rates, the only obvious alternative explanation being a greater positive selective value for variants of this type in these tribal populations. Following the same arguments laid out in the preceding paragraph, these findings would imply a total locus rate in these populations of 7×10^{-5}/locus/generation, a very much higher rate than usually postulated. Because of the many factors that enter into these indirect estimates, the error of the estimate is difficult to calculate. The possibility of significant regional differences in human mutation rates raises many tantalizing questions. Earlier I commented concerning the occurrence of natural mutagens in the environment; there are reasons to suspect that this cannot be the entire explanation. The resolution of the reason(s) for these two differing estimates will be of great interest.

Thus far, our considerations with reference to biochemical events have been directed largely toward what are usually termed micromutations. However, the evidence from genetic fine structure analysis, such as the data on the hemoglobin complexes mentioned earlier, or on the striking similarities in amino acid sequences in certain enzymes and hormones with dissimilar functions (e.g., trypsin, chymotrypsin, elastase, the haptoglobin heavy chain, and certain of the blood clotting factors [62], or lipoamide dehydrogenase and glutathione reductase [63]) leaves little doubt concerning the occurrence of larger events in the past. We may term these macromutations, the most common form of which would be duplication (or loss?) of a functional structural gene. The evolutionary potential of these events is becoming increasingly apparent (cf. [64–68]). The relatively high frequency of persons lacking one or even both α-coding sequences in many groups plus the demonstration that

TABLE 2

ESTIMATES OF MUTATION RATES IN THREE DIFFERENT GROUPS OF TRIBAL POPULATIONS, BASED ON AN INDIRECT APPROACH

		ESTIMATE ($\times 10^{-5}$/locus/generation) AND METHOD				
GROUPING	TRIBES (N)	Kimura-Ohta	Nei	Rothman-Adams	AVERAGE	INVESTIGATOR
Amerindians	12	1.43	1.69	1.71	1.61	Neel and Rothman [58]
Australian aborigines	?*	.61	.28	1.29	.73	Bhatia, Blake, and Kirk [59]
New Guinea	55	.44	.24	.56	.41	Bhatia et al. [60]

*The blurring of tribal lines with acculturations renders a precise tribal accounting difficult.

triplicated α-globin loci are not uncommon in humans [69] provide examples of "clean" duplication and deletion, presumably resulting from unequal crossovers outside the structural gene sequence. Given the identical nature of the two α-globin structural genes, Zimmer et al. [70] have argued for the "rapid" duplication and loss of the genes coding for these chains, as the basis for the "concerted" evolution they manifest. The various hemoglobin Lepore's and anti-Lepore's (hemoglobins characterized by polypeptides which are δ-β chain "hybrids"; reviewed in [2]) may, on the other hand, be presumed to result from unequal crossovers involving DNA within the structural gene sequence. The discovery in the mouse of a (nonfunctional) α-globin-like gene that has lost both intervening sequences raises the possibility that mRNA can become chromosomally integrated or in some other fashion act as a template at the time of DNA strand duplication [71]. Finally, Jones et al. [72] have recently described a form of $\delta\beta$-thalassemia, which on fine structure analysis is characterized by an inversion of the entire sequence between the $^A\gamma$ and δ-globin genes as well as deletions of approximately 1,000 and 7,500 nucleotide pairs, and have suggested a model of intrachromosomal crossing-over capable of explaining the findings. Whether or not the model is correct, the event is one of the clearest examples to date of the potentiality inherent in "macromutation." There is no doubt that the same duplication events that have been so important in evolution also create "accident-prone" chromosomes, a point established many years ago by Sturtevant's [73] and Bridge's [74] elucidation of the Bar mutation in *Drosophila* (see also [75]). Thus, the data from the hemoglobin loci, plus data beginning to emerge from other fine structure analysis, suggest duplication-deletion mutations are more common than we had thought, but recognition of this fact has been delayed because these tend to result in quantitative changes not so easily detected as the qualitative changes with which genetics usually deals.

With the current great interest in the extent to which transposons, mutator genes, and hybrid dysgenesis phenomena contribute to spontaneous mutation rates, speculation concerning the role of such mechanisms in human mutation is inevitable. It will be difficult to obtain evidence of these phenomena for human germinal mutation but easier for somatic mutation, assuming comparable phenomena in the two types of cells. "Cancer families" are an obvious starting point in the search. There is, perhaps, another approach to this quest. Some years ago, on one of our expeditions to the South American rain forest, we did cytogenetic studies on presumably normal Indians, in an effort to establish baselines for chromosomal damage [76]. To our surprise, about one in 250 cells appeared as in figure 3. This is as extreme damage as produced by any known mutagenic treatment. In the absence of any other such reports, many of our colleagues have been skeptical of the validity of this "finding." Recently, however, Dr. Awa, responsible for the cytogenetic studies in Hiroshima and Nagasaki, has observed similar cells in the control material of that study, although with a much lower frequency than in the Indians [77]. Now, cells like this will self-destruct at the next cell division, but it is not difficult to imagine lesser degrees of genetic reorganization than that implied by these preparations, occurring in germinal tissue, which would be transmissible and be accompanied by multiple changes in our DNA sequences, with the potentiality for quantum rather than slow genetic change. Our karyotypic findings could be interpreted as a correlation with the higher mutation rates we have postulated for Amerindians. At the time we felt the most probable cause of our findings was some recent viral exposure. We very badly need data concerning the randomness of the distribution of these cells within individuals. Cairns [78] has speculated that the activation of transposons or similar elements in human somatic cells may be a major factor in the genetic changes responsible for oncogenesis; the cells I have just described may lend cytological support to that suggestion. Later in this presentation I will return to the relevance of these cells to problems of human evolution.

On the basis of observations of the kinds of phenotypes resulting from

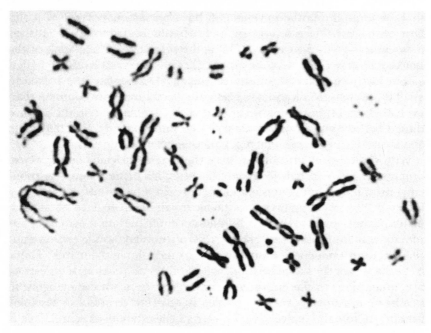

FIG. 3.—Complex metaphase from the cultured leucocytes of a 25-year-old Yanomama male. Multiple dicentrics as well as a tricentric chromosome can be seen, in addition to many free fragments.

mutation in man and other animals, the impact of mutation is considered to be generally deleterious. Evolution is thought to proceed in the face of this generally erosive phenomenon only because natural selection sifts out the relatively few favorable mutations and ultimately discards the rest. It must be admitted, however, that most of the approaches to the study of mutation have been geared to the detection of its ill effects rather than its good or potentially favorable effects. Thus, it seems clear that no study in the past has been oriented to the detection of small duplication events, whose importance in evolution we have already stressed.

From the foregoing brief review, it is apparent we are still in the early stages of defining with any precision the frequency and phenotypic impact of mutation on our species. Nevertheless, stimulated by the need to develop a frame of reference with respect to the medical implications of an increase in mutation rates from exposure to radiation or chemical mutagens, a number of committees have struggled to estimate what amount of disease is maintained in a human population by mutation. For instance, the United Nations Scientific Committee on the Effects of Atomic Radiation in its 1977 report [79] suggested, with many qualifications, that in a cohort of live-born infants approximately 60 per 1,000 could expect to develop at some time in their lives serious disease caused by mutation pressure. Approximate though this figure is (cf. [80]), it is easy to understand why most geneticists are uncomfortable with the thought that the mutation rates of civilized societies might undergo an increase.

Just how susceptible is our genetic material to external perturbation? Would that we really knew. My colleagues and I have, for some 35 years now, been involved in an effort to evaluate that question in the population of highest known risk to an increased mutation rate, the children of the survivors of Hiroshima and Nagasaki. Those studies were at first morphological in nature but in the past 10 years, paralleling the evolution of the field, they have taken a decided biochemical and cytological turn [81]. Using individualized estimates of the gonadal radiation sustained by the survivors, only available in the past several years, plus assumptions concerning the mutational contribution to our baseline phenomena, we have utilized the results of this reanalysis to estimate, for the first time, the genetic doubling dose or radiation for our species, that is, the amount of radiation that will produce the same frequency of mutations as occurs spontaneously each generation. This process results in a first *preliminary* estimate of an average gametic doubling dose of 250 rem [53]. (Note this is a revision upward of the estimate of Schull, Otake, and Neel [81].)

This new estimate, which should incorporate the frequency of both micro- and macromutation as defined earlier, has important implications for how we view the human radiation hazard. Up until the present we have been largely guided by the results of experiments with the mouse, for which the doubling dose of acute radiation for a variety of end points averages 40 rem. If our arguments, which we view as conservative, hold up, then our species is less sensitive than the mouse experiments would suggest by a factor of, perhaps, 6. Two clear implications of this study are (1) that the genetic risks of radiation, while real, have been somewhat overstated by most of the select committees considering the question and (2) that extrapolation to the human situation from the results of animal experimentation must be approached with some caution.

Why There Are Not More Mutations: Repair

Given the potentialities which we recognized earlier for mishaps, large or small, within the DNA, intuitively it seems remarkable that measured mutation rates, and the estimated frequency of serious genetic abnormalities maintained by the presence of mutation, are as low as they are. After all, assuming in the female 23 cell divisions between the newly fertilized zygote and the germ cells it would release, and in the male on average 380 (see Vogel and Rathenberg [37]), and also assuming that 500 nucleotide pairs are required to specify the average protein, then for such a protein a mutation rate of 10^{-5} would imply only one realized failure in exact replication in roughly 10 billion nucleotide replications! Although this simplistic calculation is undoubtedly incorrect in detail, it does establish an order of magnitude for the apparent fidelity of DNA replication. The reason for this fidelity, in part at least, lies in the existence of extremely sophisticated and diverse mechanisms for the repair of damaged DNA. Today a rapidly expanding literature recognizes at least eight more or less distinct kinds of repair (recent reviews in [82–85]). These repair strategies have been elucidated for the most part in microbial systems but, increasingly, similar repair systems are being recognized in higher eukaryotes, including humans. They go by such descriptive names as photoreactivation repair; nucleotide excision repair, which may be of either the "short patch" or "long patch" type; repair by simple excision of a misplaced purine or pyrimidine base; ligation of strand breaks in the DNA; and postreplication repair, occurring just after the defect in the DNA has been reproduced in the course of cell division. Perhaps the best understood type of repair is the excision of thymidine dimers in bacterial DNA following ultraviolet radiation; under optimum conditions this is approximately 95 percent effective at a dose of 10 ergs/mm² [86, 87]. The enzymes responsible for these various repair strategies, for the most part to be envisioned as hovering over the replication fork, may overlap in part with that other very complex assemblage of enzymes responsible for the replication process itself (reviewed in [88]).

Especially fascinating is the evidence for repair at a time substantially removed from the insult. Generoso et al. [89] have reported that when male mice treated with any of four chemical mutagens were mated 1–9 days after treatment to females of four different stocks, there were apparent differences in the frequency of dominant lethal effects in the resulting embryos. This was attributed to different rates of repair, in the fertilized eggs of the various stocks, of genetic lesions induced 1–9 days previously. Brandriff and Pederson [90] observed significantly increased unscheduled DNA synthesis after fertilization in male pronuclei resulting from sperm irradiation with ultraviolet some 3 hours previously. Both these examples of delayed repair involve the rather special case of the newly fertilized zygote. Of greater generality, but more ambiguous in interpretation, is the well-known observation of Russell [91] concerning the failure to recover mutations in increased numbers in those offspring of irradiated female mice conceived more than 7 weeks posttreatment. Since selective elimination of all oocytes containing induced mutations seems unlikely as the entire interpretation, one must either postulate a remarkable resistance of the immature oocyte to the primary genetic effects of radiation or a repair process continuing long after the insult. Many—perhaps the majority—of radiation-induced mutations are thought to be deletions of varying amounts of DNA, but those that are viable as homozygotes are presumably for the most part the result of nucleotide substitutions in the code; the repair of this array of potential genetic damage so long after the insult suggests a remarkable sensitivity to (or "memory" for) the original template, or else a process of directed gene conversion at meiosis.

The evidence for other important repair mechanisms during meiosis, presumably because of the close proximity of a normal template if a gene has been damaged, has recently been reviewed by Bernstein [92]. One wonders if this may be a primary mechanism for protecting against deficiency mutations; a contrast of somatic and germinal cell mutations in this respect would be important. These facts, plus the possible role of the testis-blood barrier in protecting germinal tissue against mutagens (reviewed in Fawcett [93]), suggest we must be extremely careful in extrapolating from studies on mutation rates in somatic cells to events in germinal cells. Given the rapidity with which our knowledge of the versatility of repair systems is expanding, it is not inconceivable that certain types of genetic events which appeared as mutations in one generation might be repaired at meiosis in a subsequent generation.

To complicate even further the complexity of repair, there is evidence that in human cells some repair is a two-step process, during which there is a "rearrangement" of the position of damaged sites, from regions relatively sensitive to a staphylococcal nuclease to regions relatively insensitive (reviewed in Lieberman et al. [94]). This phenomenon admits of diverse interpretations, ranging from actual physical transposition of damaged sites within the DNA strand to temporary dissolution of nucleosome structure following physical damage to the DNA strand, but, at the very least, this introduces new complexities into the repair process.

The first human disease found to be due to an inherited defect in one of these repair systems is xeroderma pigmentosum, in which there is an inability to excise and repair cyclobutane pyrimidine dimers induced by ultraviolet radiation [95, 96]. In the short interlude of the 13 years since this discovery, there have probably been more papers written in the United States about this disease than there are cases. Through complementation studies in cell culture, Hashem and colleagues [97] have identified at least eight different types of xeroderma pigmentosum. At the very least, this indicates some eight different points in the DNA strand coding for this repair enzyme at which the mutation in question can occur, and at the most, it may indicate that there are eight different loci at which mutation resulting in this repair defect may result. If the latter, then the fact that mutation at any one of these loci can disrupt the repair process indicates that the gene products act sequentially or coordinately in repair, and is further testimony to the complexity of the phenomenon.

But these replication and repair mechanisms, as well as those responsible for editing DNA transcripts discussed earlier, are also under genetic control and subject to mutation. Thus, at the very same time that we begin to visualize the complexity that keeps the rest of the complexity in order, now we have to begin to think about a further complexity that keeps the editing and repair complexity ready to cope with the basic complexity. This pyramiding of corrective measures can go on and on. We will surely agree that the emerging sophistication of the genetic repair "system" is no less than—and indeed, a necessary counterpart to—the emerging sophistication of our genetic architecture discussed earlier.

A new level of complexity for repair mechanisms is suggested by the previously noted remarkable similarity of the two γ-globin genes. On the basis of complete nucleotide sequences for the two γ-globin genes, Slightom, Blechel, and Smithies [98] suggest intergenic "conversions" occur in the germ line between such duplicated segments. This clearly has potential implications for the repair of one of these segments from a template supplied by the other. On the other hand, the primate differences in γ-chain composition suggest the conversion mechanism has not compromised evolutionary change. Data have also been recently presented suggesting, on the basis of studies of the mouse immunoglobulin $\gamma 2a$ and $\gamma 2b$ chain genes, that exons can be exchanged between genes in a multigenic family [99]. Given some sensing device for the integrity of an exon, this too can be regarded as a potential repair mechanism, but a kind of repair which has the attributes of intracellular selection.

The potential for a quite different type of repair stems from recent studies of ribosomal gene variants [100, 101]. In humans, genes coding for ribosomal RNA are distributed among "nucleolus organizers" located on five pairs of chromosomes—there are perhaps 150–200 discrete loci involved—with, then, an average of 30–40 of these ribosomal genes in the nucleolus organizer of each of the chromosomes involved. Genetic polymorphisms characterized by susceptibility to the action of specific restriction-site endonucleases have now been identified at three different sites, among these 150–200 genes. Different nucleolus organizers can contain the same restriction-site variant; this strongly suggests the occurrence of genetic exchanges among ribosomal gene clusters on nonhomologous chromosomes, regulated in such a way that the polymorphism persists. Such a system as this has obvious potential for the "repair" of a defective allele.

Earlier I mentioned that only one mutation altering electrophoretic mobility had been detected in 815,803 locus tests in humans. While we have been pursuing the biochemical approach in humans, Tobari and Kojima [102] and Mukai and Cockerham [56] (see also Voelker, Schaffer, and Mukai [103]) have been pursuing a similar approach in *Drosophila*. The use of defined polypeptides now for the first time renders cross-species comparisons of mutation rates meaningful. The estimate of the frequency of mutation resulting in electrophoretic variants from these two papers is $.25 \times 10^{-5}$/locus/generation, on the basis of 2,328,212 tests. Because of a relatively high rate of chromosomal aberrations in their experimental lines, Yamaguchi and Mukai [104] have expressed the concern that the observed rate may be somewhat elevated because of a mutator gene in the stock. Be this as it may, the limits we can place on the human rate suggest that the human rate is basically the same as that of *Drosophila*, in spite of the fact that our average generation time is roughly 200 times longer than that of *Drosophila* living in the wild state, during all of which time we are exposed to both natural and man-made mutagens, and our body temperature averages at least $10°$ C higher than that of *Drosophila*. Thus, if our repair systems were comparable to those of *Drosophila*, we might anticipate on a per locus basis a human mutation rate several hundred times greater than observed for that organism, although some of the force of that argument is diminished by the fact that a disproportionate share of mutation apparently occurs at the time of the meiotic divisions (see [105]), of which, like *Drosophila,* we have just two.

These data suggest that an essential feature of the evolution of an animal with so long a life span and so low a reproductive potential as ourselves has been the development of a much more sophisticated repertoire of repair mechanisms than most animals possess. Recently evidence for better repair mechanisms in long-lived species has emerged from comparative studies of *Mus musculus* and *Peromyscus leucopus,* and humans and other primates. The two mouse species are about the same size, but *Peromyscus* has a life span 2.5 times greater than *Mus.* Ultraviolet-induced unscheduled DNA synthesis in fibroblast cultures was 2.5 times greater for the *Peromyscus* cells. A determination of repair patch size by bromouracil photolysis following irradiation revealed that patch sizes were the same for the two species but the number of repaired sites was 2.2 times greater for *Peromyscus* [106]. With respect to primates, studies similar to the above, also involving fibroblasts, reveal a striking correlation between maximum life span and unscheduled DNA synthesis over a range of 10 years (*Saguinus*) to 90 years (*Homo*) [107]. Such studies are only in their early stages, but if other repair systems yield similar results, a basis could be provided for a better understanding of species differences in mutation rates.

Viewed in this light, it is not surprising that our first, tentative estimate of the genetic doubling dose of radiation for man is some six times greater than that for the mouse. The importance of establishing this ratio as accurately as possible to the issue of extrapolation from animal experiments to man is such that our Japanese colleagues and ourselves propose to continue the studies in Japan as long as feasible. An obvious supplement to such studies involves experimental manipulation of human and nonhuman somatic cells in tissue culture, but at present extrapolation to germinal rates from either cell type would, as noted earlier, be limited by the possibility that germinal tissue has evolved better (and different!) repair mechanisms than somatic tissue. Finally, I close this section with the obvious thought that in our efforts to protect the human genetic material from damage from environmental mutagens, efforts to minimize mutagenic exposures should be supplemented by exploration of the possibility of facilitating the action of repair enzymes.

Chance, the Other Threat to DNA

Thus far we have been principally concerned with intracellular aspects of human genetics—the threat to the genetic architecture posed by mutation and the cells' devices for coping. Let us now move to the organismic level. All of us who have taught Mendelian genetics have impressed upon our students the chanciness of what D. C. Rife some years ago referred to as the "dice of destiny," but I suspect only those of us working in population genetics appreciate how really chancy it all is. With the early deterministic formulations of population genetics, assuming as they did populations of infinite size, Fisher's classic calculations of some 50 years ago [108] demonstrated that even a gene conferring a 1 percent advantage on its possessor would survive and presumably replace the normal allele, in only two out of 100 introductions. But it was the elaboration during the past generation of a genetic calculus appropriate to finite, generally small populations (i.e., a more realistic model) that made it clear just how very vulnerable an adapted/evolving genotype was to the operation of chance, especially in the small, subdivided populations that characterized our ancestors.

Although it is convenient to formulate the problem in terms of fixation, in fact very few mutations ever go to fixation, simply because over the necessary time period, additional mutations at that locus will have been introduced into the population. Let us rather think in terms of the probability of loss of a mutation. Kimura and Ohta [109] have developed a population model which relates mean time of extinction of a mutation of neutral phenotypic value to population size: the smaller the population, the shorter the average survival time in generations. In their model, the assumptions are: a diploid population with overlapping generations; stable size and age distribution; and monogamy, with a reproductive expectancy of two children per couple and a Poisson distribution of those children. This latter assumption results in some 10 percent of couples being childless: this fraction is the equivalent of prereproductive death or sterility in this model population. Amerindian populations depart in many ways from these assumptions. In a simulation based on four interrelated villages of Yanomama Amerindians, we estimated that the mean survival time for a "neutral" mutant which is eventually lost ($\bar{t}_0$), introduced into an adult, was only 5.6 generations [51]. (It is necessary to employ the figure for introduction at maturity, since the census of rare variants is conducted on the adult generation.) By contrast, when the "variance effective" population number was used as the basis for calculations employing the Kimura-Crow formulation, the estimate was 11.7 generations. Although the simulation was less than perfect, we feel that the general magnitude of the difference is valid, reflecting the enhanced role for chance in a polygynous society with the higher mortality schedules of this population (15–20 percent infanticide and 15–20 percent pre-adult deaths).

The role of chance in small, subdivided human populations is most dramatically demonstrated when we consider how little assigning a positive selective value to an allele alters its fate. Thompson and Neel [110] have considered how selection would alter the fate of 280 alleles carried by young adults aged 10–19 in a population with a Yanomama-type structure. A branching-chain process was projected for 16 generations, where the mean number of replicates per allele (M) ranged from 1.00 to 1.15. Even at the highest value of M, corresponding roughly to a 15 percent selective advantage for the allele carrier, three-quarters of these alleles have still been lost at the end of 16 generations.

A second example of the enhanced role of chance in genetic phenomena in tribal populations stems from the manner of population extension. This is generally from the fission of preexisting villages, along lineal lines [111]. Thus, at the time a new village arises—a potential source of a new tribe—not only is a very small part of the tribal gene pool sampled, but that sampling is quite structured. We have recently shown that this structuring reduces the "effective" size of the resulting villages, as measured by the number of independent genotypes present in the villages, by one-half [112], thus increasing the role for stochastic factors in establishing gene frequencies.

We conclude from this that such was the force of chance during our pretribal and tribal days that the majority of favorable mutations, even those with what the population geneticist would consider relatively high positive selective values, were rather quickly lost and, conversely, deleterious mutations could increase in numbers in striking fashion and even go to fixation (see [113, 114]). On the other hand, of course, chance very rarely can result in genetically "good things," but even if chance gave rise to a genetically superior group of individuals, it would not ensure their perpetuation. This requires the operation of selection, to a consideration of which we now turn.

TABLE 3

THE FATE AFTER 16 GENERATIONS OF 280 ALLELES PRESENT IN INDIVIDUALS AGED 10–19 YEARS IN A POPULATION WITH THE DEMOGRAPHIC STRUCTURE OF THE YANOMAMA INDIANS

COPIES (N)*	SIMULATION		PREDICTED BY BRANCHING-CHAIN TREATMENT			
	Run 1	Run 2	$M = 1$	$M = 1.05$	$M = 1.1$	$M = 1.15$
0	247	231	256.0	243.6	228.5	211.3
1–50	28	40	23.7	34.6	44.8	50.5
51–100	4	9	.3	1.7	5.7	13.4
>100	1	0	.0	.1	1.0	4.8

NOTE.—Figures were determined by a computerized simulation program and by powering a branching-chain process the requisite number of generations, after Thompson and Neel [110]. The computer simulation program is based on a population expanding at the time of our study at a rate of approximately 25 percent per generation. The symbol M denotes the average number of copies with which the allele replaces itself each generation, and is roughly equivalent to a selection coefficient.

*This refers to the total number of copies of the original allele still present in the population at the completion of 16 generations.

Selection, the Deus ex Machina

It is clear from the foregoing that the operation of chance plus the potentiality for mutational change in the DNA—this latter despite a marvelous array of repair strategies—are powerfully disruptive threats to genetic adaptation. And yet, in the face of both of these erosive forces, life forms have not only survived, but evolved, a fact we customarily attribute to the guiding hand of natural selection. Only now is it becoming apparent how much human evolution demands of natural selection. Let us return briefly to some of the points made concerning the α- and β-globin gene clusters at the outset of this presentation. The preservation of so much of that structure for millions of years against the assault of the random mutation process, whose ill effects are so well documented in the hematology clinics of the world, leaves absolutely no doubt of the operation of strong genetic censoring mechanisms, some of which we can only dimly visualize. For instance, if the postulate by Zimmer and colleagues [70] of rapid duplication and loss of α-globin genes is correct, what then prevents the number of α-globin genes from slowly drifting upward? And, given the demonstrated frequency with which the exons are involved in partial or total loss due to unequal crossing-over, what in the world protects the pseudogenes from the same fate and ultimate elimination? Yet our ignorance of how the total system works is such that one can either argue that the apparently more rapid evolution of the introns and/or pseudogenes is due to the fact that as junk they are relatively free from selective constraints [115, 116] or, in diametric opposition, that the introns and/or pseudogenes are the regulatory "handles" within the region, and their rapid evolution reflects the cell's need for finely tuned control mechanisms [117, 118]. The demonstration, mentioned earlier, of thalassemia-like syndromes due to variation in introns is in my opinion rendering the first viewpoint increasingly untenable.

Selection is traditionally visualized as operating through differential survival or fertility with respect to genotype. But whereas the role of repair in offsetting potential mutational events can readily be demonstrated in human cells, its population counterpart, positive selection, choosing among those mutations which do occur, is almost an act of faith for humans. We have by now a multitude of examples of negative selection in humans—simply inherited diseases diminishing the probability of survival and/or reproduction. On the other hand, the paucity of examples of positive selection, in the sense of the superiority of a defined genotype over what is currently the norm, is impressive. This is not surprising, since if the superiority is great enough to be unambiguously demonstrated, then, unless a balanced polymorphism is involved, the allele associated with the selective advantage should in short order, in the evolutionary sense, replace the standard allele. The clearest example of selection-at-work in humans is of course provided by the Hgb^s allele, responsible for sickle-cell hemoglobin. A variety of lines of clinical evidence make it clear that the individual with the sickle-cell trait has a greater resistance to falciparum malaria than the individual without the trait. The results of studies in areas where falciparum malaria is hyperendemic, however, have been rather variable with respect to the exact survival and/or reproductive advantage conferred by the trait (reviewed in [119]); so that even for this presumably extremely solid example of a balanced polymorphism it has been difficult to develop precise estimates of selective differentials. The next best example of an adaptive trait in man is probably the resistance to vivax malaria conferred by the $FyFy$ genotype, a genotype encountered predominantly in blacks of West African origin (cf. [120]). The basis for this genetic resistance seems to be loss of a red-cell receptor site for the merozoite state of the parasite. Presumably, the receptor site is a normal component of cell membranes in persons of North European extraction, in whom it has other functions than waiting to be the vehicle for infection by a passing merozoite. We must assume that the loss of this receptor has some physiological consequence, and yet we have no insight into what the disadvantage of the $FyFy$ genotype is in a nonmalarious environment. Thus, both of our best examples of "positive" selection involve the loss of a normal attribute.

It is critical to efforts to conceptualize the operation of selection upon populations with the genetic structure of our species that there be a clear assessment of the amount of genetic variation in the population and its phenotypic implications. Until some 30 years ago, our knowledge of human genetic diversity was largely based on traits obvious to physical examination. The recent proliferation of biochemical techniques, and especially the introduction of electrophoresis as a convenient means of demonstrating certain types of differences among proteins, together with the blood-typing techniques stimulated by the increased use of blood transfusions, dramatically altered our view of the storehouse of genetic variation contained within human populations. Based largely on the aforementioned techniques, it has been estimated that genetic polymorphisms characterized by electrophoretic variants are to be encountered at some 20–25 percent of loci coding for soluble proteins, with the average human heterozygous at 6.3 percent of all these loci [121].

Although there have been obvious reasons why this estimate might be biased upward, the first solid data on the potential magnitude of the bias results from the lower levels of heterozygosity—roughly 1 percent— revealed by two-dimensional polyacrylamide gel electrophoresis of lysates of specific human cell types [122–124]. That the two-dimensional gel technology can detect known human protein variants has been demonstrated [125], but its efficiency relative to one-dimensional electrophoresis remains to be established. Accordingly, while there may be differences in the levels of heterozygosity characterizing the types of proteins revealed by two-dimensional electrophoresis and those enzymes thus far studied in one-dimensional systems, it is far too soon to say whether the overall heterozygosity level for the 50,000 or so loci encoding for polypeptides will be closer to the higher or lower value given above. It must in these considerations be borne in mind how much genetic variation is not being identified by electrophoresis. For instance, our laboratory, employing rather conservative criteria, has recently suggested that the frequency of individuals heterozygous for non-electrophoretic variants resulting in severe loss of enzyme activity (i.e., nulls) in a series of nine enzymes found in erythrocytes, is 2.4/1,000 determinations, and that heterozygotes for variants altering the thermostability of the same enzymes (but not their activity nor electrophoretic behavior) have a frequency of 3.8/1,000 determinations [45, 46]. One of the null variants which was encountered, of triosephosophate isomerase, achieves polymorphic proportions in American blacks. In the same material, rare (nonpolymorphic) electrophoretic variants had a frequency of 1.0/1,000. Given the conservative criteria employed, we have suggested that the total level of heterozygosity for rare, nonpolymorphic variants was roughly 1 percent, a value to be added to the usual heterozygosity index. All things considered, one is entitled to suspect that the frequency of heterozygosity at the biochemical level, all types of variants included, is apt to be at least the 6 percent quoted earlier.

More recently, the application of the same techniques that have defined the complexity of organization of the DNA is beginning to define the variation at that level. On the basis of the very early findings with a few probes of defined kilobase length [98, 126–128], it can be calculated that, with respect to the 2×10^9 nucleotide pairs in human DNA, there is something of the incredible order of 10^7 DNA polymorphisms, involving various rearrangements of the DNA, waiting to be discovered! If these are uniformly distributed between exons and other DNA, and if we assume the exons are responsible for 50,000 proteins each encoded on average by 1,000 nucleotides, then 250,000 of these polymorphisms should be in exons, roughly 5 per exon, not all, of course, finding phenotypic expression. This is a "result" to be approached with great caution (cf. [126]), but even if the distribution of restriction-site polymorphisms was so nonrandom that they were only one-tenth as frequent in exons as in the remaining DNA, the data remain suggestive of high heterozygosity indices.

The ultimate source of all this variation is, of course, the mutational process. It was H. J. Muller who most effectively among recent geneticists directed attention to the practical (as well as theoretical) implications of this process for humans (cf. esp. [129]). In 1956, Morton,

Crow, and Muller formalized in precise mathematical terms the concept of a genetic load imposed by mutation [130]. A year later Haldane [131, p. 520] wrote: "The unit process of evolution, the substitution of one allele by another, if carried out by natural selection based on juvenile deaths, usually involves a number of deaths equal to about 10 or 20 times the number in a generation, always exceeding this number, and perhaps rarely being 100 times this number." These two papers had a profound effect upon the geneticist's response to the deluge of genetic variation being revealed by the new biochemical techniques. With these formulations, it became increasingly difficult to accommodate within the selective process, as usually formulated, the amount of variation being recognized. A particularly lucid treatment of the dilemma is John Edwards's recent exposition [132] of the demands placed upon selection by the extensive linkage disequilibrium at the HLA locus, which region occupies only about one-thousandth of the human genome.

One response to the unfolding situation was the suggestion by Kimura [133] that a considerable proportion of the genetic variation encountered in populations, and even going to fixation in the form of nucleotide substitutions, must be neutral in its selective implications. At no point in the resulting polemic have the "neutralists" made specific claims as to the proportion of mutation resulting in neutral variants, but one senses that the emphasis has been gradually shifting as the evidence has unfolded [115]. Again the hemoglobin polypeptides have yielded important evidence on a critical point in population genetics, namely, just what "majority" might amount to: Boyer et al. [134] on the basis of cross-species comparisons suggest that at least 95 percent of all nonsynonymous mutations involving the β-globin polypeptide are functionally unacceptable in the homozygous state, and that 70 percent of synonymous mutations are adaptively undesirable. Thus, even if a high fraction of nucleotide substitutions actually encountered were selectively neutral, the responsible mutations are a distinct minority of those which must have occurred.

The concept of neutral variants and nucleotide substitutions has been of great heuristic value in forcing careful consideration of the question of the nature of the indirect evidence for selection. There may well be some nucleotide substitutions in evolution which are neutral in their selective implications. However, while one can envision mutations which are selectively neutral in a well-buffered organism, eventually continuing mutation should challenge these homeostatic buffers, and further mutations would no longer be neutral. This argument, originally directed toward mutation affecting minor biochemical traits, can of course be extended to the many nucleotide variants, mentioned above, now being directly detected. I suspect that even the protagonists of the "neutralist" viewpoint would agree, especially in the light of the developments I have been reviewing, that the demonstration—were that possible—that as many as 50 percent of nucleotide substitutions in exons were neutral would have little impact on the major problem of how selection operates to eliminate deleterious mutations and produce evolutionary adaptations.

In this latter regard, probably the chief development of the last 2 decades has been the concept of truncate selection, by virtue of which it was recognized that the death of a single individual could remove a number of deleterious genes from a population simultaneously [135–138]. This was a significant step toward reconciling the apparent need for genetic deaths created by the genetic load concept with the relatively low fecundity and prereproductive mortality of an animal like ourselves. One may, however, question whether even truncate selection can accommodate the demands upon selection created by the complex genetic architecture we now recognize, especially if we visualize selection as for the most part operating on alleles altering fitness with the traditional 1 or 2 percent.

The increase in gamete DNA as we ascend the evolutionary scale and, more recently, the biochemical evidence for gene duplication force us to think that from the standpoint of evolution the more important targets for selection are the macroevents best described as gene duplications, or occasionally more profound changes. The hemoglobin loci may be unreliable indicators of the frequency of such events because of the existence of duplication at both loci, but such duplication may be more common than we realize and the hemoglobin loci are, as previously noted, suggesting that duplication-deletion events may be relatively common. Some 18 years ago, in discussing the occurrence of minor hemoglobin components, such as the δ-polypeptide, I wrote: "This thought immediately gives rise to questions concerning the frequency in the mammalian genome of suppressed genetic information which periodically, as a result of mutation in the suppressing system, is exposed to the selective sieve, this presumably representing a mechanism for the evolution of 'new' proteins. Critical approaches to this question will almost certainly have to turn to experimental organisms" [139, p. 914]. The question raised in that first sentence can now be approached directly—how often will the coding sequences for proteins with marked homologies, such as, for example, lipoamide dehydrogenase and glutathione reductase, be found to be linked—but the second sentence surely failed to anticipate that our own species would be the subject of choice for the study of that question.

Since most of human evolution occurred under tribal and pretribal conditions, a principal focus of our fieldwork among Amerindians in recent years has been an effort to discern those aspects of the culture which might provide the selective process with the necessary leverage.

One of the most lasting impressions of that fieldwork was of the differing genetic roles of males and females [140]. In most tribal societies, especially hunting-gathering-limited agriculture societies where mobility is important, women space their pregnancies by a variety of means that limit them to about one birth every 3 or 4 years. Infertility is rare. The women are oriented toward a culturally imposed reproductive uniformity. This does not imply that a more capable woman does not on average raise more children to adulthood, but it does limit the opportunity for differential fertility based on genetic attributes. By contrast, in many of these cultures the males are engaged in a potentially lethal competition for a position of leadership from late childhood on. The man who emerges as the village headman usually acquires an extra wife or two for his efforts. He must be presumed to be genetically above the average. Our studies among the Yanomama show that the headman leaves twice as many children as the average male surviving to adulthood.

The issue of differential fertility has also been approached through computer simulation. Figure 4 depicts the number of grandchildren born, in a simulation program, to a cohort of Yanomama males aged 0–9. The potential for positive selection operating through headmen is the chief mechanism we have been able to define tending to offset the

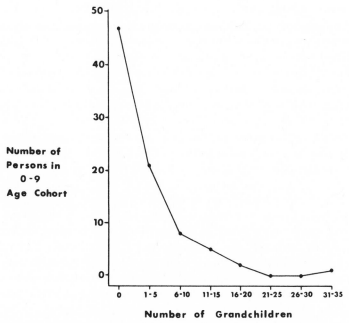

Fig. 4.—The number of grandchildren produced in a computer-simulation program by a cohort of 86 Yanomama males aged 0–9 years, after Neel [140].

erosive effects of mutation and chance on our genetic material. On the other hand, it must be freely admitted that we simply do not know how much of the differential expectation of reproductive success shown in figure 4 has genetic implications. Incidentally, with the disappearance of the institution of headmanship, we may have lost one of the chief vectors for positive selection thus far identifiable.

In summary, it has become increasingly difficult to reconcile the unfolding organization of human DNA with traditional formulations of the operation of natural selection. Efforts to accommodate the situation by such strategies as the postulate of neutral nucleotide substitutions, truncate selection, or emphasis on the role of "headmen" as prime agents of natural selection, valuable though they have been, at this point seem only partially to meet the emerging theoretical needs of population genetics.

These Wondrous Times

It is commonplace these days to assert that the biological revolution in which we find ourselves is of equal magnitude for our concept of self as were the Copernican and Darwinian revolutions. Already the current revolution has progressed from a first naive flush of enthusiasm to a sober realization of the complexity of what we are revealing. A fitting analogy is with the physicist, making the transition from the old mechanistic formulations to the relativistic concepts that dominate current thought. It is perhaps one of the great intellectual convergences of all human thought that so shortly after Jacques Monod [141] could write *Le Hasard et la nécessité,* the physicist-cosmologist Wheeler [142, p. 395] would ask: "But as for the molecules themselves, the particles of which they are made and the fields that couple them, is it conceivable that they too derive their way of action, their structure and even their existence from multitudinous accidents?"

It should at this point in this essay be clear that I am rejecting the concept that any substantial portion of our DNA is redundant, a carryover from the past, or that any substantial portion of the perturbations to which it is subject are neutral, meaningless in nature. In viewing the DNA which is in excess of the needs we can visualize for it as "junk," a hodgepodge of genes rendered useless by evolutionary advance, we have been engaging in an exercise of considerable arrogance. Because we could not immediately see a function for it, it obviously had none. In adopting this position, we turned our back on a well-established principle of evolution, the loss of unneeded organs. There are so many ways to delete segments of DNA that it is difficult to believe that any substantial fraction of what we have is just being passively carried along. While no one will argue that there is no junk DNA, recent discoveries leave less and less room for it. It is to me a sobering thought that we can ingeniously bypass all those complexities to move the human genes for insulin and somatotropin about even with our imperfect comprehension of how the system is maintained.

As we attempt to understand this complexity better, we must be prepared to rethink both the impact of mutation and the manner of operation of natural selection. With respect to mutation, I have already hinted that a higher proportion of the total spectrum may have a favorable impact, especially in the long run, than current wisdom would hold. Testing this hypothesis is much more difficult than testing the hypothesis that some mutation has obvious deleterious effects. Data are particularly needed on the frequency of occurrence of small duplications. With respect to selection, much more attention will in the future be directed to other avenues of natural selection than the traditional mortality and fertility differentials among live-born offspring. Three neglected areas in the study of selection stand out.

1. It now appears that at least half of all human zygotes are lost prior to term [143, 144]; Austin [143] in particular has emphasized the possibility that most of these losses have a genetic basis. A striking development of the past 15 years has been the recognition that at least 5–6 percent of recognized pregnancies are characterized by chromosomal abnormalities [145, 146]. Most of these chromosomal mutations are characterized by aneuploidy and result in early fetal loss, but Jacobs [147] estimates that structural chromosomal rearrangements consistent with survival arise with a mutation rate of 1×10^{-3}/generation. The meiotic quadrille is not so precise as we had thought. Mohrenweiser and Neel [46] have suggested that as many as 3 percent of conceptuses may be homozygous for null variants. Since those nulls which have been recognized postnatally are usually associated with ill health, and since geneticists have not established such a high frequency in newborn infants, we have suggested that, like the chromosomal variants, many of these are eliminated in utero. But only a bare beginning has been made on our understanding of the genetic bases for fetal loss, especially very early loss. Recently Beer et al. [148] have reported that women with consecutive spontaneous abortions of unknown etiology shared HLA antigens associated with the A, B, and D/DR loci with their spouses significantly more often than was the case for a control group. In the past, geneticists have been concerned with the effects on pregnancy wastage of genetic differences between fetus and mother, as exemplified by Rh disease. This new finding suggests that for some genetic systems, fetal-maternal similarities may lead to early pregnancy losses, and illustrates both the subtlety and potentiality for in utero selection. The human uterus, which has often seemed almost capricious in its rejection of its contents, may, in fact, frequently be responding to a genetic imperative of which we have been quite unaware.

2. There are other possible inconspicuous possibilities for the selection which appears necessary to our genetic survival. Of the 30,000 oocytes in the ovaries of a newborn human female, only some 250–300 will realize the possibility of fertilization, and of the 300,000,000 sperm in a male ejaculate, only one can be successful in achieving fertilization. Is it possible that there is more genetic selection involved in which eggs and which sperm are "successful" than we think to be the case? There is, to be sure, abundant evidence to suggest that sperm can function even when lacking one or several chromosomes of the normal haploid complement (cf. esp. [149]), but such observations do not preclude the possibility of selection based on positive attributes determined by the paternal genotype.

3. Some aspects of the maintenance of the structure of the globin complex (as an example of a gene cluster) or of the ribosomal DNA gene complex can best be explained if the "corrections" which must be postulated are preferential for a particular DNA sequence. This would correspond to genomic (as opposed to phenotypic) "selection," by mechanisms yet unknown. To the extent that these corrections are guided by coexisting duplications, then these duplications are not only the material for future evolution but an insurance policy for the present.

To pursue a different line of thought concerning insufficiently appreciated possible modes of selection: Some paleontologists have long been impressed by the evidence for bursts of speciation resulting in *clades* (Eldridge and Gould [150]; see also [151]) especially, in recent years, stressing the role of "sudden" allopatric speciation (as contrasted to phyletic gradualism), a possibility also raised, with different terminology, by geneticists such as de Vries [152] and Goldschmidt [153]. De Vries's position was weakened when the "saltations" he observed in Oenothera were found to have a rather special cytological basis. Goldschmidt argued on the basis of his experiments with *Lymantria* and *Drosophila* that mutation involving rate-controlling processes, perhaps clustered, might have profound consequences, creating "hopeful monsters," some few of which might prove adaptive. The response to Goldschmidt's suggestion was at best unenthusiastic, as was the initial response, 10 years later, to McClintock's first description of bursts of mutation in maize resulting from the presence of what we now term "transposons" [154]; today we consider the possibility that in natural populations of the higher eukaryotes, the majority of mutation may occur in bursts under the influence of transposons, mutator genes, and "hybrid dysgenesis" (reviewed in [155–158]). In the phenomenon termed hybrid dysgenesis in *Drosophila,* there is also an increase in crossing-over, which, like mutation, could promote new gene complexes.

With respect to the human line of descent, at this congress Dutrillaux will review data suggestive of such a *clade* at the time of origin of the great apes, some 4,000,000 years ago, and Tobias, with specific reference to man, will discuss fossil evidence consistent with a burst of diversification some 2,500,000 years ago (but see [159]). Just as the vast majority of mutations, even if favorable, are doomed to be lost, so, we may assume, are all traces of the vast majority of clustered genetic events which might have served as the basis of allopatric speciation or subspeciation. Thus, at this early stage in our study of population genetics, we have no data on the frequency of such clustered events, which perhaps should not be regarded as unique phenomena as much as the extremes along a spectrum of nonrandomly distributed mutational events, which less pronounced "clustering" in general contributes to evolution in a much less striking fashion than the concept of *clades* evokes.

Earlier I referred to the manner in which Yanomama villages fission along lineal lines, a process which with successive repetitions can lead to marked differentiation on the basis of chance alone. But in theory this also sets the stage for what Eshel [160] has termed "clone selection," wherein it is the survival probability of the clone rather than the individual that is critical; under certain conditions (substitute "lineage" for "clone"), this is consonant with strong selection and rapid evolution in human populations. A localized increase in mutation and crossing-over (equal and unequal) under the influence of one of these "mutator" phenomena in an isolated group, a group which by chance has allele frequencies "receptive" to this increase in mutation, followed by clonal-type selection, offers an interesting alternative to the prevailing view of the accumulation of variation through an almost clocklike introduction of mutation followed by selection over the entire range of the species. The complexly damaged cells we saw in the Yanomama may represent another type of evidence of clustered mutation, this in somatic cells, but evidence for the germinal tissue counterpart may be the rare children with complex chromosome rearrangements (reviewed in Pai [161]). Our clinical sieve is increasingly tuned to detecting the ill effects of mutational events such as these children represent—but how would we recognize a mutation of potentially adaptive significance?

It would appear that the mathematics of gene substitution might be substantially altered if changes at the DNA level were clustered in time and space. Such clustering, especially if it involved genetic elements linked on the same chromosome, implies, at least initially, fewer phenotypes than genetic changes. Selection would, thus, operate on a greater number of potential gene substitutions simultaneously than would be the case with evenly spaced mutational events. Some rather rigorous mathematical treatments will be required to explore just how greatly the combination of clustered mutational events, increased crossing-over, and clonal selection might alter the evolutionary strategy of moving from one "adaptive peak" to another.

I would not like my sense of awe and wonder, as our genetic makeup becomes better understood, to be seen as a retreat into mysticism. It is, rather, a realization that our genetic organization and the way in which it evolved is a far more complicated subject than seemed the case during the period of scientific euphoria that followed the elucidation of the structure of DNA and the cracking of the genetic code. A genetic organization that has been evolving 4 billion years can scarcely be expected to have yielded to our assault in the short 30-year period since these seminal developments. Perhaps the principles that govern this organization will, in retrospect, seem as simple as those governing the code, but for now that thought can only be an article of faith. Currently there is in the air the sense of the dawn of a new day in population genetics—but there is also considerable confusion as to which way to turn to catch the sunrise. We are a privileged group at a privileged time in the development of scientific thought and understanding—privileged not in the arrogant use of power, but privileged to witness and contribute to an unfolding story whose final implications we can only dimly foresee.

REFERENCES

1. McKusick, V. A. The anatomy of the human genome. *J. Hered.* 71:370–391, 1980.
2. Weatherall, D. J., and Clegg, J. B. Recent developments in the molecular genetics of human hemoglobin. *Cell* 16:467–479, 1979.
3. Winter, W. P.; Hanash, S. M.; and Rucknagel, D. L. Genetic mechanisms contributing to the expression of the human hemoglobin loci. *Adv. Hum. Genet.* 9:229–291, 1979.
4. Efstratiadis, A., et al. The structure and evolution of the human β-globin gene family. *Cell* 21:653–668, 1980.
5. Leder, P. The organization and expression of cloned globin genes. *Harvey Lect.* 74:81–100, 1980.
6. Symposium on Recombinant DNA. *Science* 209:1317–1435, 1980.
7. Maniatis, T.; Fritsch, E. F.; Lauer, J.; et al. The molecular genetics of human hemoglobins. *Annu. Rev. Genet.* 14:145–178, 1980.
8. Orkin, S. H., and Nathan, D. G. The molecular genetics of thalassemia. *Adv. Hum. Genet.* 11:233–280, 1981.
9. Jensen, E. O., et al. The structure of a chromosomal leghemoglobin gene from soybean. *Nature* 291:677–679, 1981.
10. Kinniburgh, A. J., and Ross, J. Processing of the mouse β-globin MRNA precursor: at least two cleavage-ligation reactions are necessary to excise the larger intervening sequence. *Cell* 17:915–921, 1979.
11. Spritz, R. A., et al. Base substitution in an intervening sequence of a β+-thalassemic human globin gene. *Proc. Natl. Acad. Sci. USA* 78:2455–2459, 1981.
12. Orkin, S. H.; Gott, S. C.; and Hechtman, R. L. Mutation in an intervening sequence splice junction in man. *Proc. Natl. Acad. Sci. USA* 78:5041–5045, 1981.
13. Modiano, G.; Battistazi, G.; and Motulsky, A. G. Nonrandom patterns

of codon usage and of nucleotide substitutions in human α- and β-globin genes: an evolutionary strategy reducing the rate of mutations with drastic effects? *Proc. Natl. Acad. Sci. USA* 78:1110–1114, 1981.

14. GRANTHAM, R. Workings of the genetic code. *Trends Biochem. Sci.* (December), pp. 327–331, 1980.
15. SHIMIZU, A., et al. Ordering of mouse immunoglobulin heavy chain genes by molecular cloning. *Nature* 289:149–153, 1981.
16. BODMER, W. F. Gene clusters, genome organization, and complex phenotypes. When the sequence is known, what will it mean? *Am. J. Hum. Genet.* 33:664–682, 1981.
17. NATHENSON, S. G., et al. Structure of H-2 major histocompatibility complex products: recent studies on the H-2K^b MHC mutants. In *Current Trends in Histocompatibility*, edited by S. FERRONE and R. REISFELD. New York: Plenum, 1980.
18. NAIRN, R.; NATHENSON, S. G.; and COLIGAN, J. E. Isolation, characterization, and amino acid sequence studies of the cyanogen bromide fragments of the H-2D^d glycoprotein. *Eur. J. Immunol.* 10:495–503, 1980.
19. NAIRN, R.; NATHENSON, S. G.; and COLIGAN, J. E. Amino acid sequence of cyanogen bromide fragments CN-C (residues 24–98) of the mouse histocompatibility antigen H-2D^d. A comparison of the amino-terminal 100 residues of H-2D^d, D^b, K^d, and K^b reveals discrete areas of diversity. *Biochemistry*, in press.
20. NAIRN, R.; YAMAGA, K.; and NATHENSON, S. G. Biochemistry of the gene products from murine MHC mutants. *Annu. Rev. Genet.* 14:241–277, 1980.
21. KLEIN, J. H-2 mutations: their genetics and effect on immune functions. *Adv. Immunol.* 26:55–146, 1978.
22. STEINMETZ, M., et al. Three cDNA clones encoding mouse transplantation antigens: homology to immunoglobulin genes. *Cell* 24:125–134, 1981.
23. ZIMMERMAN, M.; MUMFORD, R. A.; and STEINER, D. F. (eds.). Precursor Processing in the Biosynthesis of Proteins. *Ann. N.Y. Acad. Sci.* 343:449, 1980.
24. AKIL, H., and WATSON, S. J. Beta-endorphin and biosynthetically related peptides in the central nervous system. In *Handbook of Psychopharmacology*, vol. 17, edited by L. IVERSON, S. IVERSON, and S. SNYDER. New York: Plenum, in press.
25. ABRAHAMSON, S., et al. Uniformity of radiation-induced mutation rates among different species. *Nature* 245:460–462, 1973.
26. SANKARANARAYANAN, K. Evaluation and re-evaluation of genetic radiation hazards in man. III. Other relative data and risk assessment. *Mutat. Res.* 35:387–414, 1976.
27. SCHALET, A. P., and SANKARANARAYANAN, K. Evaluation and re-evaluation of genetic radiation hazards in man. I. Interspecific comparison of estimates of mutation rates. *Mutat. Res.* 35:341–379, 1976.
28. DRAKE, J. W., and BALTZ, R. H. The biochemistry of mutagenesis. *Annu. Rev. Biochem.* 45:11–37, 1976.
29. LINDAHL, T. DNA repair enzymes acting on spontaneous lesions in DNA. In *DNA Repair Processes*, edited by W. W. NICHOLS and D. G. MURPHY. Miami: Symposia Specialists, 1977.
30. BALTZ, R. H.; BINGHAM, P. M.; and DRAKE, J. W. Heat mutagenesis in bacteriophage T4: the transition pathway. *Proc. Natl. Acad. Sci. USA* 73:1269–1273, 1976.
31. CLARK, A. M. Naturally occurring mutagens. *Mutat. Res.* 32:361–374, 1976.

32. NAGAO, M., and SUGIMURA, T. Environmental mutagens and carcinogens. *Annu. Rev. Genet.* 12:117–159, 1978.
33. HARNDEN, D. G. Viruses, chromosomes, and tumours: the interaction between viruses and chromosomes. In *Chromosomes and Cancer,* edited by J. GERMAN. New York: Wiley, 1974.
34. TAYLOR, J. H.; WOODS, P. S.; and HUGHES, W. L. The organization and duplication of chromosomes as revealed by autoradiographic studies using nitrium-labelled thymidine. *Proc. Natl. Acad. Sci. USA* 70:3395–3399, 1957.
35. SHAFER, D. A. Replicative by-pass model of sister chromatid exchanges, implications for Bloom's syndrome and Fanconi's anemia. *Hum. Genet.* 39:177–190, 1977.
36. CARTER, N. D., et al. Phosphoglucomutase polymorphism detected by isoelectric focusing: gene frequencies, evolution and linkage. *Ann. Hum. Biol.* 6:221–230, 1979.
37. VOGEL, F., and RATHENBERG, R. Spontaneous mutation in man. In *Advances in Human Genetics,* vol. 5, edited by H. HARRIS and K. HIRSCHHORN. New York: Plenum, 1975.
38. GARDNER, R. J. M. A new estimate of the achondroplasia mutation rate. *Clin. Genet.* 11:31–38, 1977.
39. OBERKLAID, F., et al. Achondroplasia and hypochondroplasia. *J. Med. Genet.* 16:140–146, 1979.
40. STEVENSON, A. C., and KERR, C. B. On the distribution of frequencies of mutation to genes determining harmful traits in man. *Mutat. Res.* 4:339–352, 1967.
41. CAVALLI-SFORZA, L. L., and BODMER, W. *The Genetics of Human Populations.* San Francisco: W. H. Freeman, 1971.
42. YASUDA, N. An average mutation rate in man. *Jpn. J. Hum. Genet.* 18:279–287, 1973.
43. NEEL, J. V. Some problems in the estimation of spontaneous mutation rates in animals and man. In *Effects of Radiation on Human Heredity.* Geneva: World Health Organization, 1957.
44. ADAMS, J. G., III, et al. Hemoglobin Indianapolis (β112[G14]arginine). *J. Clin. Invest.* 63:931–938, 1979.
45. MOHRENWEISER, H. W. Frequency of enzyme deficiency variants in erythrocytes of newborn infants. *Proc. Natl. Acad. Sci. USA* 78:5046–5050, 1981.
46. MOHRENWEISER, H. W., and NEEL, J. V. Frequency of thermostability variants and estimation of the total "rare" variant frequency in human populations. *Proc. Natl. Acad. Sci. USA* 78:5729–5733, 1981.
47. ROTHMAN, E. D.; NEEL, J. V.; and HOPPE, F. M. Assigning a probability for paternity in apparent cases of mutation. *Am. J. Hum. Genet.* 33:617–628, 1981.
48. KIMURA, M., and OHTA, T. The average number of generations until extinction of an individual mutant gene in a finite population. *Genetics* 63:701–709, 1969.
49. NEI, M. Estimation of mutation rates from rare protein variants. *Am. J. Hum. Genet.* 29:225–232, 1977.
50. ROTHMAN, E. D., and ADAMS, J. Estimation of expected number of rare alleles of a locus and calculation of mutation rate. *Proc. Natl. Acad. Sci. USA* 75:5094–5098, 1978.
51. LI, F.; NEEL, J. V.; and ROTHMAN, E. A second study of the survival of a neutral mutant in a simulated Amerindian population. *Am. Nat.* 112:83–96, 1978.

52. HARRIS, H.; HOPKINSON, D. A.; and ROBSON, E. B. The incidence of rare alleles determining electrophoretic variants: data on 43 enzyme loci in man. *Ann. Hum. Genet.* 37:237–253, 1974.

53. SATOH, C., et al. Genetic effects of atomic bombs. *Proceedings of the Sixth International Congress of Human Genetics,* in press.

54. NEEL, J. V.; MOHRENWEISER, H. W.; and MEISLER, M. H. Rate of spontaneous mutation at human loci encoding protein structure. *Proc. Natl. Acad. Sci. USA* 77:6037–6041, 1980.

55. ATLAND, K. Monitoring for changing mutation rates using blood samples submitted for PKU screening. *Proceedings of the Sixth International Congress of Human Genetics,* in press.

56. MUKAI, T., and COCKERHAM, C. C. Spontaneous mutation rates at enzyme loci in *Drosophila melanogaster. Proc. Natl. Acad. Sci. USA* 74:2514–2517, 1977.

57. NEEL, J. V. In quest of better ways to study human mutation rates. In *Human Mutation: Biological and Population Aspects,* edited by E. HOOK. New York: Academic Press, in press.

58. NEEL, J. V., and ROTHMAN, E. D. Indirect estimates of mutation rates in tribal Amerindians. *Proc. Natl. Acad. Sci. USA* 75:5585–5588, 1978.

59. BHATIA, K. K.; BLAKE, N. M.; and KIRK, R. L. The frequency of electrophoretic variants in Australian aborigines and indirect estimates of mutation rate. *Am. J. Hum. Genet.* 31:731–740, 1979.

60. BHATIA, K. K., et al. The frequency of private electrophoretic variants and indirect estimates of mutation rate in Papua New Guinea. *Am. J. Hum. Genet.* 33:112–122, 1981.

61. NEEL, J. V., and ROTHMAN, E. D. Is there a difference between human populations in the rate with which mutation produces electrophoretic variants? *Proc. Natl. Acad. Sci. USA* 78:3108–3112, 1981.

62. HEWETT-EMMETT, D.; CZELUSNIAK, J.; and GOODMAN, M. The evolutionary relationships of the enzymes involved in blood coagulation and hemostasis. *Ann. N.Y. Acad. Sci.* 370:511–527, 1981.

63. WILLIAMS, C. H.; ARSCOTT, L. D.; and SCHULZ, G. E. Amino acid sequence homology between pig heart lipoamide dehydrogenase and human erythrocyte glutathione reductase. *Proc. Natl. Acad. Sci. USA,* in press.

64. HOROWITZ, M. H. On the evolution of biochemical synthesis. *Proc. Natl. Acad. Sci. USA* 31:153–157, 1945.

65. LEWIS, E. B. Pseudoallelism and gene evolution. *Cold Spring Harbor Symp. Quant. Biol.* 16:159–174, 1951.

66. OHNO, S. *Evolution by Gene Duplication.* New York: Springer, 1970.

67. ZUCKERKANDL, E. The appearance of new structures and functions in problems during evolution. *J. Mol. Evol.* 7:1–57, 1975.

68. DAYHOFF, M. O., et al. Evolution of sequences within protein superfamilies. *Naturwissenschaften* 62:154–161, 1975.

69. GOOSSENS, M., et al. Triplicated α-globin loci in humans. *Proc. Natl. Acad. Sci. USA* 77:518–521, 1980.

70. ZIMMER, E. A., et al. Rapid duplication and loss of genes coding for the α chains of hemoglobin. *Proc. Natl. Acad. Sci. USA* 77:2158–2162, 1980.

71. NISHIOKA, Y.; LEDER, A.; and LEDER, P. Unusual α-globin-like gene that has cleanly lost both globin intervening sequences. *Proc. Natl. Acad. Sci. USA* 77:2806–2809, 1980.

72. JONES, R. W., et al. Major rearrangement in the human β-globin gene cluster. *Nature* 291:39–44, 1981.

73. STURTEVANT, A. H. The effects of unequal crossing over at the Bar locus in *Drosophila*. *Genetics* 10:117–147, 1925.
74. BRIDGES, C. B. The Bar "gene" a duplication. *Science* 83:210–211, 1936.
75. GELBART, W. M., and CHOVNICK, A. Spontaneous unequal exchange in the rosy region of *Drosophila melanogaster*. *Genetics* 92:849–859, 1979.
76. BLOOM, A. D., et al. Chromosomal breakage in leucocytes of South American Indians. *Cytogenet. Cell Genet.* 12:175–186, 1973.
77. AWA, A. A. Personal communication.
78. CAIRNS, J. The origin of human cancers. *Nature* 289:353–357, 1981.
79. UNITED NATIONS SCIENTIFIC COMMITTEE ON THE EFFECTS OF ATOMIC RADIATION. *Sources and Effects of Ionizing Radiation*. New York: United Nations, 1977.
80. NEEL, J. V. Mutation and disease in man. *Can. J. Genet. Cytol.* 20:295–306, 1978.
81. SCHULL, W. J.; OTAKE, M.; and NEEL, J. V. A reappraisal of the genetic effects of the atomic bombs: summary of a thirty-four year study. *Science* 213:1220–1227, 1981.
82. NICHOLS, W. W., and MURPHY, D. G. (eds.). *DNA Repair Processes*. Miami: Symposia Specialists, 1977.
83. HANAWALT, P. C.; FRIEDBERG, E. C.; and FOX, C. F. *DNA Repair Mechanisms*. New York: Academic Press, 1978.
84. HANAWALT, P. C., et al. DNA repair in bacteria and mammalian cells. *Annu. Rev. Biochem.* 48:783–836, 1979.
85. GENEROSO, W. M.; SHELBY, M. D.; and DE SERRES, F. J. (eds.). *Symposium on DNA Repair and Mutagenesis in Eukaryotes*. New York: Plenum, 1980.
86. HOWARD-FLANDERS, P. DNA repair. *Annu. Rev. Biochem.* 37:175–200, 1968.
87. SETLOW, R. B., and SETLOW, J. K. Effects of radiation on polynucleotides. *Annu. Rev. Biophys. Bioeng.* 1:293–346, 1972.
88. DEPAMPHILIS, M. L., and WASSARMAN, P. M. Replication of eukaryotic chromosomes: a close-up of the replication fork. *Annu. Rev. Biochem.* 49:627–666, 1980.
89. GENEROSO, W. M., et al. Genetic lesions induced by chemicals in spermatozoa and spermatids of mice are repaired in the egg. *Proc. Natl. Acad. Sci. USA* 76:435–437, 1979.
90. BRANDRIFF, B., and PEDERSEN, R. A. Repair of the ultraviolet-irradiated male genome in fertilized mouse eggs. *Science* 211:1431–1433, 1981.
91. RUSSELL, W. L. Effect of the interval between irradiation and conception on mutation frequency in female mice. *Proc. Natl. Acad. Sci. USA* 54:1552–1557, 1965.
92. BERNSTEIN, C. Why are babies young? Meiosis may prevent aging of the germ line. *Perspect. Biol. Med.* 22:539–544, 1979.
93. FAWCETT, D. W. The cell biology of gametogenesis in the male. *Perspect. Biol. Med.* 22:556–573, 1979.
94. LIEBERMAN, M. W., et al. The role of chromatin structure in DNA repair in human cells damaged with chemical carcinogens and ultraviolet radiation. In *Environmental Carcinogenesis*, edited by P. EMMELOT and E. KRIEK. Amsterdam: Elsevier–North-Holland, 1979.
95. CLEAVER, J. E. Defective repair replication of DNA in Xeroderma pigmentosum. *Nature* 218:652–656, 1968.
96. CLEAVER, J. E., and BOOTSMA, D. Xeroderma pigmentosum: biochemical genetic characterization. *Annu. Rev. Genet.* 9:19–38, 1975.
97. HASHEM, N., et al. Clinical characteristics, DNA repair, and complementa-

tion groups in xeroderma pigmentosum patients from Egypt. *Cancer Res.* 40:13–18, 1980.

98. SLIGHTOM, J. L.; BLECHEL, A. E.; and SMITHIES, O. Human fetal $^G\gamma$- and $^A\gamma$-globin genes: complete nucleotide sequences suggest that DNA can be exchanged between these duplicated genes. *Cell* 21:627–638, 1980.

99. OLLO, R., et al. Comparison of mouse immunoglobulin γ2a and γ2b chain genes suggests that exons can be exchanged between genes in a multigenic family. *Proc. Natl. Acad. Sci. USA* 78:2442–2448, 1981.

100. ARNHEIM, N., et al. Molecular evidence for genetic exchanges among ribosomal genes on nonhomologous chromosomes in man and apes. *Proc. Natl. Acad. Sci. USA* 77:7323–7327, 1980.

101. KRYSTAL, M., et al. Human nucleolus organizers on nonhomologous chromosomes can share the same ribosomal gene variants. *Proc. Natl. Acad. Sci. USA* 78:5744–5748, 1981.

102. TOBARI, Y. N., and KOJIMA, K. A study of spontaneous mutation rates at ten loci detectable by starch gel electrophoresis in *Drosophila melanogaster. Genetics* 70:397–403, 1972.

103. VOELKER, R. A.; SCHAFFER, H. E.; and MUKAI, T. Spontaneous allozyme mutations in *Drosophila melanogaster:* rate of occurrence and nature of mutants. *Genetics* 94:961–968, 1980.

104. YAMAGUCHI, O., and MUKAI, T. Variation of spontaneous occurrence rates of chromosomal aberrations in the second chromosomes of *Drosophila melanogaster. Genetics* 78:1209–1221, 1974.

105. MAGNI, G. E., and VON BORSTEL, R. C. Different rates of spontaneous mutation during mitosis and meiosis in yeast. *Genetics* 47:1097–1108, 1962.

106. HART, R. W.; SACHER, G. A.; and HOSKINS, T. L. DNA repair in a short- and a long-lived rodent species. *J. Gerontol.* 34:808–817, 1979.

107. HART, R. W., and DANIEL, F. B. Genetic stability in vitro and in vivo. In *Advances in Pathobiology,* vol. 7, edited by C. BOREK, C. M. FENOGLIO, and D. W. KING. New York: Thieme-Stratton, 1980.

108. FISHER, R. A. *The Genetical Theory of Natural Selection.* Oxford: Clarendon, 1930.

109. KIMURA, M., and OHTA, T. *Theoretical Aspects of Population Genetics.* Princeton, N.J.: Princeton Univ. Press, 1971.

110. THOMPSON, E. A., and NEEL, J. V. The probability of founder effect in a tribal population. *Proc. Natl. Acad. Sci. USA* 75:1442–1445, 1978.

111. NEEL, J. V. The genetic structure of primitive human populations. *Jpn. J. Hum. Genet.* 12:1–16, 1967.

112. SMOUSE, P. E.; VITZHUM, V. J.; and NEEL, J. V. The impact of random and lineal fission on the genetic divergence of small human groups: a case study among the Yanomama. *Genetics* 98:179–197, 1981.

113. OHTA, T. Slightly deleterious mutant substitutions in evolution. *Nature* 246:96–97, 1973.

114. OHTA, T. Role of very slightly deleterious mutations in molecular evolution and polymorphism. *Theor. Popul. Biol.* 10:254–275, 1976.

115. LI, W.-H.; GOJOBORI, T.; and NEI, M. Pseudogenes as a paradigm of neutral evolution. *Nature* 292:237–239, 1981.

116. MIYATA, T., and HAYASHIDA, H. Extraordinarily high evolutionary rate of pseudogenes: evidence for the presence of selective pressure against changes between synonymous codons. *Proc. Natl. Acad. Sci. USA* 78:5739–5743, 1981.

117. VANIN, E. F., et al. A mouse α-globin-related pseudogene lacking intervening sequences. *Nature* 286:222–226, 1980.

118. APIRION, D. Intervening sequences: how did they come about and why are they there? *Perspect. Biol. Med.,* in press.
119. RUCKNAGEL, D. L., and NEEL, J. V. The hemoglobinopathies. In *Progress in Medical Genetics,* edited by A. G. STEINBERG. New York: Grune & Stratton, 1961.
120. MILLER, L. H., et al. The resistance factor to *Plasmodium vivax* in blacks: the duffy—blood-group genotype Fy-Fy. *N. Engl. J. Med.* 295:302–304, 1976.
121. HARRIS, H. *The Principles of Human Biochemical Genetics,* 3d rev. ed. Amsterdam: Elsevier/North-Holland Biomedical Press, 1980.
122. MCCONKEY, E. H.; TAYLOR, B. J.; and PHAN, D. Human heterozygosity: a new estimate. *Proc. Natl. Acad. Sci. USA* 76:6500–6504, 1979.
123. WALTON, K. E.; STYER, D.; and GRUENSTEIN, E. I. Genetic polymorphism in normal human fibroblasts as analyzed by two-dimensional polyacrylamide gel electrophoresis. *J. Biol. Chem.* 254:7951–7960, 1979.
124. SMITH, S. C.; RACINE, R. R.; and LANGLEY, C. H. Lack of genic variation in the abundant proteins of human kidney. *Genetics* 96:967–974, 1980.
125. WANNER, L. A.; NEEL, J. V.; and MEISLER, M. H. Separation of allelic variants by two-dimensional electrophoresis. *Am. J. Hum. Genet.,* 34:209–215.
126. JEFFREYS, A. F. DNA sequence variants in the $^G\gamma$, $^A\gamma$-, δ-, and β-globin genes of man. *Cell* 18:1–10, 1979.
127. TUAN, D., et al. Restriction endonuclease mapping of the human γ globin gene loci. *Nucleic Acids Res.* 6:2519–2544, 1979.
128. EWENS, W. J.; SPIELMAN, R. S.; and HARRIS, H. Estimation of genetic variation at the DNA level from restriction endonuclease data. *Proc. Natl. Acad. Sci. USA* 78:3748–3750, 1981.
129. MULLER, H. J. Our load of mutations. *Am. J. Hum. Genet.* 2:111–176, 1950.
130. MORTON, N. E.; CROW, J. F.; and MULLER, H. J. An estimate of the mutational damage in man from data on consanguineous marriages. *Proc. Natl. Acad. Sci. USA* 42:855–863, 1956.
131. HALDANE, J. B. S. The cost of natural selection. *J. Genet.* 55:511–524, 1957.
132. EDWARDS, J. H. Allelic association in man. In *Population Structure and Genetic Disorders,* edited by A. W. ERIKSSON, H. K. FORSIUS, H. R. NEVANLINNA, et al. New York: Academic Press, 1980.
133. KIMURA, M. Genetic variability maintained in a finite population due to mutational production of neutral and nearly neutral isoalleles. *Genet. Res.* 11:247–269, 1968.
134. BOYER, S. H., et al. The proportion of all point mutations which are unacceptable: an estimate based on hemoglobin amino acid and nucleotide sequences. *Can. J. Genet. Cytol.* 20:111–137, 1978.
135. MILKMAN, R. D. Heterosis as a major cause of heterozygosity in nature. *Genetics* 55:493–495, 1967.
136. KING, J. L. Continuously distributed factors affecting fitness. *Genetics* 55:483–492, 1967.
137. SVED, J. A.; REED, T. E.; and BODMER, W. F. The number of balanced polymorphisms that can be maintained in a natural population. *Genetics* 55:469–481, 1967.
138. CROW, J. F., and KIMURA, M. Efficiency of truncate selection. *Proc. Natl. Acad. Sci. USA* 76:396–399, 1979.
139. NEEL, J. V. Some currently neglected aspects of human biochemical genetics. In *Genetics Today, Proceedings of the Eleventh International Congress on Genetics.* Oxford: Pergamon, 1964.
140. NEEL, J. V. On being headman. *Perspect. Biol. Med.* 23:277–294, 1980.

141. MONOD, J. *Le Hasard et la nécessité.* Paris: Éditions du Seuil, 1980.
142. WHEELER, J. A. Fontiers of time. In *Problems in the Foundations of Physics,* Proceedings of the International School of Physics "Enrico Fermi," course 72, edited by G. TORALDO DI FRANCIA. Amsterdam: North-Holland, 1979.
143. AUSTIN, C. R. Pregnancy losses and birth defects. In *Reproduction in Mammals,* vol. 2, edited by C. R. AUSTIN and R. V. SHORT. Cambridge: Cambridge Univ. Press, 1972.
144. ROBERTS, C. J., and LOWE, C. R. Where have all the conceptions gone? *Lancet* 1:498–499, 1975.
145. JACOBS, P. A. The load due to chromosome abnormalities in man. In *The Role of Natural Selection in Human Evolution,* edited by F. M. SALZANO. Amsterdam: North-Holland, 1975.
146. CARR, D. H., and GEDEON, M. M. Population cytogenetics of human abortuses. In *Population Cytogenetics,* edited by E. B. HOOK and I. H. PORTER. New York: Academic Press, 1977.
147. JACOBS, P. A. Mutation rates of structural chromosome rearrangements in man. *Am. J. Hum. Genet.* 33:44–54, 1981.
148. BEER, A. E., et al. Major histocompatibility complex antigens, maternal and paternal immune responses, and chronic habitual abortions in humans. *Am. J. Obstet. Gynecol.,* in press.
149. FORD, C. E. Gross genome imbalance in mouse spermatozoa: does it influence the capacity to fertilize? In *Proceedings of the International Symposium on the Genetics of the Spermatozoon,* edited by R. A. BEATTY and S. GLUECKSOHN-WAELSCH. Edinburgh and New York: By the editors.
150. ELDRIDGE, N., and GOULD, S. J. Punctuated equilibria: an alternative to phyletic gradualism. In *Models in Paleobiology,* edited by T. J. M. SCHOPF. San Francisco: Freeman, Cooper, 1972.
151. GOULD, S. J. *The Panda's Thumb.* New York: Norton, 1980.
152. DE VRIES, H. *Die Mutationstheorie,* Bds. 1 and 2. Leipzig, 1901, 1903.
153. GOLDSCHMIDT, R. *The Material Basis of Evolution.* New Haven, Conn.: Yale Univ. Press, 1940.
154. McCLINTOCK, B. Induction of instability at selected loci in maize. *Genetics* 38:579–599, 1953.
155. THOMPSON, J. N., JR., and WOODRUFF, R. C. Mutator genes—pacemakers of evolution. *Nature* 274:317–321, 1978.
156. THOMPSON, J. N., and WOODRUFF, R. C. Increased mutation in crosses between geographically separated strains of *Drosophila melanogaster. Proc. Natl. Acad. Sci. USA* 77:1059–1062, 1980.
157. SVED, J. A. The "Hybrid Dysgenesis" syndrome in *Drosophila melanogaster. Bioscience* 29:659–664, 1979.
158. GREEN, M. M. Transposable elements in *Drosophila* and other *Diptera. Annu. Rev. Genet.* 14:109–120, 1980.
159. CRONIN, J. E., et al. Tempo and mode in hominid evolution. *Nature* 292:113–122, 1981.
160. ESCHEL, I. Clone selection and the evolution of modifying features. *Theor. Popul. Biol.* 4:196–208, 1973.
161. PAI, G. S., et al. Complex chromosome rearrangements: report of a new case and literature review. *Clin. Genet.* 18:436–444, 1980.

ON LUMPERS AND SPLITTERS, OR THE NOSOLOGY OF GENETIC DISEASE

VICTOR A. McKUSICK, M.D. *

As Knut Faber pointed out in his *Nosography* [1], Mendelism contributed much to nosology (the study of the classification of disease) by focusing attention on specific entities. Bacteriology, with its similar emphasis on specific etiology and specific entities, had a comparable effect. One need only recall that a century ago, symptoms such as jaundice, dropsy, and anemia were viewed as entities, in much of medicine at least, to realize the nosologic contributions of Mendelism and bacteriology. But the main object here is to review some contemporary problems in the nosology of genetic disease.

The two leading principles in genetic nosology are pleiotropism and genetic heterogeneity. "Pleiotropism" means multiple effects of a single etiologic factor, for example, a single gene in the genetical use to which the term is usually put. Pleiotropism is the usual, but not the sole, basis for hereditary syndromes. Linkage, that is close situation on the same chromosome of genes for the separate manifestations, is a theoretically unsatisfactory and as yet unproved explanation for Mendelizing syndromes. Because of crossing over, which in time separates even closely situated loci, linkage produces no permanent association of traits in a population.

"Genetic heterogeneity" refers to the existence of two or more fundamentally distinct entities with essentially one and the same phenotype. Nosologists in all fields tend to be either "lumpers" or "splitters" (Fig. 1).[1]

* Professor of medicine, The Johns Hopkins University School of Medicine, Baltimore, Maryland 21205. This paper was presented in part at the First Conference on the Clinical Delineation of Birth Defects, The Johns Hopkins Hospital, May 20, 1968. An article on this subject appears in *Clinical Delineation of Birth Defects*, Vol. V, No. 1, January, 1969.

[1] The resemblance of the splitter to the revered Civil War president is a clue to my bias. The cartoons also indicate that splitting is harder work than lumping.

Fig. 1

Fig. 2.—Little People of America convention, Baltimore, July 21, 1968

Psychologists tell us that we find it easier to recognize similarities than differences. Hence a natural tendency to lumping exists. However, geneticists are forced to be splitters because of their recurrent encounters with genetic heterogeneity in recent years.

The principle of genetic heterogeneity is illustrated by a consideration of Figure 2, a photograph of the national annual meeting of an organization of dwarfs and midgets. Clearly, if one were to undertake a study of the genetics of short stature, or of the physiologic defects underlying short stature, or of other aspects including even the psychologic and sociologic impact of short stature, separation of separate categories would be necessary. In this absurdly obvious example, genetic heterogeneity is evident even to the casual observer, and the experienced eye can pick out more than a dozen distinct entities. (Of course, heterogeneity can also involve non-genetic bases for the phenotype, so called phenocopies.)

Pleiotropism is "many from one"—multiple phenotypic features from one etiologic factor, one gene. Genetic heterogeneity is "one from many" —one and the same or almost the same phenotype from several different etiologic factors.

In medical genetics, awareness of pleiotropism and genetic heterogeneity have developed particularly in the last twenty years, the second a bit later than the first. Looking back on my own work in medical genetics, I recognize that pleiotropism was a leading concern in its earlier stages when I began working, for example, with the Marfan syndrome. Latterly, genetic heterogeneity has become increasingly the focus—for example, in studies of the genetic mucopolysaccharidoses and the separation of homocystinuria from the Marfan syndrome. Successive editions of *Heritable Disorders of Connective Tissue* [2] illustrate this trend in nosology.

In an earlier period, medical genetics suffered from excessive and improper splitting, which was at least partly inadvertent, arising as it did from specialization in medicine. Seeing cases of one and the same entity, physicians in different specialties were concerned mainly with features falling within their particular purview and often failed to recognize that the feature of particular interest to them was merely part of a syndrome. Examples are angioid streaks of the ophthalmologist and pseudoxanthoma elasticum of the dermatologist. Thus, medical geneticists have been, and continue to be, lumpers to the extent that they pull together the pleiotropic manifestations of genetic syndromes. (Indeed, medical geneticists can be the generalists of modern medicine.) As it was put by Jonathan Hutchinson (1828–1913), a pioneer syndromologist [3], "We must analyze, and seek to interpret partnerships in disease."

The Genetic Entity

What constitutes an entity? How does one identify genetic heterogeneity? Few would quibble with the statement that the phenotype resulting primarily from a specific and unitary factor is an entity. Thus, delineation of genetic entities is on surest ground if a fundamental biochemical defect or chromosomal fault is identified. Trisomy D, trisomy E, and *cri-du-chat* syndromes are now recognizable with a high order of accuracy on clinical grounds alone. Demonstration of a specific etiopathogenetic basis permits separation of groups of cases in which the frequency of clinical features can be determined and with which new cases can be compared, like an "unknown" with a "standard." One can then make use of Bayesian methods by which individual manifestations are given weightings according to their frequency in the specific entity on the one hand, and in persons without the specific entity on the other hand. Of course, Down's syndrome was diagnosable with considerable accuracy before the era of chromosomology, and Walker's scoring method for use in connection with dermatoglyphic diagnosis was already developed, but the specificity of the diagnostic technique was improved by working from karyotypically proven cases. Greater difficulty is encountered in delineating entities among congenital disorders when the etiology is not known or when a unitary biochemical defect has not been shown and, consequently, unitary etiopathogenesis is not certain. Such is the problem, for example, with the Cornelia de Lange syndrome, the Turner phenotype with normal karyotype, Rubinstein's syndrome, Prader-Willi syndrome, etc. In these, overdiagnosis is a risk; unless the clinical features are unusually distinctive, criteria are likely to be either too rigorous or too inclusive and a happy middle ground is not easily attained. When one does not have a specifically diagnostic laboratory test or other diagnostic means independent of the clinical phenotype, one cannot list the frequency of occurrence of specific features without falling into circular reasoning. (Small chromosomal aberrations beyond the limits of resolution of presently existing cytogenetic methods seem likely as the basis of many syndromes which show rather remarkable case-to-case reproducibility but which do not show sufficiently strong familial aggregation to be considered Mendelian and do occasionally show detectable chromosomal changes. But until, by improved methods, small aberrations are actually demonstrated, such is only speculation.) At the stage when homocystinuria had not yet been separated from the Marfan syndrome, what was the usefulness of tabulating precise percentage figures for the frequency of various clinical features? Even today the lack of a specific diagnostic test in the Marfan syndrome impedes delineation at the mild end of the spectrum of severity, and the possibility of residual heterogeneity limits the value of quantitaive statements on the percentage frequency of components.

How Does One Identify Genetic Heterogeneity?

The methods for recognizing genetic heterogeneity, the existence of separate entities in what has been considered a single phenotype, are mainly three: clinical, genetic, and biochemical.

CLINICAL METHODS

Differences in the phenotype are the most treacherous basis for decision on genetic heterogeneity. After all, similarity of phenotype is what leads to mistaken impressions of homogeneity (or unity) in the first place. Furthermore, genetic disorders, particularly those inherited as autosomal dominants, vary widely in severity, presumably in large part because of effects of modifier genes. Some aspects of a syndrome may be missing in the individual case. For example, among sibs who inherited the Marfan gene from a parent, one may have all the cardinal features of the syndrome but another may not show ectopia lentis. On the average, 50 per cent of the genes of sibs are identical by descent. Differences in the other half of the genome account for differences in expression of the single major gene, just as placing a pleiotropic gene on different genetic backgrounds in mice can result in suppression of some aspects of the syndrome produced by that gene [4].

Clinical differences in phenotype can come to the support of other methods for recognizing genetic heterogeneity. For example, when pedigree pattern suggested the existence of an X-linked form of mucopolysaccharidosis (Hunter syndrome), analysis of phenotype in comparison with the autosomal recessive type showed several phenotypic differences, particularly absence of corneal clouding [2].

Analysis of phenotype can be most reliably used to delineate an entity when an extensive kindred with many persons affected with an autosomal dominant is available for study, or a deme[2] (really an extended kindred) with many cases of an autosomal recessive. If the disorder is rare (as are most Mendelizing disorders), then one can be confident that the identical gene is responsible for all cases in a given kindred or in a given deme. The range of variability in the phenotype of one entity can be more securely determined than is possible on the basis of a large number of randomly ascertained families, each with only a few affected members. In the latter situation circularity enters because of the criteria by which families were selected in the first place.

Let me illustrate with example of dominants. The view that familial multiple polyposis of the colon and Gardner's syndrome are separate entities has been challenged [6] by surgeons, who, focusing mainly on individual cases, feel that they represent a spectrum with no cutoff point between two distinct entities. The strongest support for distinctness comes from large family studies: The most extensively involved family of multiple colonic polyposis yet studied is, to my knowledge, that of Asman and Pierce [7] with about ninety affected persons, not one of whom had extra bowel tumors of the type seen in Gardner's syndrome. The most extensively involved family with Gardner's syndrome is, to my knowledge, the one originally studied by Eldon Gardner. More than 20 persons [8] in this kindred have the entity, although at the time of study a few seemed to have only polyps and others seemed to have only extra-intestinal tumors.

Another dramatic example is provided by families with amyloid neuropathy. While he was a research fellow with me from 1965 to 1967, Mahloudji [9] studied a form of amyloid neuropathy which was initially ascertained in eleven seemingly unrelated kindreds living mainly in Maryland and in neighboring states. Genealogic sleuthing showed that all eleven kindreds were descended from a couple who had immigrated from Germany in the eighteenth century. The giant kindred contained at least 145 persons who were affected according to reliable information or who must have had the gene because of their position in the genealogy. Personal study of over fifty affected persons from this single kindred leaves no doubt that the disorder is distinct from the amyloid neuropathy observed in equally numerous cases in Portugal by Andrade [10]. Table 1

[2] "Deme" is defined by George P. Murdock [5] as a local endogamous community, or consanguineal kin group.

reviews the phenotypic differences of the two types which are given either an eponymous or a geographic name; the ailment in Mahloudji's enormous kindred appears to be identical with that studied in Indiana by Rukavina and his colleagues [11] in a kindred of Swiss extraction.

The range of variability of the ordinarily very rare Ellis–van Creveld syndrome [12] ("six-fingered dwarfs") has been gauged better from analysis of the sixty-one cases which have up to now been recognized in a single inbred group, the Lancaster County (Pa.) Amish. The fact, for example, that one-third die before age six months cannot be determined from analysis of reported cases (which also number about sixty) because of biases of ascertainment. Sometimes a syndrome apparently inherited

TABLE 1

DIFFERENTIAL FEATURES OF TWO TYPES OF
HEREDITARY AMYLOID NEUROPATHY

Characteristic	Type I of Andrade (Portuguese type)	Type II of Rukavina (Indiana type)
Age of onset..........	3d decade	5th decade
Site of onset..........	Feet	Hands
G.I. complaints........	Common	Rare
Impotence and sphincter disturbance..........	Common	Rare
Foot ulcers..........	Common	Rare
Duration of illness......	4–12 years	14–40+ years

as a recessive is observed in two or more sibs. Especially if the parents are related, the possibility comes up that the affected children fell heir to two unrelated recessives rather than the disorders being a pleiotropic syndrome produced by a single mutant gene. (Close linkage of two recessives would increase the simulation of pleiotropism.) If a large inbred group has many individuals showing the coincidence of the two manifestations, single-gene etiology is strongly supported. Cartilage-hair hypoplasia [13] has been delineated by the fact that over eighty Amish persons (by present count) have the associated hair and skeletal abnormality. Similarly, for a dominant such as the Peutz-Jeghers syndrome, in which the coincidence of such different features as melanin spots and intestinal polyposis is difficult to explain on physiologic grounds, the study of an extensively affected kindred (e.g., reference [14]) lends strong support to its single-gene basis. If separate genes were responsible for the two features, they would become separated at some point in a large kindred (even if closely linked) through the process of genetic crossing-over. This is, then, essentially a genetic approach to definition of a syndromal entity—which brings us to this aspect.

GENETIC METHODS

Genetic methods for identifying heterogeneity include mode of inheritance, allelism test, and linkage study. A number of phenotypes are known which is different families exhibit autosomal dominant, autosomal recessive, and X-linked patterns of inheritance. No one could consider the autosomal and X-linked forms anything but separate entities since the mode of inheritance indicates that the genes are located in different parts of the genome. The distinctness of the autosomal dominant and autosomal recessive forms is likewise defensible in many instances. I have had an opportunity to study an X-linked spastic paraplegia [15] and an autosomal dominant form [16]. Autosomal recessive spastic paraplegia, with and without other features, has been observed by my colleagues and me [17, 18] and by others. Retinitis pigmentosa and peroneal muscular atrophy were examples of triple mode of inheritance that Allan [19] cited when he pointed out the principle that is sometimes called Lenz's law: The autosomal dominant form is usually the mildest and the autosomal recessive form the most severe, with the X-linked recessive form occupying an intermediate position as to severity.

When two individuals homozygous at the same locus have children, all their children are likewise homozygous at that locus. Deaf-mutes frequently marry, and in some such families all the children are also deaf-mutes; the parents must have an allelic form of deafness. But in others, although both parents seem from the evidence in their family trees to have a recessively inherited form of deaf-mutism and although the disorder in the parents is phenotypically indistinguishable, all children have normal hearing. Presumably the parents are homozygous at separate loci; they have non-allelic forms of recessive congenital deafness. Chung, Robinson, and Morton [20] concluded that homozygosity at any one of many loci may result in congenital deafness of a phenotypically indistinguishable type—a large amount of genetic heterogeneity. I find this conclusion plausible because usually, when one studies an inbred group for specific recessives, one finds deaf-mutism whatever else may or may not be present. We [21] have observed fourteen cases of deaf-mutism among the Amish of Lancaster County and thirty-eight cases of deaf-mutism among the conservative Mennonites of the same county. In one instance, an Amish deaf-mute married a Mennonite deaf-mute. All three of their children hear normally, a fact which suggests non-allelic recessive deafness in the two religious isolates. Among the cases of isolated growth-hormone deficiency, we have identified two genetically distinct, although phenotypically indistinguishable, types by the fact that an affected man and woman had two normal children [22]. Heterogeneity in recessive albinism [23] and in recessive congenital amaurosis, or retinal aplasia [24], has been demonstrated by the same approach.

Linkage, the demonstration by family studies that two genes are reasonably close together on the same chromosome pair, can demonstrate heterogeneity. The gene locus for one form of elliptocytosis is closely situated to the Rh blood-group locus. On the other hand, a form of elliptocytosis as yet phenotypically indistinguishable is determined by a gene at a locus which shows no linkage with the Rh locus [25]. Even if other methods had not revealed the heterogeneity in X-linked hemophilia (hemophilias A and B), sufficiently extensive linkage studies might have uncovered the heterogeneity: The hemophilia A locus is closely linked to that for glucose-6-phosphate dehydrogenase and those for color blindness, but the hemophila B locus is so far removed that independent assortment occurs through crossing over [26].

BIOCHEMICAL METHODS

Demonstration of biochemical differences is in man the most critical and definitive means of establishing genetic heterogeneity. The biochemical changes observed may be useful even though moderately far removed from primary gene action. For example, the pattern of urinary excretion of mucopolysaccharide in the mucopolysaccharidoses helps distinguish several types [2] (e.g., types III and IV from each other and from types I and II), although some are distinguished with difficulty on this ground (e.g., types I and II). Differentiation is surer when a biochemical defect closer to primary gene action is demonstrable (e.g., an enzyme defect). Firm delineation of multiple forms of glycogen storage disease [27] and of hereditary non-spherocytic hemolytic anemia [27] is now possible on the basis of demonstrated defects in different enzymes.

Even when deficiency in the activity of a specific enzyme is demonstrable, residual heterogeneity may exist. This has been shown to be the case with non-spherocytic hemolytic anemia due to glucose-6-phosphate dehydrogenase deficiency [27] and that due to pyruvate kinase deficiency [28]. It has also been shown in galactosemia [29] and in phenylketonuria (or at least phenylalaninemia) [30]. It is suspected or tentatively demonstrated in homocystinuria and several other conditions. This heterogeneity can be allelic; mutation may have occurred at different points in the cistron so that amino acid substitution occurs at different places in the enzyme polypeptide, resulting in several forms of mutant enzyme each with a deficiency of enzyme activity but with different reaction characteristics and different degrees of deficiency. Following the precedent of cystathioninuria [31] and suggestive differences in response to pyridoxine in homocystinuria, one might think that some mutations in enzymes affect mainly the combination or collaboration of the enzyme with the cofactor. The net effect of either type of mutation would be deficiency of enzyme activity.

Thus, three main approaches are available for delineating separate genetic entities by uncovering genetic heterogeneity—clinical, genetic, and biochemical. The complementarity of hemophilias A and B—blood from patients with hemophilia A corrects the coagulation defect in patients with hemophilia B, and vice versa—is a physiological demonstration of heterogeneity which, I suppose, can be included in the clinical method. Morphologic methods used, for instance, in sorting out genetic disorders of the central nervous system, can perhaps be viewed also as a special type of clinical analysis. Cell-culture techniques are supplementing in a powerful manner both morphologic and biochemical approaches to the unraveling of heterogeneity (e.g., reference [32]).

Practical Difficulties in Entity Recognition

The practical problems associated with recognition of genetic entities are considerable. The entity may have a certain Gestalt which, although unmistakable, is scarcely definable on the basis of a single case, and it may be difficult for even the most skilled word artist to convey the features to others who may have seen identical cases. In this field a photograph can be worth a thousand words. (The Williams facies of the hypercalcemia—supravalvar stenosis—mental retardation syndrome is a good example.) The entities are individually rare. A single case may not impress; it may take two to arouse suspicion that a discrete entity is involved, and three may be necessary to convince. Of course, if two or more sibs are identically affected, this helps, but with the small families now the rule, a majority of families will have only one affected child (if the condition is an autosomal recessive), and of course in those conditions which are sublethal dominants, all or almost all cases result from fresh mutation and are sporadic.

Joseph Warkany of Cincinnati points out to me that among the several hundred thousand patients in institutions for the mentally retarded in this country many as yet undelineated syndromes exist, each with phenotypically more or less striking features, but, because each is an isolated case, the physician does not know what to make of it. The problem is a logistic one, how to get together the two or three or more cases which may exist in widely separated institutions. Once an entity is described on the basis of two or three cases, other cases tend to be found rather quickly. Many persons had seen a case of microcephaly, beaked nose and broad thumbs and great toes, but had not known what to make of it until Rubinstein and Taybi put the syndrome together [33].

Inbred Groups and Nosology

The usefulness of inbred groups in defining the phenotypic limits of an entity because of the high probability that the same gene underlies all cases has already been discussed. Furthermore, delineation of "new," that is, hitherto unrecognized, genetic entities with recessive inheritance is enhanced in inbred groups. At least three factors account for increased visibility of recessives in such groups: (1) Inbreeding increases the number of homozygotes. (2) These groups usually have large families. (3) The "groupness" increases conspicuousness of disorders.

Contrary to a prevalent misconception—that inbreeding causes a build-up of "bad" genes in a population, consanguinity per se does not change the frequency of *genes* in a population. It does change the proportion of *genotypes*, increasing the number of homozygous individuals at the expense of heterozygotes.

Intuitively one can appreciate the fact that the greater the number of children that are produced by two parents heterozygous for the same recessive gene, the greater is the chance that two or more will be affected with a given entity. This is shown mathematically in Table 2. Of families with two heterozygous parents, ascertainable because at least one affected child has been born, 86 per cent will have only one child affected if there are only two children in the family. Of families with three children, 73 per cent will have only one child affected. Four-child families will in

62 per cent of instances have only one child affected. Not until six-child families are reached will more than half the families have more than one affected child. When we see in a sibship a single case of an unusual clinical picture which seems to represent an entity, we cannot be as certain as we can be if at least two sibs are identically affected.

The "groupness" increases conspicuousness. A deme is effectively a large kindred, so that the considerations are similar to those just mentioned. But, in addition, sociologic uniqueness of the group increases visibility. The occurrence of multiple cases of an unusual clinical picture in

TABLE 2

WHEN SIBSHIPS AFFECTED BY AN AUTOSOMAL RECESSIVE DISORDER ARE ASCERTAINED THROUGH THE PRESENCE OF AT LEAST ONE AFFECTED CHILD,* WHAT PROPORTION OF ASCERTAINABLE SIBSHIPS HAVE ONLY ONE CHILD AFFECTED?

1-sib families	100%
2-sib families	86
3-sib families	73
4-sib families	62
5-sib families	52
6-sib families	43
7-sib families	36
8-sib families	30
9-sib families	24
10-sib families	20
11-sib families	16
12-sib families	13

* It is further assumed that ascertainment is complete or at least unbiased so that a random selection of affected sibships is achieved.

a sociologic distinctive group like the Amish is likely to impress the observer even though few or none of the cases are in sibs. Thalassemia major occurred in the Mediterranean basin in large numbers of cases, but the definitive description [34, 35] came from a pediatric hematologist in Detroit, Michigan, who could not help being impressed with the unusual disorder which occurred always in children of Mediterranean extraction.

The Naming of Genetic Entities

Once recognized, entities present problems in naming—and names are important. A syndrome has "arrived" if it has a name. An unfortunate consequence of naming can be the mistaken impression that we understand the condition. With few congenital disorders does the state of knowledge permit a designation based on specific etiology or pathogenetic mechanism, as in the thalidomide syndrome, the rubella syndrome, and the trisomy-18 syndrome. Eponyms have their usefulness as mere labels, with no prejudice as to the nature of the basic defect, but they place a severe strain on the memory. Both physicians' names (e.g., most eponyms) and patients' names (e.g., Hartnup's disease) are used. I prefer to say *the* Marfan syndrome (rather than Marfan's syndrome) because it makes it clear that the surname is merely a tag. After all, Marfan described only the skeletal features.

Use of one facet of the syndrome as a name for the whole has obvious risks, since that feature may occur as an isolated anomaly or may be a part of other syndromes, and, on the other hand, it may be missing in some cases of the particular syndrome. For these reasons "arachnodactyly" is a poor designation for the Marfan syndrome. Of course, if the component manifestation selected for designating the whole is a striking, unique, and invariable feature of the syndrome, no problem arises. Examples are focal dermal hypoplasia, incontinentia pigmenti, osteogenesis imperfecta, and chondrodystrophia calcificans congenita, although even these are not immune to nosologic, and consequent terminologic problems.

Ideally, the designation should help one remember the features of a syndrome. Such is the virtue of the naming system which has been used especially by the Gorlin school of syndromologists [36]: oro-facio-digital, oculo-auriculo-vertebral, oculo-dento-digital, and so on. It creates no major problem that the trilogic is appropriate to more than one entity; one delineated after the first one is simply called "number 2" (as Rimoin and Edgerton [37] have done for the OFD syndromes) or given an entirely different name. When their names are reduced to initials such as OFD, OAV, ODD, and FDH (in the Gorlin system of nomenclature), congenital malformation syndromes come to sound like governmental agencies—and perhaps often have other similarities!

Beginning with the glycogen storage diseases and following with the mucopolysaccharidoses, numbering of disorders in a particular group of entities has become one way to cope with the nomenclature problem. It has proven useful to have eponyms to go along with the numbers. Thus, glycogenoses I–VI carry eponyms von Gierke, Pompe, Forbes, Andersen, McArdle, and Hers, respectively. Mucopolysaccharidoses I–VI carry eponyms Hurler, Hunter, Sanfilippo, Morquio, Scheie, and Maroteaux-Lamy, respectively. In general, it is easier to remember the eponym than the number. The eponym conjures up a mental picture of the case(s) the man described. Geographic designations have occasionally been used —for example, the Portuguese and Indiana varieties of amyloidosis (Table 1) and the malformation syndromes called Amsterdam (Cornelia de Lange syndrome) and Rostock types.

Although a few simple guidelines for the naming of genetic entities can be laid down, usage in the long run dictates the preferred designation.

The Numerical Status of Genetic Nosology

The catalogues of Mendelian traits [27] which, with computer aids, I maintain on a continuously updated basis permit an estimate of how many traits are known in man. The numbers presumably relate to separate loci; variant hemoglobins with a change in the β polypeptide chain

TABLE 3

NUMBER OF MUTANT PHENOTYPES ACCORDING
TO MODE OF INHERITANCE

TYPE	a. MAN		b. MOUSE Green (1967) [39]
	Verschner (1959) [38]	McKusick (1968) [27]	
Autosomal dominant..	285	344 (+449)*	99
Autosomal recessive...	89	280 (+349)	207
X-linked.............	38	68 (+55)	12

* The figures in parentheses indicate the number of additional traits for which the particular mode of inheritance has been suggested but not proven.

are, for example, listed as one item. Some allelic traits may be separately listed, but these must be more than counterbalanced by others where non-allelic heterogeneity has not been recognized. The number of firmly established traits are shown in Table 3, as well as the number of traits included only provisionally, either because the particular mode of inheritance is not considered proven or because the separateness from other listed entities is not certain. The well-established loci in man, numbering about 700, probably represent only a fraction of the whole. Man may have as many as 100 times as many genes in all, perhaps even more than that [40].

Probably, proportionately, many more recessives than dominants remain to be discovered. The predominance of known dominants over recessives in man is an anomalous situation in comparison with other species, for example, the laboratory mouse, where more recessives than dominants are known (see Table 3, *b*). The difference is mainly due to differences in mating patterns. Most *visible* mutations (mutations with effects evident without refined methods for studying the phenotype) are probably recessive. In man such a mutant gene can arise and be lost—either by chance or because of some disadvantage even in the hetero-zygote—without meeting up with itself in a homozygote. Or if a homozygote does occur, it has a strong likelihood of being a sporadic case because of the small size of human families (again see Table 2), and the isolated case may escape recognition as a distinct entity. In mice, on the other hand, because of brother-sister and other close matings, as well as the larger number of offspring, a recessive mutation is likely to come more promptly to attention.

Summary

In genetic nosology both lumping and splitting have a place: lumping in connection with pleiotropism; splitting in connection with genetic heterogeneity.

REFERENCES

1. K. H. FABER. Nosography. 2d ed. New York: Hoeber, 1930.
2. V. A. McKusick. Heritable disorders of connective tissue. 1st, 2d, and 3d eds. St. Louis: Mosby, 1956, 1960, and 1966.
3. J. HUTCHINSON. Arch. Surg. (London), 4:361, 1893.
4. L. C. DUNN and D. R. CHARLES. Genetics 22:14, 1937.
5. G. P. MURDOCK. Social structure. New York: Macmillan, 1948.
6. W. G. SMITH. Dis. Colon, Rectum, January-February, 1968.
7. H. B. ASMAN and E. R. PIERCE. Unpublished observations.
8. E. J. GARDNER. Personal communication, 1968.
9. M. MAHLOUDJI, R. D. TEASDALL, J. J. ADAMKIEWICZ, W. H. HARTMANN, P. A. LAMBIRD, and V. A. McKusick. Medicine, 48:1, 1968.
10. C. ANDRADE. Brain 75:408–427, 1952.
11. J. G. RUKAVINA, W. D. BLOCK, E. C. JACKSON, H. F. FALLS, J. H. CAREY, and A. C. CURTIS. Medicine 35:239–334, 1956.
12. V. A. McKusick, J. A. EGELAND, R. ELDRIDGE, and D. E. KRUSEN. Bull. Hopkins Hosp. 115:306–336, 1964.
13. V. A. McKusick, R. ELDRIDGE, J. A. HOSTETLER, J. A. EGELAND, and U. RUANGWIT. Bull. Hopkins Hosp. 116:285–326, 1965.

14. A. J. McAllister, N. F. Hicken, R. G. Latimer, and V. R. Condon. Amer. J. Surg. 114:839–843, 1967.
15. A. W. Johnston and V. A. McKusick. Amer. J. Hum. Genet. 14:83–94, 1962.
16. V. A. McKusick. Cold Spring Harbor Symp. Quant. Biol. 39:99–114, 1964.
17. H. E. Cross and V. A. McKusick. Arch. Neurol. 16:1–13, 1967.
18. ———. Arch. Neurol. 16:473–485, 1967.
19. W. Allan. Arch. Intern. Med. 63:1123–1131, 1939.
20. C. S. Chung, O. W. Robinson, and N. E. Morton. Ann. Hum. Genet. 23:357–366, 1959.
21. M. C. Mengel, B. W. Konigsmark, and V. A. McKusick. EENT Monthly (in press).
22. D. L. Rimoin, T. J. Merimee, and V. A. McKusick. Unpublished observations.
23. P. D. Trevor-Roper. Brit. J. Ophthal. 36:107–110, 1952; Proc. Roy. Soc. Med. 56:21–24, 1963.
24. P. J. Waardenburg and J. Schappert-Kimmijser. In: P. J. Waardenburg, A. Franceschetti, and D. Klein (eds.). Genetics and Ophthalmology, vol. 2. Springfield, Ill.: Thomas, 1959 only, 1963.
25. N. E. Morton. Amer. J. Hum. Genet. 8:80–96, 1956.
26. V. A. McKusick. On the X chromosome of man. Washington: American Institute of Biological Sciences, 1964.
27. ———. Mendelian inheritance in man. Catalogs of autosomal dominant, autosomal recessive and X-linked phenotypes. 2d ed. Baltimore: Johns Hopkins Press, 1968.
28. R. H. Bigley and R. D. Koler. Ann. Human Genet. 31:383–388, 1968.
29. D. Y.-Y. Hsia. Metabolism 16:419–437, 1967.
30. V. H. Auerbach, A. M. DiGeorge, and G. G. Carpenter. In: W. L. Nyhan (ed.), Amino acid metabolism and genetic variation, pp. 11–68. New York: McGraw-Hill, 1967.
31. G. W. Frimpter. Science 149:1095–1096, 1965.
32. B. S. Danes and A. G. Bearn. Lancet 1:241–243, 1967.
33. J. H. Rubinstein and H. Taybi. Amer. J. Dis. Child. 105:588–608, 1963.
34. T. B. Cooley and D. Lee. Trans. Amer. Pediat. Soc. 37:29, 1928.
35. T. B. Cooley. Amer. J. Dis. Child. 43:705, 1932.
36. R. J. Gorlin and J. J. Pindborg. Syndromes of the head and neck. New York: McGraw-Hill, 1964.
37. D. L. Rimoin and M. T. Edgerton. J. Pediat. 71:94–102, 1967.
38. O. Fr. v. Verschner. Genetik des Menschen: Lehrbuch der Humangenetik. Munich and Berlin: Urban & Schwarzenberg, 1959.
39. M. Green. Personal communication, August, 1967.
40. F. Vogel. Nature, 201:847, 1964.

ON SOME TRENDS IN UNDERSTANDING
THE GENETICS OF MAN

*JAMES V. NEEL and WILLIAM J. SCHULL**

Introduction

The objectives and directions of scientific investigation in the United States are currently coming under a scrutiny such as they have perhaps not enjoyed since the nineteenth century [1]. Efforts to meet some of the very proper questions being raised can proceed from either of two viewpoints, the very general—which runs the risk of lacking the substance on which to base a solid philosophy—and the more specific—which quickly can assume the form of special pleading. In this presentation, despite the risks, we will attempt to be rather specific about a particular area, on the thesis that from such communications, if properly executed, comes the stuff on which to base the discussions which ultimately lead to a reasonable consensus. But the hope of a consensus with these present thoughts and emphases was not a primary consideration in the preparation of this communication.

The emergence in the past twenty years of the study of human genetics as a strong and rapidly moving discipline is surely one of the significant currents in modern biology and medicine. This article will attempt to appraise some trends which suggest that the future of the study of human genetics holds some directions and emphases which have thus far been insufficiently appreciated. The implications of these developments for

* Department of Human Genetics, University of Michigan Medical School, Ann Arbor. Publication of this paper in *Perspectives* has been supported in part by contributions from the United Health Foundations, Inc., and The National Foundation.

In a wide-ranging review of this nature, detailed referencing is impossible; we have mentioned only key articles or reviews. To those whose important contributions are not specifically named, we apologize. We do wish to acknowledge the extent to which the preparation of this article has profited from discussions with Drs. Walter Bodmer, George Brewer, James Crow, Th. Dobzhansky, Henry Gershowitz, Robert Krooth, T. E. Reed, Donald Rucknagel, C. R. Shaw, Margery Shaw, Donald Shreffler, Charles Sing, C. Stern, John Sved, and Richard Tashian.

human welfare cannot be honestly appraised at the moment but may be considerable. Many of the points to be mentioned are "in the air" to an extent that the authors claim no great originality but hope to have combined the fragments in a worthwhile fashion. The presentation is aimed at the scientific community rather than the professional geneticist and, space limitations being what they are, will of course, out of the mass of material available, emphasize those facts which best illustrate our general thesis.

For the purposes of this discussion, we shall recognize the following fields within the study of human genetics: biochemical, chromosomal, somatic cell, clinical, population, and formal (mathematical). These areas are not clearly demarcated, as illustrated, for example, by the union of biochemical and clinical genetics in the study of such a disorder as cystic fibrosis of the pancreas. We shall not touch upon the field of somatic cell genetics at all, but in developing our theme will first refer briefly to selected developments in four of the remaining five fields, reserving comments on some developments in mathematical genetics until later.

The most impressive recent developments have been in the biochemical and cytological aspects of human genetics. The wide-scale employment of newly developed, inexpensive, and easily operated apparatus to screen for abnormalities in protein molecules has revealed an unsuspected wealth of inherited biochemical variations, some of which have now been characterized down to the ultimate unit of change in a protein, the amino acid. Many of these variants occur with a frequency such that the genetic systems of which they are a component must be termed "polymorphic," that is, upward of 2 per cent of the population is heterozygous at this locus, a frequency which, as pointed out by Ford [2], is not likely to be maintained by mutation. Because of the importance which polymorphisms will assume in what follows, it seems necessary to digress long enough to establish certain facts concerning their genetic properties. Consider a locus at which a gene A may undergo mutation to a variety of alleles, of which we will consider only one, A'. If A' when homozygous results in a phenotype which is superior in some respect to that associated with A, then it will in time replace A in the population (except in the case where heterozygote AA' is at a disadvantage). Since, however, the phenotypic effect associated with each established gene in the genome reflects the action of a long period of natural selection, it is clear from both theory and observation that, unless the environment is changing rapidly, only rarely will a new mutation when homozygous be associated with a selective advantage. Given a mutation rate of the order of 1×10^{-5} per gene per generation, a commonly assumed frequency,[1] then if the heterozygote does not differ from normal, at genetic equilibrium the frequency of the homozygous mutant phenotype will be determined by its selective disadvantage. If, for instance, the mutant phenotype regularly fails to reproduce, because of early death or sterility, then equilibrium frequency for the phenotype will be 1×10^{-5}, (i.e., equal the mutation rate). If the reproductive performance is half-normal, the equilibrium frequency is 2×10^{-5}. In a randomly mating population, in the first instance, the frequency of heterozygotes is approximately six per 1,000 and in the second instance, nine per 1,000. Even when the mutant phenotype has a 90 per cent survival value, the frequency of heterozygotes is only twenty per 1,000. Whenever, then, more than 2 per cent of a population is heterozygous for a given gene, one must consider four principal possibilities, namely, (1) that the mutant type is near normal in its properties (neutral mutation), (2) that the heterozygote enjoys some selective advantage over normal that offsets the gene loss because of the impaired performance of the homozygous mutant (balanced polymorphism), (3) that a new gene with favorable effects is in the process of replacing its predecessor, or, finally, (4) that genotypes which now appear selectively equivalent, at other gene frequencies are not equivalent (frequency-dependent selection; see below). Other possibilities exist, such as introduction of the gene through migration from another population (which only shifts the problem to the parent population), or "positive" selection in one sex but "negative" in the other (on a priori grounds, rare), or, under certain circumstances, reproductive compensation without heterozygote advantage, but these are the four that require primary consideration. One of the very important properties of a balanced polymorphism is that once established it does not depend on recurrent mutation for the maintenance of genetic diversity, while the decrease in population fitness because of an increased mutation rate may be less, and more transient, in a balanced polymorphism than in a "classical" system where selection favors a homozygous type.

Quite a number of the polymorphisms involving biochemical traits

---

[1] Frequencies of this order are based on the detection of gross phenotypic change or loss of function. The rate of mutation resulting in specific amino acid substitutions in protein molecules is of course much lower. However, although equilibrium values are altered, the general argument remains the same for such "specific" mutations.

which have been discovered recently, such as those affecting human hemoglobin, or the deficiency in erythrocytic glucose-6-phosphate dehydrogenase, or the abnormal acetyl transferase originally discovered through its inability to inactivate isonicotinic acid hydrazide, arose from disease-oriented studies. The recognition of most of the approximately ten human blood group systems in which variation occurs in polymorphic proportions, and within which the differences also ultimately are biochemical, likewise stems from disease-oriented studies (i.e., investigation of hemolytic disease of the newborn or transfusion reactions). This orientation to disease in the discovery of new systems has made difficult generalizations concerning the relative frequency of polymorphic systems among all (or a random sample of) genetic loci at which structural genes occur. For instance, in the blood group systems, where discovery so often depends on an individual elaborating an antibody in response to an antigenic stimulus, it would appear that the serological incompatibility which is prerequisite to antibody formation, arising as it does through maternal-fetal or transfusion incompatibility, would be present more frequently in polymorphic than non-polymorphic systems, and so create a bias toward the discovery of the former.

However, with the recognition of the fruitfulness of screening in uncovering new variation, many laboratories have turned to non-disease-oriented studies. One of the more active groups engaged in such screening operations has reported that, among ten human enzymes chosen simply for convenience of study, three were found polymorphic (one of these due to polymorphism at two genetic loci) in an English population [3]. Lewontin [4], basing his argument on an asymptotic loss of the bias toward discovering polymorphisms in the blood group systems as the number of loci discovered increases, has suggested that the proportion of loci for which an English population is polymorphic is about one-third and that the probable heterozygosity per individual is about 0.16. For reasons mentioned above, this approach might be expected to yield an upper limit to the proportion of polymorphic loci. Studies with experimental organisms are yielding similar results. Shaw [5] reports variation in six out of eight dehydrogenases studied in the deer mouse, *Peromyscus*, the criterion for selection being primarily the existence of convenient test systems. Similar studies on *Drosophila* reveal that of eighteen loci associated with proteins identified by electrophoresis, again selected for ease of identification, seven were found subject to inherited variations in frequencies readily justifying the term "polymorphic" [6].

The various biochemical techniques employed in the analysis of variation will not, of course, reveal all the alternative forms of a given protein. Thus, amino acid substitutions in a polypeptide which do not involve a change in the net charge of the polypeptide will not usually be detected; more than half of the theoretically possible substitutions are of this type. Likewise undetected will be minor variations in genetic control mechanisms, in consequence of which greater or lesser amounts of a given polypeptide are formed. There must thus be a great deal of currently hidden genetic variation, of which the same proportion may occur in polymorphic systems as in the case of the more easily detected variation. Assuming man has at least 100,000 loci encoding for protein structure, even a very conservative projection from these findings calls for human diversity on a hitherto undreamed-of scale.

Equally explosive have been the developments in human cytogenetics. Again the key has been technical innovation, in this instance the appearance of simple, convenient, and inexpensive procedures for short-term cell culture. At least three different syndromes resulting from a supernumerary autosome are well established (trisomy 13, trisomy 18, and trisomy 21). Although no viable conceptus lacking an autosome has yet been recognized, there are at least three syndromes resulting from the partial deletion of an autosome (partial deletion of short arm of 5, short arm of 18, long arm of 18). Variations in the number of X chromosomes from one to five, and variations in the number of Y chromosomes from zero to three, and almost all possible combinations of these numerical variations in the X and Y, have been described. Some 20 per cent of spontaneously aborted foetuses are found to have abnormal chromosome numbers. And, currently, the counterpart at the chromosomal level of the genetically controlled biochemical polymorphisms is beginning to emerge, as normal-appearing persons are recognized to possess and transmit minor but persistent chromosomal anomalies, in the form of over- or undersized satellites, constrictions, or differences in length (reviewed in [7, 8]). Perhaps the most clearly documented of these minor variations are the differences, both racial and familial, in the length of the Y chromosome [9]. It is difficult to believe that these variations are all related simply to chromosomal expansion and contraction; surely, in at least some instances, qualitative differences are involved. Thus, human cytogenetics is beginning to yield evidence for a morphological counterpart to the biochemical variation being recognized but with the important additional inference that differences large enough to be seen with the light microscope should involve rather large blocks of genetic material.

In clinical genetics, two trends stand out, namely, the recognition of an increasing number of diseases, usually quite rare, which can be attributed to a simple biochemical defect, and the parallel recognition that the genetic predispositions to many of the much more common disorders—diabetes, gout, hypertension, disturbed lipid metabolism (including coronary heart disease), schizophrenia, and some congenital malformations—are less simply inherited than once thought. While the details remain obscure, these latter diseases increasingly appear to have a complex, multifactorial basis, that is, result from the interaction of a number of genes (and not always the same genes) at different genetic loci, these genetic determinants usually acting in concert with non-genetic factors of various types. At present this trend toward multifactorial formulations is based as much on a recognition of the many biochemical events contributing to the disorder (each potentially under independent genetic control) and the continuous and bell-shaped (or quasi-continuous) distribution of the trait as on formal genetic analysis. Indeed, the formal analysis of multifactorial inheritance remains just as difficult and unsatisfactory in man as in all other organisms. It is worth emphasizing that, since in multifactorial systems genetic effects at the phenotypic level are largely additive, there is a continuum in the manifestations of the trait, with the precise dividing line between "normal" and "abnormal" arbitrary.

Although the term "multifactorial inheritance" conjures up the image of many loci, it is important to keep in mind that with only two alleles at each of two loci there are nine possible genotypes which, in the absence of dominance and, if subject to environmental modifications of their expression, can give rise to a reasonable approximation to a continuity of phenotype and, depending on gene frequencies and mode of gene action, to a bell-shaped frequency distribution. Thus, the term "multifactorial" is not inconsistent with a continuing quest to identify the effects of specific genes in the etiology of a given disease, and, in fact, genetic studies of many of these diseases will progress in proportion to the ability to identify the various genetic components by appropriate studies.

The problem is well illustrated by our current knowledge of the genetic basis of diabetes mellitus (reviewed in [10]). Here is a common, extremely important disease formerly thought to have a simple genetic basis but now increasingly regarded as the result of a complex interaction between environmental and genetic factors, with, in consequence, a spectrum of susceptibility to this disorder. Its ultimate cause remains a mystery. With the demonstration that there may be an *excess* of circulating insulin at some stage in the prandial cycle in many persons predisposed to diabetes, the early view of diabetes as resulting from a simple lack of insulin is yielding to a much more complex formulation. Currently, attention is being directed to the possibility of such diverse primary etiologies as an initial "overproduction" of insulin, an overresponsiveness of an insulin antagonist system, an abnormality of the small blood vessels, an abnormal insulin molecule, or endocrine imbalance in which the pituitary and adrenal play important roles. The future will surely bring clarification of which one or, more probably, what combination of these and other defects render the individual susceptible to diabetes mellitus, and with this clarification the way will be open to definitive genetic studies. In passing, we note that the incidence of diabetes mellitus is such that if there are genes at three or four different loci which contribute to the diabetic susceptibility, they would appear to have individual frequencies such that they must be regarded as polymorphisms.

A similar situation is emerging with respect to the genetics of gout and hyperuricemia (reviewed in [11]). Formerly thought to be a discrete, qualitative departure from "normality" under the control of a single gene, it now becomes increasingly probable that there is a unimodal continuum of susceptibility, as measured by blood uric acid levels, with the entity known as gout, a clinical manifestation of the tail of a continuous, somewhat skewed distribution. Such a finding is best explained by multifactorial inheritance, interacting with strong environmental determinants. As in the instance of diabetes, the nature of the ultimate defect(s) is still unclear, with such diverse possibilities under investigation as a primary overproduction of uric acid, an abnormally high resorption rate in the renal tubule during the absorptive phase for urate, or a decreased tubular secretion during the secretory phase for urate. On an intuitive basis, one anticipates variability in each of these various aspects of urate metabolism, with the individuals most susceptible to hyperuricemia and gout usually resulting from particular combinations of the variations, although occasionally single genes may appear to play key roles in urate metabolism [12]. And if the variability in each step of urate metabolism is under the control of relatively few genes, again, as in the case of diabetes, one is forced to ascribe to these genes frequencies which classify them as polymorphisms.

As a final example in this section, we might consider the problem of histocompatibility, so vital to organ transplantation. The evidence seems clear that in both mouse and man there are several loci (in addition to the ABO locus) at which genes responsible for histoincompatibility occur but that one of these loci (H-2 in the mouse; HLA in man) is far more important than the others. In man, at least fifteen different alleles may occur at this major locus, apparently with frequencies that make them polymorphisms [13, 14]. The persistence of the histocompatibility antigens over long periods in cell lines maintained in tissue culture [15], while other cellular antigens are lost, suggests these antigens are associated with vital cell functions, the nature of which (and the reason for their maintenance in multiple forms) being unknown at present.

Turning now to population genetics, as each new polymorphism has been recognized in a given ethnic group, at once an effort has been made to test for the occurrence of this polymorphism in other ethnic groups. Although some polymorphisms, such as those associated with certain hemoglobinopathies and the glucose-6-phosphate dehydrogenase deficiency, show rather marked ethnic restrictions, in general it seems clear that polymorphisms occur to about the same extent in all populations, although from one population to the next the frequency of the various alleles in a system, and even the nature of the alleles present, varies significantly.

Many of the populations studied—Australian Aborigines, South African Bushmen, Pacific Island groups, some tribes of American Indians—answer to the description "remote," "small," "isolated," and "inbred." These are the conditions under which genetic variability should be diminished. In fact, however, the specific polymorphisms are usually found to persist in such groups. An exact evaluation of the total amount of genetic variability present in such a group is difficult, but in at least one population in which an effort was made to derive a rather complete picture, the Xavante Indians of the Brazilian Mato Grosso [16], the total amount of genetic variation was not greatly diminished from that to be observed in such a cosmopolitan population as Hamburg, Germany.

A Central Problem of Human Genetics

Thus, developments in four of the principal fields of human genetics have directed attention to the frequency, the ubiquity, and the persistence of the genetic polymorphisms. To put it very simply, *they have now reached such numbers and new ones are being recognized at such a rate, that the understanding of their significance is unquestionably a focal problem of modern biomedicine.* For the sake of perspective, it should be emphasized that the fact of intraspecific diversity has long been recognized. From the earliest records to the present, man has commented on his diversity, and in this generation the amount of genetically determined variation in both plants and animals, particularly hidden variability, has been a favorite theme of such distinguished contemporary biologists as Dobzhansky, Mayr, and Stebbins. What is new is our recognition of its potential extent and our ability to reduce this to the primary manifestation of genetic differences, namely, variation in proteins and related substances. In so designating the "central nature" of this area to human genetics, we have no intention of downgrading the importance of such other areas of research as gene-protein relationships, genetic control mechanisms, "somatic cell" genetics, or chemical mutagenesis. But from the standpoint of both theory and potential benefit to mankind, this area has the character of a foundation. Thus, concern about the long-range effects on the species of increased mutation rates will be strongly tempered by knowledge of the extent of "buffering" polymorphisms, and the potential role of manipulation of genetic mechanisms for the amelioration of human disease is inversely proportional to the extent to which important diseases have their basis in multifactorial inheritance.

Now, as noted earlier, in its simplest form a genetic polymorphism is usually defined as the occurrence in a population of more than 2 per cent of individuals heterozygous for a pair of alleles. Some years ago, Fisher [17] argued that even in the case of a newly mutated gene with a desirable phenotypic effect there was only a small probability (approximately two times the selective advantage) it would establish itself in a population, while genes with neutral or undesirable phenotypic effects had an almost negligible chance of becoming firmly established. Furthermore, a gene which might have had a function in the past but had now lost the function would relatively soon (in an evolutionary sense) be lost from populations of small or moderate size through random genetic drift. Yet, as mentioned earlier, not only are similar polymorphisms found in groups that have been long separated, but the frequencies of the component genotypes of a system may be remarkably similar in these groups. Accordingly, the working hypothesis in approaching the polymorphisms would seem to be that the majority of them are functional, for the most part of two types, namely, (1) transient, in which one allele is either moving to a new equilibrium value or is in the process of replacing another, or (2) balanced, in which there is a proportional loss of A and A' genes because the heterozygote is superior to both homozygotes. (We will not get involved in the complexities of polymorphisms involving three or more alleles; they follow the same principle, but it is possible to have stability even when a heterozygote is less fit than one of the homozygotes.) The occurrence of most of the recognized polymorphisms in a variety of widely separated primitive or recently primitive groups of different ethnotypes renders it unlikely that very many are transient, unless the rate of gene substitution is so slow that the polymorphism, derived from a common ancestor of great antiquity, is in all groups progressing toward genetic fixation of one of its components at the same slow rate. But, in either case, the selection which must be assumed responsible for the polymorphism implies an impaired performance, in terms of survival or reproduction, of certain genotypes. If, for instance, the polymorphism is maintained through differential mortality then in a "steady state" equilibrium, relatively more AA and $A'A'$ individuals die prior to or during the reproductive period than AA', the precise relative death rates determining the frequencies in the population in question of the A and A' genes. This loss is often referred to as the "load" imposed by the polymorphism.

As noted above, current genetic theory is strongly deterministic with respect to the polymorphisms. The fact is, however, that thus far few evidences of this determinism have been forthcoming either at the physiological or statistical levels (see below). This is particularly true with regard to variations in enzymes and hormones. To take an extreme example, the A chain of porcine insulin is identical to that of man, whereas that of bovine insulin differs by two amino acids and that of ovine, by three [18, 19], yet all three types of insulin seem to have the same effectiveness in experimental animals and in man [20]. Indeed, even much greater variations in the structure of this molecule seem not to be associated with very significant differences in molecular activity (reviewed in [21]). On the other hand, whichever the "ancestral" type may be, the gene in the other species responsible for the amino acid substitution has completely replaced the ancestral gene; this would seem to imply a selective differential, which is not detected by bio-assay techniques.

Clearly, unlikely as it seems at present, the logic of the situation demands that we do not abandon the possibility that some of the polymorphisms are not and never have been maintained by selection, that is, that the genotypes AA, AA', and $A'A'$ are selectively equivalent. Possible alternatives to this determinism (discussion in [5]) include genetic drift acting on a neutral gene, or the accumulation of a specific gene thru the simple pressure of repetitive mutation (which would be very specific in type). Unfortunately (from the standpoint of the experimentalist), given a failure in the present to demonstrate any selective effect associated with a specific gene, the converse, the "demonstration" of "everlasting neutrality," will be impossible. We must also be prepared for the possibility, stressed by Motulsky ([22]; see also [23]), that a substantial fraction of the human polymorphisms were at one time in the (recent) history of our species established by selective forces now no longer operative. Now, the ubiquity of specific polymorphisms suggests a widespread function in primitive man. The loss of function must have occurred in relatively recent times. Thus, this point of view implies a very significant recent change in the nature of natural selection, with the implication that man may be in a period of rapid evolution. A "positive" demonstration of this very important possibility, which many biologists will feel demands a greater change in the nature of natural selection than seems probable, might issue from proper studies on primitive man, but the problem of assembling a sufficiently large sample may be insurmountable (see below). Some insight might, of course, accrue from the mathematical exploration of the limiting cases of specific models (see, e.g. [24]). It will be difficult to satisfy the strict determinist except with positive evidence, but a failure of the appropriate studies to provide adequate evidence for the postulated selection for numerous polymorphisms would have such an impact on biological thought that the "negative" studies would be well justified.

Returning now to the concept of the "load" of a polymorphism, although there are a number of hints of selective differentials associated with specific genotypes of specific polymorphic systems (several to be mentioned below), for only one polymorphism—that created by the high frequency of the gene for sickle cell hemoglobin in some populations—do we at present have a reasonable understanding of how the load is balanced. In this instance, the "load," created by sickle cell anemia, is in regions of *P. falciparum* endemicity apparently balanced by the greater-than-normal resistance of the heterozygote to death from malaria. However, even with respect to this much-quoted case, there remain important gaps in our knowledge, such as the extent (if any) to which the heterozygote is handicapped in regions where falciparum malaria does not occur. Nor is the mechanism by which heterozygosity confers protection to malaria understood.

For the past twenty years, thinking about genetic loads has for the most part been in terms of specific genetic loci rather than in terms of the total organismic phenotype, which is the product of many loci. To a large extent, this stems from a classic paper by Muller [25], entitled "Our Load of Mutations." In this paper, the concept was developed that each deleterious mutation in time resulted in one genetic death because sooner or later the carrier of that mutation was eliminated thru natural selection. Muller's treatment of the consequences of mutation was essentially additive, that is, in visualizing the cost of an increased mutation rate in human morbidity and mortality, he added the consequences at the various separate loci under consideration. Man seemed especially vulnerable to the effects of an increased mutation rate because he not only did not have the "reproductive reserve" by which other species might compensate for increased numbers of defective offspring but possessed a system of ethics which compelled him to try to preserve the defectives. It was, of course, recognized that by chance some individuals would carry more deleterious genes than others, but the question of a differential action of selection on those who carried the largest number of deleterious mutations seems not to have received explicit treatment. Mutation rates per gamete as large as 1.0 seemed untenable because they in effect implied that at equilibrium there were no survivors to maintain the species. Consciously or not, during the past twenty years, this concept of additivity has pervaded much of the thinking about the cost of maintaining the polymorphisms. As the number of polymorphisms has multiplied, it has become apparent that with this additivity formulation our species was about to run out of "load space" [26]. As recently as 1966, following their demonstration of the amount of polymorphism in *Drosophila*, Lewontin and Hubby [27] wrote: "We then have a dilemma. If we postulate weak selective forces, we cannot explain the observed variation in natural populations unless we invoke much larger mutation and migration rates than are now considered reasonable. If we postulate strong selection, we must assume an intolerable load of differential selection in the population."

A notable exception to this locus-by-locus approach has been an argument advanced by Morton, Crow, and Muller in 1956 [28]. Their objective was to estimate the relative contribution to the load of a population, as revealed by inbreeding, of loci at which genetic variability was maintained by a balance of selection and recurrent mutation as contrasted to loci at which variability reflected a balance of opposing selective forces, that is, the polymorphisms. They demonstrated that, granted certain assumptions, the ratio of the load of a completely inbred population[2] to an outbred one would be large, say greater than twenty or so, if the majority of the concealed variability exposed upon inbreeding was maintained by a balance between selection and recurrent mutation; whereas if a substantial portion of the load was ascribable to the existence of genes with heterozygous advantage, the ratio would be small. We have summarized the relevant data elsewhere [29]; suffice it here to say that the studies on man and other animals do not reveal ratios which, even under the aforementioned argument, may be unambiguously construed as supporting the view that recurrent mutation is the primary mechanism through which genetic variability is stabilized. Be this as it may, these stimulating arguments not only precipitated a great deal of thought and renewed investigative effort but a drum-fire of objections and criticisms directed at the legitimacy of the underlying assumptions and certain practical difficulties of interpretation of inbreeding data. In particular, with their assumption of the non-synergistic (i.e., additive) action of deleterious genes, there were marked limitations on the number of polymorphisms a species such as man, with its relatively low reproductive potential, could maintain. Of late, the controversy over this formulation has subsided in a miasma of objection and counterobjection (see [29]) to a point where some have been tempted to assert that "load is dead."

But within the past year there have been developments with an important bearing on the conflict between the limited number of polymorphisms which the mortality structure of human populations could main-

[2] The concept of a "completely inbred population" (coefficient of inbreeding equal to 1.0) is, of course, with respect to man, a statistical abstraction but convenient to the argument.

tain under the previous (additive and non-synergistic) mathematical treatments of the subject and the mounting evidence from the laboratory concerning the frequency of polymorphisms [30–32]. Over the years, one of the most telling arguments in support of the notion that the bulk of genetic variability is maintained through a balance of selective forces and recurrent mutation and migration has been the apparently prohibitive cost of maintaining an equivalent variability by polymorphic means. Though implicit, it was perhaps not generally appreciated that this argument assumes that the viability of an individual is *directly* determined by a simple combination of selective values, locus-by-locus, such that the combination determines the probability of survival. It has been argued that under these circumstances the mean population fitness cannot be much lower than the fitness of the most fit (see, e.g. [24]). The latter is now known not to be necessarily true, and cogent and biologically meaningful alternatives to the action of selection posited above have been advanced. Sved and his colleagues ([30]; see also [31, 32]) have shown that the magnitude of the fitness of the most fit is of quite limited importance in determining the selective values at individual loci. Alternatively stated, they find that "individuals with extreme selective values are so rare as to play little part in determining the variance in selective values." They (and King [31]) have also shown that, if selection acts in such a way that the probability of survival is not directly determined by the individual selective values, that is, if there is a threshold effect, then the cost of maintaining a large number of heterotic loci is by no means necessarily prohibitive. This would be true, for example, if the selective values merely contributed to the vigor with which an individual competed for restricted resources. Under these circumstances, the probability of survival would not be directly related to one's genotype but would vary as a function of the composition of the population, the environment (availability of the restricted resources), and stochastic elements. This same reformulation will almost certainly encourage another round of thinking concerning the theoretical upper limits to the rate of spontaneous mutation, and the impact on both individuals and populations of increases in mutation rates. Doubtless, much remains to be done to refine these new formulations, but the way is clearly open to a more realistic approach to the problem of genetic variation in populations. *For an understanding of genetic man, we must, fully conscious of the difficulties ahead, turn increasingly to the study of interactions between genetic sys-*

tems, that is, to the genetics of the whole organism and its relations to its complex environment.

The "system" which regulates the expression of the various polymorphisms in an individual, and the interactions of individuals, is one which we can only dimly begin to visualize. However, it seems clear not only that selection may be quite specific for particular combinations of genes but also that the polymorphisms must sometimes exhibit a functional interdependence, that is, that both positive and negative synergism probably occur. One simple example is already at hand, involving peptic ulcer and the ABO blood groups and the secretor trait [33]. The incidence of duodenal ulceration is increased in individuals who are blood type O, and also in individuals who are non-secretors of the ABO blood group substances. These two characteristics are inherited independently of one another. If we take the incidence of duodenal ulceration in persons who are secretors and not of blood type O as unity, then the incidence in secretors who are type O is 1.4; in non-secretors who are not O, 1.8; and in non-secretors who are type O, 2.7. This latter figure contrasts to the relative incidence of 2.2 to be expected with simple additivity. For a recent treatment of the complexities of interaction possible with only two two-allele polymorphic systems (and review of pertinent literature), the reader is referred to references [34] and [35].

Recently, Kojima and Yarbrough [36] have described a polymorphism involving a *Drosophila* enzyme in which at gene frequencies below equilibrium (gene frequency of 0.30 for the *F* allele) selection appeared to favor phenotypes in which this gene was present, but at higher frequencies, selection was apparently for phenotypes lacking the gene. They interpret these findings, in the face of apparent selective neutrality at equilibrium, as evidence of frequency-dependent selection. The experimental conditions would seem to minimize the possibility that the effect is due to linkage of a second gene with the marker, but other explanations of the "effect," save frequency-dependent selection, remain. Such a situation is not yet known in man, and the difficulties in demonstrating the action of selection on specific genetic systems being what they are for man (see below), may never be demonstrated. But possibilities, admittedly somewhat fanciful, come to mind. Let us assume a genetic component to aggressive behavior. A primitive, polygynous society in which each male was highly aggressive might so decimate itself in the struggle for leadership that it was non-viable. At high frequencies of aggressiveness, the non-aggressive male who could keep aloof from the sanguine struggle might stand a better chance of survival and reproduction than the more aggressive, with his chance a function of the amount of aggressiveness in the group. But at low frequencies of this same trait, the aggressive male who assumed leadership (and multiple wives) would be the object of positive selection, and the more passive the group, the greater his reproductive potential. There would thus be an intermediate frequency at which this phenotype (and its genetic basis) would tend to be stabilized. It is in this connection noteworthy that in *Drosophila* several investigators have reported that in mixed populations a minority genotype appears to mate more frequently than expected on the basis of relative frequency [37, 38]. *We must recognize the probability that the genetic regulatory systems to which populations are subject are no less complex and finely balanced than those at the cellular level so brilliantly elucidated by the molecular biologist. What is emerging is a much more subtle and sophisticated view of the genetic integrity of populations than those earlier concepts which saw man delicately poised between input from mutation and output from selection.*

Approaches to Elucidating the Functions of the Polymorphisms

Having stated what is emerging as one—if not *the*—central problem to be attacked in human (and population) genetics (as we see it), we now turn to our second major thesis, namely, that the solution of this problem will proceed primarily (but not entirely) from studies which skilfully combine aspects of biochemical, clinical, population, and statistical genetics, and whose scope and complexity will greatly exceed the studies of the past. We will list the principal approaches which now present themselves—mindful that we are almost certainly overlooking some of the more significant and that new approaches may at any moment outmode those which today look promising.

Here is perhaps the best place to emphasize that the strategy of the polymorphisms will differ from the "strong inference" strategy of the physical sciences, or even from some branches of biology, such as molecular genetics. Such is the complexity of the situation that even with all the aid the computer can provide (see below), we will for some time have to be satisfied with constantly refining our estimates of "truth" rather than in framing clear tests of hypotheses, tests which result in a distinction between truth and non-truth (cf. [39]). Otherwise stated, starting from almost complete ignorance of the functions and interactions of the polymorphisms, we may expect to converge by successive approximations on the desired answers; but it is unrealistic to predicate our strategy on brilliant breakthroughs which will suddenly illumine all.

1. *Intensive studies of the functions of the genetically variant proteins.*—For many of the proteins subject to genetic variation, their physiological significance is still poorly understood. For instance, is the function of haptoglobin solely to bind and transport free hemoglobin, and is the function of the transferrins limited to the transport of iron? With better knowledge of function, those aspects of the biology of the organism upon which a genetic variant might impinge could be seen more clearly.

2. *Studies of polymorphisms in relation to specific phenotypes (usually disease entities).*—A considerable effort has already gone into this type of investigation (reviewed in [40–44]). The approach, briefly, is to type individuals with a wide variety of phenotypes, normal or diseased, for as many of the polymorphisms as possible. The question of the proper controls is critical—the use of siblings has much to recommend it but is especially laborious. We have earlier in this paper mentioned the two relationships most firmly established thus far, between the ABO blood type and secretor type and susceptibility to peptic ulcer, and between the sickle cell gene and susceptibility to malaria. There seems also to be a real relationship between the ABO blood type and susceptibility to gastric carcinoma, type A individuals being more prone. It is worth noting that, since peptic ulcer appears to be primarily a disease of civilization and gastric carcinoma a disease of the old age which relatively few reached until recently—and which in any event is largely postreproductive—these specific predispositions may have had relatively little to do with establishing present-day gene frequencies, but the example is there. Furthermore, with the control of malaria, the selective differentials which have established the sickle cell polymorphism should be sharply altered. Thus, for the three polymorphisms for which disease relationships are best established, there is evidence of changing selective pressures. For any of the polymorphisms, the possibility exists that, even when one is successful in detecting an action of selection, it is not the same selective pressure which resulted in the establishment of the polymorphism.

It must be recognized with respect to this and several of the approaches to follow, that, under the newer formulations, the detection of the function of a specific gene substitution, or the estimation of the selective value of a specific polymorphism at a specific locus, now appears more difficult. This is because the value (or disadvantage) of possessing a certain gene (or combination of several genes) may be offset by other aspects of the total genome. It has, of course, long been recognized that the expression of a particular gene is related to the total genotype; what is new is the complexity of this total genotype, with the expanding possibilities for offsetting or complicating the role of a specific gene. With the many possibilities for interaction, there may be a mutually stabilizing influence of many polymorphisms on one another. The practical consequence of this to the investigator is the difficulty in detecting the influence of a specific polymorphism and the strong possibility of drawing a blank in even a large study.

3. *The establishment of large cohorts whose members, typed for as many polymorphisms as possible, are followed from birth to death.*—This approach, very simply, involves a search for the differential elimination of certain genotypes from the sample. Such cohorts, which should include the earliest products of conception with which it is reasonable to work, must be organized in a variety of cultures and environments. The search is for evidence that selection is operating on the tail of a distribution based on many different genetic systems. One is thus looking for the elimination of phenotypes characterized by complex combinations of genes. The statistical problems engendered by this approach are formidable, since in effect one engages in multiple-variables analysis in which the higher interaction terms assume a special significance, and as is well known, the interpretation of these higher-order interactions is especially difficult.

4. *Statistical studies of associations of polymorphisms in ethnically homogeneous populations, especially among older age groups.*—Again, the observations should be planned for as wide a variety of environments as possible. This approach is easier than the preceding two but lacks their specificity, since instead of looking for evidence of selection operating on the tails of a distribution, now one in effect looks at the distribution after various "tails" in an *n*-dimensional space have been selected against, and attempts to reconstruct the nature of these "tails." Unfortunately, with human populations in their present state of rapid transition, it will be difficult to define a population in which the very old and very young are drawn from comparable gene pools.

5. *Studies of inbreeding effects.*—Ten years ago, interests in these studies ran high for reasons previously stated, but with the realization that the Morton-Crow-Muller formulation could not yield clear-cut insights (summary in [29]), interest in the results of inbreeding has waned. The newer developments to which reference has been made again suggest, however, that such studies can set upper limits for the importance of the polymorphisms. Very preliminary calculations based on the results of inbreeding studies in man suggest that, even with the new formulations, the number of polymorphisms which could be maintained by selection is well below the number estimated to be present on the basis of the surveys mentioned earlier. Rigorous interpretation of these studies still hinges nonetheless on the immediacy which obtains between the relative, constituent selective values associated with an individual's genotype and the probability of survival of that individual. Particularly relevant, therefore, would seem to be further studies of mortality and morbidity directed toward disclosure of the probable manner of action of selection through disease, especially disease of a microbial origin. It has been argued that microbial diseases may have evolved to a point such "that only individuals below a certain point on the scale of relative selective values would be selected against" [30]. If this obtains, a considerable component of the total amount of selection may be of a more subtle nature than has heretofore been generally supposed.

6. *The search for distortions of genetic ratios.*—Assuming random union of egg and sperm at the time of fertilization—an assumption that may not always be entirely justified—then departures from the expected proportions of offspring from the various types of matings possible in specific systems, as measured at various stages in the life cycle, signify the action of selection. Perhaps the best documented example of this is the deficiency of blood type A individuals from marriages between type O women and men heterozygous for the genes responsible for the A and O traits. The most comprehensive treatments of this subject estimate that, in Caucasian populations, maternal-fetal incompatibility with respect to the ABO system reduces fertility by 6.3 per cent and causes the elimination of 9.4 per cent of incompatible zygotes [45]. Similar estimates have been made for Japanese populations [46]. Several extensive recent studies in widely separated parts of the world suggest that these are overestimates (or that selective pressures are diminishing) [23, 47–49]. There are large differences be-

tween the results of different investigations, and the question of the "true" effect is far from settled [44]. This particular system is so pivotal to our thinking that a major effort to clarify the situation seems justified. Although the involvement of the ABO locus in survival and fertility seems firmly established, in principle if not in magnitude, the status of other possible distortions in genetic ratios is quite controversial. While this type of approach, like (3) and (4), can seldom be expected to yield clues to specific functions of polymorphisms, it at least should serve to identify those polymorphisms currently playing important roles in survival and reproduction.

7. *Comparisons of polymorphisms between related species.*—With respect to physiologically identical proteins in related species, will some tend to be stable but others polymorphic, the latter suggesting responses to types of selection common to related species? In an anthropocentric essay such as this, the comparison of greatest interest involves man and the other primates. At the moment, there are several examples of similar polymorphisms in related species, one of the best being the polymorphism of the serum transferrins in man (reviewed in [50]), the macaque (reviewed in [51]), and the chimpanzee [52]. Hemoglobin polymorphism is probably even more extensive in various macaque species (reviewed in [53]) than in man but has not yet been reported in the chimpanzee. There are also cases of relatively invariant proteins, such as the erythrocyte carboxylic esterases of man and other primates [54]. On the other hand, erythrocyte acid phosphatase, so commonly polymorphic in man [55], has as yet only rarely been observed to be polymorphic in other primates [56, 57]. Also, electrophoretic variants of red-cell carbonic anhydrase I, which are rarely observed in human populations, have been found to occur in polymorphic proportions in several primate species [57, 58]. This general area is only now being explored; it is obviously too soon for generalizations. Either outcome (species similarities in polymorphisms in homologous proteins vs. absence of correspondence in closely related species) provides interesting opportunities for speculation. In particular, a general absence of correspondence would appear to indicate either a remarkable species specificity in the operation of selective factors, a conclusion which, in view of the correspondence of important biological attributes in higher primates, is difficult to accept or else a "randomness" (non-functionality) of the polymorphisms which, in view of their specificity and frequency, is equally difficult to accept.

8. *Studies of experimental animal populations.*—Animal experimentation will be a necessary complement to the foregoing. Chromosomal polymorphism has been recognized for some thirty years in *Drosophila*, and observations on both free-living and caged populations have demonstrated the responsiveness of particular chromosomal types to environmental change [59, 60]. Now an abundance of biochemical polymorphisms is also being recognized in *Drosophila* [6]. Biochemical and other polymorphisms are also being discovered at an increasing rate in a variety of infrahuman mammals (reviewed in [61, 62]). Although most of these have been discovered in laboratory or domesticated strains, it is clear they are also present in wild strains (e.g., of *Peromyscus* [63] and *Mus* (reviewed in [64]); see also preceding section). The H-2 locus in the mouse, with its many alleles might be especially suitable experimental material for the study of the mechanism maintaining a mammalian polymorphism. A cohort of mice or *Drosophila* of size 100,000, or parallel cohorts under different environments, can be followed at a fraction of the cost involved in following a human cohort of this size. Obvious experiments range from a simple study of the characteristics, especially reproductive, of individuals of diverse genotypes, to the synthesis of experimental populations differing in the representation and frequency of the polymorphisms, such populations to be followed for an indefinite number of generations under conditions more nearly simulating a state of nature than the usual laboratory. Other experimental approaches to the question of the survival value of various gene combinations could involve inbreeding experiments, wherein now for the first time, by following "marked" loci, a meaningful study of the decay of heterozygosity with inbreeding (and inferences concerning the selective forces opposing the decay) becomes possible. Observations of this type may be expected to provide models of interaction and selection. Furthermore, it seems clear that out of these controlled situations will emerge insights which may provide entirely new ways of looking at some of the human polymorphisms. But while, then, we would be extremely strong proponents of extensive investigations with experimental material, we feel it necessary to emphasize that no amount of animal work can substitute for the necessary studies on men.

9. *Comparisons of the relative variability within defined populations of specific polymorphic systems.*—In a balanced polymorphism, the tendency of selection to maintain an equilibrium gene frequency is opposed by the stochas-

tic events associated with the constitution of each generation from the preceding. The stronger the stabilizing selection, the less dispersive the role chance can play. Given a series of relatively small populations, where dispersive factors are more effective than in large populations, then polymorphisms with respect to which the gene frequencies cluster relatively tightly around the over-all mean are more apt to be subject to strong selection than those for which the means are more variable. The identification of the more stable polymorphisms could suggest systems for which it will be especially profitable to search for the action of selection, by various of the approaches already enumerated. In man, comparisons of related tribal populations, such as of American Indians or New Guinea Melanesians or Australian Aborigines, should be undertaken when sufficient data are at hand. An important reservation to this approach is that relatively recently arisen polymorphisms, such as may be represented by the hemoglobin S and C polymorphisms, may, despite being subject to strong selection, exhibit regional variability because they are still being dispersed through migration.

10. *The development of mathematical models.*—Patently, the approaches of which we speak can be fully exploited only if there exist meaningful mathematical and statistical models which can serve as the bases for estimation of parameters of interest. At the very least, there must be the means to demonstrate significant departures from predictions based on the model and to estimate the selective disadvantages presumed to underlie these departures. Somewhat more ambitiously, we should like to incorporate into our models of human evolution and population structure such evidence as may be gleaned from the studies discussed. Neither of these is a forlorn hope. There already exists a substantial and reasonably elegant theory to guide segregation analyses (see [65, 66]). Undoubtedly, further advances are not only possible but will be forthcoming as data accumulate to guide the directions of the theory.

Some hypotheses require for their testing mathematical models of population structure. Until quite recently, only two basic models existed; one of these Sewall Wright termed the "island" model and the other, the "neighborhood." Neither of these seemed particularly appropriate descriptions of the situations which must have prevailed during man's early history or even in many areas of the contemporary scene. Kimura ([67]; see also [68, 69]) has contributed a third which he has called the "stepping-stone" model. He envisages an infinite array of colonies between which migration can occur, but in any one generation an individual can move at most "one step" in any one of the possible directions uniting the colonies. This surely constitutes a further step toward reality insofar as human populations are concerned but remains some distance removed from the emerging picture of the population structure of economically primitive groups. Here one is apt to observe a series of fissions and fusions, often along family lines, of disparate size and uncertain frequency [16]. How to incorporate developments of this nature into models of population structure is not clear, although, as we shall see, computers hold promise in this respect. One can predict substantial progress in this direction, if progress proves proportional to the interest which has converged on these issues.

Studies of as many of these various types as feasible carried out in the surviving primitive populations would be of special significance, since it is under these conditions that many of the polymorphisms arose and many present gene frequencies may have been established. However, it is by no means clear that at this cultural level sufficiently large samples can be assembled for definitive results. Furthermore, the relatively high rate of "non-paternity" in many such groups could invalidate efforts at segregation analysis. Finally, as long as these groups remain primitive, medical documentation of the desired accuracy will be difficult to obtain. Be all this as it may, ours is the last generation of scientists to have the opportunity to attempt to determine whether in these relatively simple cultures it will be possible to identify evolutionary forces obscured or altered by the complexity of modern civilization.

The extent to which the necessary large-scale studies can be related to the present-day (or modified) provisions for the collection of vital and health statistics (such as the U.S. Census, the National Health Survey, various state agencies, or even the records of voluntary health insurance agencies) is not clear. While such tie-ins would extend the violations of personal privacy which some already find intolerable, a proper educational campaign should go far to eliminate these reservations. In the extreme, one can visualize a cohort the members of which, once defined and suitably classified, would not again be personally approached, input into the study being entirely through access to a variety of types of records maintained by various organizations.

Some Comments on the Probable Magnitude of the Effort Involved

The scale of effort which some of these studies require will almost certainly exceed all previous efforts in human genetics. Some notion of the scope which it may be necessary to generate is afforded by the following two simple illustrations (see also [50, 70]):

1. Suppose one wished to determine whether there was a direct selective advantage, as reflected in diminished mortality in the first forty-five years of life, associated with being of blood group A as contrasted with blood group O. Simply and more generally stated, we wish to estimate and test the significance of the difference between two proportions. The crux is to construct an adequate test, that is, one which gives a reasonable prior probability of detecting a difference of the magnitude of interest, if, in fact, such a difference obtains. Intuitively, one grasps that the requisite samples of observations will be related to the proportion of individuals succumbing in the first forty-five years of life who are of blood group O (the "standard") and the magnitude of the difference between this value and the proportion of persons of blood group A (the "experimental" group) succumbing in the same period. If one is prepared to specify the minimum difference of interest and acceptable frequencies of errors, both with respect to erroneous rejection of the null hypothesis when it is true (the so-called Type I error) and its mistaken acceptance when it is false (the Type II error), then the size of the samples needed to estimate the two proportions can be calculated [see 71]. Let us presume that some 10 per cent of type O individuals die before their forty-fifth birthdays (for convenience, we ignore possible sex differences). Given Type I and Type II error frequencies of 5 and 10 per cent, respectively, the detection of a 10 per cent shift in mortality, that is, a change from 10 to 9 per cent, can be shown to require two samples of approximately 16,000 individuals each; the detection of a 1 per cent shift requires samples of more than 1,500,000 persons. The samples needed to demonstrate other types of differences for this as well as other "standards," given the same error frequencies, can be read from Figure 1.

2. A second example of sample-size requirements is as follows: Modern segregation analysis is, in a sense, concerned with the discrepancy between the theoretical frequency of an event under a particular genetic and sampling model and the "observed" frequency, and the use of this discrepancy as a basis for the estimation of a series of parameters presumably relatable

to selection, different mechanisms of origin of the event, etc. Unfortunately, the operating characteristics of the procedures by which these latter parameters are estimated are not known, and hence the powers of the methods, in the statistical sense, are conjectural. It seems reasonable to assume, however, that they are in some fashion a function of the precision with which the "true" frequency can be estimated. If it is assumed that the variability in estimates of the "true" frequency is solely binomial in origin,

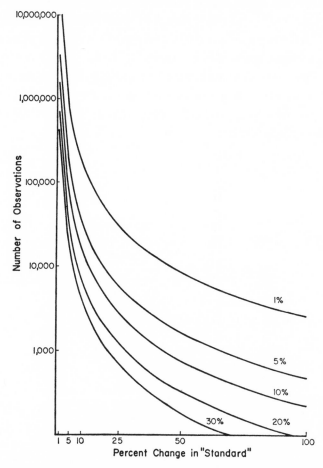

Fig. 1.—The number of observations required to compare two proportions, a "standard" and an "experimental," if it is assumed that Type I errors will occur five times in 100 judgments and Type II errors one time in ten. The "experimental" proportion, the abscissa, is expressed as the per cent change in the "standard," where the latter may be 1, 5, 10, 20, or 30 per cent. The ordinate represents the size of the two samples required to estimate the "standard" as well as the "experimental" proportion.

irrespective of the basis of the discrepancy between the "true" value and the one predicted by the null model, then one can determine, albeit crudely, the sample sizes needed. Though different criteria can be invoked, one could require, for example, that the error in the estimate of the "true" frequency not exceed half the selective differential of interest. Thus, if one studied only matings of AB × O individuals with a view toward using the distortion of the expected segregation ratio among their issue, namely, 1 : 1, as a basis for estimating the selective differentials which might obtain, and if there was reason to believe this differential was but 1 per cent of the expected proportion of type A individuals, then a sample of 40,000 observations would be needed to attain the stated precision. If the variability in the estimate contains a non-binomial element in addition to the binomial, even a larger sample would, of course, be needed. Figure 2 sets out the sample sizes needed for other selective differentials under the conditions outlined above.

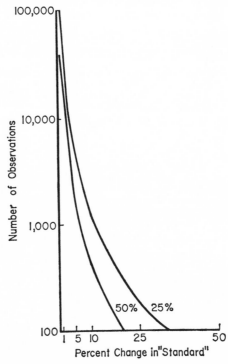

Fig. 2.—The number of observations needed to demonstrate statistical significance of the departure of an estimated proportion from its prior theoretical value. The departure is expressed as the per cent change from a "standard," in one instance, the ratio 1 : 1, and in the other, 1 : 3.

After perusal of these two figures, we believe the conclusion to be inescapable that to date none of the so-called negative studies on selection and genetic traits has been adequate to detect (or exclude) even relatively large selective advantages which might be associated with certain specific genotypes. Conversely, some of the "positive" studies imply effects so large as to be inherently improbable. In this connection, it is well to re-emphasize (see [72]) how small *systematic* errors, such as can creep into blood typings (discussion in [73]), can vitiate even the largest study.

To be adequate a study must usually involve observations on the viability and fertility of, not just thousands of observations, but hundreds of thousands, and quite possibly millions. Assuming the necessary sample homogeneity can be achieved, can our society afford studies of this magnitude? This question is perhaps best phrased in a different context—at this juncture in human history can the society which can invest enormous sums to put a man on the moon afford, when the world is being so rapidly reshaped, not to make at least a comparable effort to understand the genetic nature of the human species? How else do we acquire that ultimate understanding, that baseline, which permits evaluation of the impact of rapidly changing environments on populations, or the deep understanding of the individual on which successful therapy or, better, prophylaxis is ultimately based?

The Role of the Computers in These Developments

The implementation of this strategy will rely heavily on modern computer technology. Indeed, it could be said that without the computers the above-outlined strategies would be impossible. Furthermore, the proper use of the computers introduces a new element into the strategy, an element to be discussed in due course in this section. Though computational aids of a sort have existed for tens of centuries—witness the antiquity of the abacus or soroban—and the basic concepts upon which modern computers operate have been known for at least 130 years, the technological innovations in this area of the past two decades beggar imagination. Less than twenty-five years ago, Howard Aiken, in concert with IBM, developed the Mark II, the first machine actually capable of performing all of the tasks which Charles Babbage, a century earlier, had clearly foreseen as essential to an "analytical machine." This device filled rooms, relied heavily upon mechanical rather than electrical principles, required tenths of seconds to complete a single numeric operation, and, because of its reliance upon electronic tubes, was both error prone and sensitive to its environment. In the time required for the Mark II to perform eight additions, contemporary machines, the CDC 6600 for example, can perform 3,-000,000. Today's machines are surprisingly compact by yesterday's standards, can tolerate much wider variation in temperature and humidity, compute in nanoseconds, and are virtually errorless. In fact, so uncommon are errors (the mean error-free time, ignoring mechanical breakdowns, is more than forty hours on the newer computers) that to these devices can now be delegated a wide variety of functions only dimly perceived a decade or so ago.

Modern computers will undoubtedly contribute to the resolution of the problem under discussion in numerous ways, some presently unforeseen, but among these ways will be through their use, in real time,[3] to monitor and control processing equipment, to analyze data, to link records, and to simulate situations too complex to be handled by more conventional means. The bases for this assertion are briefly as follows:

The computer as a processor-controller.—We have indicated that studies to identify and estimate the magnitude of the selective forces maintaining the human polymorphisms must and will exceed the scope of the studies to which we have all become accustomed. Thousands, tens of thousands, and in some instances hundreds of thousands of specimens will need to be processed. Clearly, efforts of these dimensions are prohibitively expensive, if not impossible, if each specimen must be prepared and processed by hand. But the latter is no longer necessary, for as industry has amply demonstrated, computers can be used for real-time data acquisition, analysis, and control. That is to say, it is presently possible, for example, for certain computers to accept the information generated by an amino acid analyzer, convert this information to digital form, process the latter in some specified manner, evaluate and record the results, and if need be, exert supervisory control over the amino acid analyzer itself. Even relatively modestly sized computers, such as the IBM 1800, could monitor several amino acid analyzers. The savings in processing time thus possible are substantial; one laboratory which is presently using a computer-controlled automatic amino acid analysis by ion-exchange chromatography

[3] Real time describes the performance of some computation relevant to an event while that event is in progress with a view toward possibly influencing the course of the event.

followed by colorimetric determinations of the separated amino acids estimates a savings of two man hours per analysis in a procedure formerly requiring five to six hours [74]. The savings in time in a gas chromatographic analysis are even more striking, for a relatively larger proportion of effort is invested in the calculation of the contribution of the various amino acids to the protein under analysis. A two and one-half hour procedure becomes a half-hour one.

At the moment, without further major technological innovations, it is possible to automate the detection and measurement of any protein which can be or could be assayed by spectrophotometric, colorimetric, or immuno-chemical methods. And, in fact, a number of proteins are even now being automatically assayed in connection with certain screening projects. Means do not as yet exist for automating electrophoretic assays in the manner here described, but considerable energy is being expended in this direction, and undoubtedly the barriers which presently exist will soon be resolved. It seems possible, in the light of the foregoing, that in another decade or two the means will be at hand to assay an individual's phenotype automatically for a considerable number of polymorphic systems and from a single specimen.

The computer as an aid to data analysis.—Computers because of their capacity to recycle almost endlessly can obtain solutions to equations which are otherwise numerically totally intractable. Modern segregation analysis or the contemporary approaches to the estimation of genetic distances and phylogenetic relationships (cf. [75–78]) would be virtually impossible without a computer, and this might also be said of many of the other mathematical and statistical developments to which brief reference has already been made. The large-scale digital computer has in truth revolutionized numerical analysis. Out of this revolution have come new statistical techniques, but no less important, perhaps, a broader use of a variety of analytical possibilities previously denied all but the most persevering of investigators. Parts of multivariate theory, for instance, are more than thirty years old, yet the calculations involved in their application have appeared so laborious that these procedures have seen only limited use until very recently. Many of these procedures are now routinely provided with the "software" which accompanies a computer, and while it is a commonplace to deplore the use of statistical techniques which the investigator himself may poorly understand, it remains to be established that such use does more harm than good. Be this as it may, one of the major contributions which the computer will bring to the problem of the polymorphism is a more insightful view into the data which may be collected.

The computer and record linking.—Various governmental, quasi-governmental, and private organizations are custodians of a wealth of information of a vital statistical and demographic nature which if appropriately assembled might have important implications with respect to the maintenance of genetic variability. The assembly, from these diverse sources, of all of the information of biological pertinence which may be available on a particular individual and the collection of individuals into family units has been and continues to be a formidable undertaking. Remarkable strides have been made, however, both in the theory of systems of retrieving this information (a description of many of these developments is to be found in [79]) and in the mechanical and electronic aids to implement these systems. The significance of these developments for human population genetics has been recognized for some years (see, e.g., [80]), and research to effect the most meaningful use of these innovations is now being vigorously pursued at several institutions, here and abroad. While all of the uses to which these data may be ultimately put cannot presently be divined, the potential, granted further efforts to upgrade the quality of the observations, seems very great indeed. However, this potential can never be realized without specific efforts to give a genetic orientation to such undertakings as the U.S. Health Interview Survey or even the U.S. Census.

The computer as a simulator.—Possibly one of the more far-reaching contributions which digital computers can make to the problem before us as well as to all of biology will come through their use as a means to simulate processes too complex to be tractable to more conventional methods of analysis. Biologists have been much slower to capitalize on this use of the computer than physicists who have, for years, used Monte Carlo and other forms of computer simulation to solve problems which arise in nuclear physics or reactor design, for example. There is a growing awareness of the possibilities of these machines in this context, however, and one now finds demographers (see, e.g. [81]), anthropologists [82, 83], geneticists (e.g. [84–86]) and communications scientists [87], to mention but a few of the disciplines interested in the biology of populations, actively engaged in the construction for the computer of models more nearly approximating reality than those for which analytical solutions exist. These constructs are quite possibly still overly naïve, but they have nonetheless occasioned a more orderly "thinking out" of the sequence of events involved in the persistence and spread of mutant genes in a population, and the relationship of population structure to these latter events. Patently, the opportunity which the computer affords to watch the passage of a population from one state to another can provide insights into the dynamics of a genetic system which no end of steady state analysis will ever elicit. Future models will unquestionably be more realistic and much more complicated, and with their increasing complexity it will be more difficult to apprehend intuitively the probable consequences of wilfully introduced perturbations of the model. But here computer-controlled display devices may come to our aid. They not only will reduce complex data to graphs and figures as desired but permit the investigator to watch, quite literally, the unfolding of his simulation and to interpose new instructions or different parameters at seemingly critical times in a particular experiment or run. Again, this form of interaction between man and model simply cannot be achieved through conventional analytical approaches.

Finally, it seems worth noting that the computer may contribute in yet another way, namely, as a powerful adjunct to the logical and axiomatic analysis of adaptive genetic systems. An extensive literature has developed in the area of automata theory since the pioneering efforts of Turing, von Neumann, and others, which merits close scrutiny as to its applicability to biology, both cellular and populational. Some progress in this direction can already be recorded [88].

The Question of Timing

Timing is all-important in the affairs of men, scientific and otherwise. Is now the time to initiate studies of the type just discussed? One may confidentially predict that, given continuation of the present level of research support, in another ten years the number of polymorphisms which could be brought to bear on a study will have increased by a factor of 4 or 5. Since these new ones are as apt to have detectable biological effects as those already discovered, and since in consequence of our rapidly changing environment many polymorphisms may have lost their functions (and hence, investigatively, represent lost causes), perhaps we should wait until more traits are at hand. Moreover, the number of variables, genetic and non-genetic, which if considered simultaneously could enter into a cohort study today would strain all but the largest computer—perhaps we should wait for the next round of computer technology.

In our opinion, the most immediately fruitful lines of attack for human populations are items (7) and (9), which offer indirect approaches to identifying functional polymorphisms. However, these completed, with or without "success," little is to be gained by delaying the initiation of major studies along the other lines mentioned in this review. There is at hand a sufficient number of polymorphisms for a very significant start on the problem—there are already numerous tantalizing hints of functions, requiring further investigations. Moreover, the genetic polymorphisms to which we have been referring have been detected in blood, saliva, and urine. With the availability of N_2 freezers, it is now possible to store specimens for a long period of time, returning to them for additional observations when in due course there has been discovered a large number of new polymorphisms, for which the specimens could then be classified. This would vitiate at least one reason for delay.

However, it seems self-evident that many polymorphisms of great significance to the species either will not be discovered until after the original specimens of any presently initiated survey are exhausted or else involve tissues it is not yet convenient to sample. How, for instance, are we to approach what is probably the most important group, those affecting the functioning of the central nervous system? As of today, this is an area of almost total ignorance, and progress is apt to be extremely slow. Thus, were large-scale studies initiated in the near future, it would have to be with full knowledge that many of the most important genetic systems were yet to be discovered—and with the hope that the experience gained in early, preliminary studies would have a high transfer value for later and better investigations.

Perspective, and the Application to Human Affairs

The foregoing obviously is not, nor was it intended to be, comprehensive. Hopefully, it has caught some of the surge of excitement and emergent problems in this aspect of human genetics. The emphases will surely be felt to be debatable by some. But in this respect, no brief review could expect to receive universal agreement.

An underlying theme in this presentation has been meant to be a concern, at this stage, on understanding rather than immediate applications. No reference has thus far been made to "genetic engineering," which in its more restricted sense refers to the purposive alteration of certain genotypes, but in the looser sense is extended to manipulation of genetic control mechanisms as well. This is by design. The many developments in "molecular genetics," when extrapolated to man, raise numerous intriguing possibilities whose active discussion is well under way [89-95]. Cautious exploration of these possibilities must continue with a humility born of man's proven ability to disturb complex biological systems through ignorance or half-knowledge. On the other hand, the individuals toward whom much of the current thinking about genetic engineering is directed will at any time be a small minority of our species. But from the standpoint of the species as a whole, the future rests for the most part with the much more numerous "normal" individuals, whose effective functioning determines to a large measure our ability to cope with our growing problems (see also [96]). Insufficient thought has been directed toward that kind of "environmental engineering" which enables man to make the most effective use of his present genotypes [97].

Three concrete examples suffice. The first is diabetes, already discussed. It appears that man's genotype has not adjusted to the change in diet and exercise patterns of recent centuries; one response to this change may be diabetes mellitus. Is diabetes a sufficient problem that we should actively seek for that combination of diet and exercise which will minimize this disease in the various susceptible genotypes we will soon be able to recognize, and then consciously structure our culture in this direction? A second example is coronary heart disease, which alone now probably removes from productivity a greater proportion of effective human effort than all the rare biochemical disorders. Again, should we consciously strive to maximize the genotypic potential of predisposed persons as identified by the appropriate studies of the future, by minimizing environmental precipitants? Third, we turn to the very difficult area of the mind. Without doubt, the greatest and most difficult challenge confronting the human geneticist is to understand the genetic basis for human behavior and mentation. Highly aggressive behavior (however culturally modified) has been a feature of most primitive human societies. Its appearance all over the world, in the most diverse possible cultures, leaves little doubt it has a genetic component. The territorial imperative is nowhere better illustrated than in our own species. We really have no idea of the extent to which the past few thousand years of civilization have modified these ancient attributes; but as we identify polymorphisms concerned with the functioning of the central nervous system, increasingly this question can be put on an objective basis. Be this as it may, society must develop outlets for this genetic aggression, perhaps individually designed on the basis of genetic knowledge, while—hopefully—awaiting the necessary evolutionary developments.

To some, this emphasis undoubtedly seems hopelessly impractical. It is much easier to think that biomedical science will find ways to patch up individuals with genetic defects (or their genes) than it is to decide that our culture has reached a point where society as a whole must make a conscious effort to develop in such a manner that the genetic potential of each individual, once determined, is realized to its fullest. In fact, these are not alternative approaches; modern medicine will continue to improve upon its treatment at all levels of the individual with genetically determined disease even as the geneticist and those similarly oriented strive for the deeper understanding from which should stem more effective prophylactic measures.

Hovering over all these considerations is the "population problem." There is a clear consensus that human reproduction should be curtailed. Any discussion of the immediate future of human genetics which fails to touch upon this has clearly skirted its responsibilities. To what extent should genetic considerations enter into decisions to limit family size? Individuals aware of genetic problems within the family clearly have the right to access to the most current knowledge possible, and as individuals are entitled to whatever decisions they see as proper. But should systematic efforts be made to discourage the reproduction of classes of individuals? Now, the rate at which genetic change occurs is, by the standards of human history, slow. We have just seen how very elementary is our knowledge of man's genetic structure. It would seem far preferable for the present to proceed upon a voluntary "quota system" based upon a clear intellectual commitment to the number of children necessary to stabilize population numbers, rather than to attempt to lay down genetic guidelines which, in the light of the inadequacy of our present knowledge, and ability to visualize the future, can only embarrass geneticists fifty years hence. For example, should the world not solve the population problem, there will be a decrease in the general level of nutrition, and quite possibly a lowering of medical control measures. Drug-resistant *P. falciparum* malaria, already a problem in some parts of the world, may spread; it would, from the population standpoint, be a mistake to have taken measures to lower the frequency of the sickle cell gene. And if the diabetic predisposition is really a "thrifty" genotype [98], unusually efficient in food utilization, it would have been a mistake to have discouraged the reproduction of diabetics if mean caloric consumption falls. The foregoing should not be taken as an escapist argument that we cannot develop an intelligent eugenic program until the genetic structure of man is *completely* understood but, rather, that at the present time clearly too little is known to permit a formulation of the necessary scope, sophistication, and soundness.

Conclusion

We have in this essay argued that of all the exciting developments in the field of human genetics in the past twenty years, the most challenging is the realization of the extent of man's concealed biochemical variability, concerning the function of which we are in almost total ignorance. The polymorphisms resulting from this variability are thought to be the genetic equivalent of biochemical "buffer" systems, in that the gene frequencies they reflect are not easily disturbed by changes in mutation rate or temporary fluctuations in selective pressure. Some of the functions of the polymorphisms undoubtedly relate to important human differences and disease susceptibilities. It is argued that, in the crucial task of understanding man's biological nature, the major efforts required to understand the polymorphisms—many enumerated in this discussion—deserve particular attention because of ramifications into all aspects of biology and medicine, and are now feasible because of developments in theory and computer technology which permit a meaningful approach to problems of this complexity.

REFERENCES

1. G. H. DANIELS. Science, **156**:1699, 1967.
2. E. B. FORD. *In:* J. S. HUXLEY (ed.). The new systematics, pp. 493–513. Oxford: Oxford Univ. Press, 1940.
3. H. HARRIS. Roy. Soc. (London), Proc., B, **164**:298, 1966.
4. R. C. LEWONTIN. Amer. J. Human Genet., **19**:681, 1967.
5. C. R. SHAW. Science, **149**:936, 1965.

6. J. L. Hubby and R. C. Lewontin. Genetics, **54**:577, 1966.

7. J. German. Texas Rep. Biol. Med., **24** (Suppl.):347, 1966.

8. R. A. Pfeiffer. Third Int. Congr. Human Genet. Proc., pp. 103–121. Chicago: Univ. Chicago Press, 1967.

9. M. M. Cohen, M.W. Shaw, and J.W. MacCluer. Cytogenetics, **5**:34, 1966.

10. J. V. Neel. Sixth Congr. Int. Diabetes Fed., Proc. (in press).

11. ———. Third Int. Symp. Pop. Stud. Rheumatic Dis. (in press).

12. J. E. Seegmiller, F. M. Rosenbloom, and W. N. Kelley. Science, **155**:1682, 1967.

13. F. H. Bach and D. B. Amos. Science, **156**:1506, 1967.

14. E. S. Curtoni, P. L. Mattiuz, and R. M. Tosi (eds.). Histocompatibility testing. Copenhagen: Munksgaard, 1967.

15. S. G. Gangal, D. J. Merchant, and D. C. Shreffler. J. Nat. Cancer Inst., **36**:1151, 1966.

16. J. V. Neel and F. M. Salzano. Amer. J. Human Genet., **19**:554, 1967.

17. R. A. Fisher. Roy. Soc. (Edinburgh), Proc., **50**:205, 1930.

18. H. Brown, F. Sanger, and R. Kitai. Biochem. J., **60**:556, 1955.

19. D. S. H. W. Nicol and L. F. Smith. Nature (London), **187**:483, 1960.

20. D. A. Scott and A. M. Fisher. Roy. Soc. (Canada), Trans., **34**(5):137, 1940.

21. L. F. Smith. Amer. J. Med., **40**:662, 1966.

22. A. G. Motulsky. Human Biol., **32**:28, 1960.

23. N. E. Morton, H. Krieger, and M. P. Mi. Amer. J. Human Genet., **18**:153, 1966.

24. M. Kimura and J. F. Crow. Genetics, **49**:725, 1964.

25. H. J. Muller. Amer. J. Human Genet., **2**:111, 1950.

26. J. V. Neel. *In:* S. J. Geerts (ed.). Genetics today, vol. **3**, pp. 913–921. Oxford: Pergamon, 1965.

27. R. C. Lewontin and J. L. Hubby. Genetics **54**:595, 1966.

28. N. E. Morton, J. F. Crow, and H. J. Muller. Nat. Acad. Sci., Proc., **42**:855, 1956.

29. W. J. Schull and J. V. Neel. The effects of inbreeding on Japanese children. New York: Harper & Row, 1965.

30. J. A. Sved, T. E. Reed, and W. F. Bodmer. Genetics, **55**:469, 1967.

31. J. L. King. Genetics, **55**:483, 1967.

32. R. D. Milkman. Genetics, **55**:493, 1967.

33. R. Doll, H. Drane, and A. C. Newell. Gut, **2**:352, 1961.

34. S. Wright. Nat. Acad. Sci., Proc., **58**:165, 1967.

35. W. F. Bodmer and J. Felsenstein. Genetics, **57**:237, 1967.

36. K. Kojima and K. M. Yarbrough. Nat. Acad. Sci., Proc., **57**:645, 1967.

37. C. Petit. Biol. France et Belg., Bull., **92**:1, 1958.

38. L. Ehrman, B. Spassy, O. Pavlovsky, and Th. Dobzhansky. Evolution, **19**:337, 1965.

39. J. V. Neel. Amer. J. Human Genet. **18**:3, 1966.

40. J. A. Fraser Roberts. Brit. Med. Bull., **15**:129, 1959.

41. C. A. Clarke. *In:* A. Steinberg (ed.). Progress in medical genetics, vol. **1**, pp. 81–99. New York: Grune & Stratton, 1961.

42. J. A. Fraser Roberts. *In:* J. V. Neel, M. W. Shaw, and W. J. Schull (eds.). Genetics and the epidemiology of chronic diseases, pp. 77–86. Washington: U.S. Government Printing Office, 1965.

43. L. H. Muschel. Bacteriol. Rev., **30**:427, 1966.

44. T. E. Reed. *In:* World population conference, vol. **2**, pp. 498–502. New York: United Nations, 1967.

45. C. S. Chung and N. E. Morton. Amer. J. Human Genet., **13**:9, 1961.

46. C. S. Chung, E. Matsunaga, and N. E. Morton. Jap. J. Human Genet., **5**:124, 1960.

47. Y. Hiraizumi. Cold Spring Harbor Symp. Quant. Biol., **29**:51, 1964.

48. H. B. Newcombe. Amer. J. Human Genet., **15**:449, 1964.

49. T. E. Reed. Amer. J. Human Genet. (in press).

50. E. R. Giblett. *In:* A. G. Steinberg and A. G. Bearn (eds.). Progress in medical genetics, vol. **2**, pp. 34–63. New York: Grune & Stratton, 1962.

51. M. Goodman, A. Kulkarni, E. Poulik, and E. Reklys. Science, **147**:884, 1965.

52. M. Goodman, W. G. Wisecup, H. H. Reynolds, and C. H. Kratochvil. Science, **157**:48, 1967.

53. N. A. Barnicot, E. R. Huehns, and C. J. Jolly. Roy. Soc. (London), Proc., B, **165**:224, 1966.

54. R. E. Tashian. Amer. J. Human Genet., **17**:257, 1965.

55. L. Beckman. Isozyme variations in man. Monographs in human genetics, vol. **1**. New York: Karger, 1966.

56. L. Y. C. Lai. Acta Genet. et Statist. Med., **17**:104, 1967.

57. R. E. Tashian. Personal communication, 1967.

58. R. E. Tashian, D. C. Shreffler, and T. B. Shows. Ann. N.Y. Acad. Sci. (in press).

59. Th. Dobzhansky. Genetics, **28**:162, 1943.

60. S. Wright and Th. Dobzhansky. Genetics, **31**:125, 1946.

61. C. Cohen (ed.). Ann. N.Y. Acad. Sci., **97**:1, 1962.

62. G. C. Ashton, D. G. Gilmour, C. A. Kiddy, and F. K. Kristjansson. Genetics, **56**:353, 1967.

63. D. I. Rasmussen. Evolution, **18**:219, 1964.

64. M. L. Petras. Evolution, **21**:259, 1967.

65. N. E. Morton. *In:* W. J. Burdette (ed.). Methodology in human genetics, pp. 17–52. San Francisco: Holden-Day, 1962.

66. E. Peritz. Ann. Human Genet., **30**:183, 1966.

67. M. Kimura. Annu. Rep. Nat. Inst. Genet., Japan, **3**:62, 1963.

68. G. Malecot. Entretiens Monaco Sci. Humaine, pp. 205–212, 1962.

69. M. Kimura and G. H. Weiss. Genetics, **49**:561, 1964.

70. W. C. Boyd. Amer. J. Phys. Anthropol., **13**:37, 1955.

71. C. Eisenhart, M. W. Hastay, and W. A. Wallis. Techniques of statistical analysis. New York: McGraw-Hill, 1947.

72. J. H. Edwards. Brit. J. Preventive Soc. Med., **11**:79, 1957.

73. J. V. NEEL, F. M. SALZANO, P. C. JUNQUEIRA, F. KEITER, and D. MAYBURY-LEWIS. Amer. J. Human Genet., **16**:52, 1964.

74. M. I. KRICHEVSKY, J. SCHWARTZ, and M. MAGE. Anal. Biochem., **12**:94, 1965.

75. L. D. SANGHVI. Amer. J. Phys. Anthropol., **11**:385, 1953.

76. L. L. CAVALLI-SFORZA and A. W. F. EDWARDS. *In:* S. J. GEERTS (ed.). Genetics today, vol. **3**, pp. 923–932. Oxford: Pergamon, 1965.

77. D. J. ROGERS, H. S. FLEMING, and G. ESTABROOK. *In:* TH. DOBZHANSKY, M. K. HECHT, and W. C. STEERE (eds.). Evolutionary biology, vol. **1**, pp. 169–196. New York: Appleton-Century-Crofts, 1967.

78. W. M. FITCH and E. MARGOLIASH. Science, **155**:279, 1967.

79. B. C. VICKERY. On retrieval system theory. 2d ed. Washington: Butterworths, 1965.

80. H. B. NEWCOMBE and P. O. W. RHYNAS. *In:* The use of vital and health statistics for genetic and radiation studies, pp. 135–153. New York: United Nations, 1962.

81. H. HYRENIUS, I. ADOLFSSON, and I. HOLMBERG. Demographic models: second report. Demographic Institute, pp. 1–36. Sweden: Univ. Goteborg, 1966.

82. A. BRUES. Amer. J. Phys. Anthropol., N.S., **21**:287, 1963.

83. P. KUNSTADTER, R. BUHLER, F. F. STEPHAN, and C. F. WESTOFF. Amer. J. Phys. Anthropol., N.S., **21**:511, 1963.

84. W. J. SCHULL and B. R. LEVIN. *In:* J. GURLAND (ed.). Stochastic models in medicine and biology, pp. 179–196. Madison: Univ. Wisconsin Press, 1964.

85. L. L. CAVALLI-SFORZA and A. W. F. EDWARDS. Amer. J. Human Genet., **19**:233, 1967.

86. J. W. MACCLUER. Amer. J. Human Genet., **19**:303, 1967.

87. R. ROSENBERG. Simulation of genetic populations with biochemical properties. Ph.D. thesis, Univ. Michigan, 1967.

88. W. R. STAHL. Perspect. Biol. Med., **7**:373, 1965.

89. TH. DOBZHANSKY. Mankind evolving. New Haven, Conn.: Yale Univ. Press, 1962.

90. G. WOLSTENHOLME (ed.). Man and his future. London: Churchill, 1963.

91. T. M. SONNEBORN (ed.). The control of human heredity and evolution. New York: Macmillan, 1965.

92. R. D. HOTCHKISS. J. Hered., **56**:197, 1965.

93. N. H. HOROWITZ. Perspect. Biol. Med., **9**:349, 1966.

94. E. L. TATUM. Perspect. Biol. Med., **10**:19, 1966.

95. J. LEDERBERG. Amer. Natur., **100**:519, 1966.

96. J. M. THODAY. Third Int. Congr. Human Genet., Proc., pp. 339–350. Chicago: Univ. Chicago Press, 1967.

97. J. V. NEEL. Harvey Lect., **56**:127, 1961.

98. ———. Amer. J. Human Genet., **14**:353, 1962.

BIRTH DEFECTS: FROM HERE TO ETERNITY*

SAMUEL J. AJL† and JOSEPH MORI‡

Introduction: The Long March

When in 1958 the National Foundation for Infantile Paralysis (now known as March of Dimes Birth Defects Foundation) preempted for its future activity a field it chose to call birth defects, it was a violent wrench in its history. What a difference from polio! There the goal was clear, attainable, attained—in 20 years. Birth defects were then still considered a neglected area of medicine. But the transition came at a time of rising interest and expanding knowledge in human genetics and teratology. It was soon evident that birth defects covered a vast, complex, amorphous territory, seemingly spread over the whole world of biomedicine. One could not be sure where birth defects ended and other things began. The frontier of birth defects is still clouded in the murky interiors of multifactorial diseases and will likely remain so for a long time.

We are faced with the fact that causes of birth defects are omnipresent in the environment and imprinted in our genes. Many birth defects arise from the interplay between a particular environment with a particular genotype. The challenge of birth defects will probably increasingly occupy the minds of the government, scientists, and the public in general. Pretty soon we may even find the phrase "birth defects" in our dictionaries, and it will be interesting to see what definition the lexicographers will come up with.

We cannot say we hope to conquer birth defects in your lifetime. We cannot offer the possibility of a universal panacea. Prevention of some birth defects is already possible, and many others will join this list, but for all genetic diseases, and for some others as well, prevention will involve one or other ways of reproductive control by individuals or families involved, a practice requiring emotion-laden decisions which on

*Lecture delivered by S. J. Ajl at the Second Triennial Basil O'Connor Starter Research Colloquium, Key Biscayne, Florida, November 15, 1979.

†Vice-president for research, March of Dimes Birth Defects Foundation, 1275 Mamaroneck Avenue, White Plains, New York 10605.

‡Medical editor, March of Dimes Birth Defects Foundation, 1275 Mamaroneck Avenue, White Plains, New York 10605.

the public scale may be subjected to cross fire by contending groups with differing ethical outlooks. Prevention in the areas of our interest is unlikely to become public health measures with legal sanction—such as mandatory vaccination. Gone is the age of innocence when the scientific community and health agencies like the National Foundation for Infantile Paralysis operated in a kind of privileged sanctuary set aside by the society for "good causes." We will be competing for support from a public made up of small and large interest groups, often at war with each other, and must tread cautiously over the shifting sands of American pluralism.

Basic research is where everything starts. When asked what good is basic research, Grant Liddle answered, what good is the baby [1]? It embodies all the potentials of the future. And basic research is also where everything comes back. The conceptual, technical, and applied developments it gives rise to lead to another series of problems to be faced by the younger investigators, just as the next generation's babies face a different world shaped by the earlier ones. The money we put into research is the fuel by which we hope to accelerate or at least maintain this helical trajectory of science.

One of the unsung virtues of a good investigator is to sense quickly and exploit vigorously other men's discoveries, including the competitors'. Similarly, one of the attributes of a vice-president for research should be to know which way the fish are running this season, or better still, to sense which way the fish may be going next season. To cultivate this power of divination we sometimes indulge in ad hoc surveys of what is going on in various areas of research germane to our programs. You will be relieved to learn that this is not going to be a systematic review of various fields. Rather, it will be hit-and-run forays into the frontiers of several fields of interest to the March of Dimes Birth Defects Foundation, in the hope of bringing back some inklings of what may lie ahead.

For surely Lou Thomas was right when he said that the major discovery of modern biology is the knowledge of our ignorance about nature. "It would have amazed the brightest minds of the 18th century enlightenment" he wrote, "to be told . . . how little we know and how bewildering seems the way ahead. It is this sudden confrontation with the depth and scope of ignorance that represents the most significant contribution of 20th century science to the human intellect" [2].

This is what makes our science forward looking: we are always staring at the horizon and thinking up ways of going beyond, knowing full well that, if we do, we will be faced with a new horizon. "It is not so bad . . ." said Thomas, "if you are totally ignorant; the hard thing is knowing in some detail, the reality of ignorance." The awareness of ignorance may be hard to bear, but it is the bread and wine of our science. If we know some things, we can imagine what we do not know, what we should know. A smattering of knowledge can define for us what Ebert called "known unknowns" [3]. This talk, then, is a travelogue in the countries of "known unknowns."

Genes

Whoever said, "In the Beginning was the Word and the Word was DNA" may find himself disputed by God or by Leslie Orgel but not by us. We subscribe to the faith that DNA is the universal alphabet with which the book of life is composed; the only immortal part of us that connects T4 bacteriophage with Max Delbruck. So we start at the Beginning—with stretches of DNA we call genes.

Now that we have an armory of cutlery with which to chop up DNA and a stable of plasmids upon which the pieces so obtained can be engrafted for mass production, we owe to bacteria the means for making the most direct approach to the sacred precinct of the mammalian genome. The first few genes obtained in this way are now on the assembly lines of molecular biology, and the first big surprise to hit us was the discovery that at least some genes are discontinuous within themselves: there are sequences of nucleotides of varying lengths within the structural gene which are not translated into the messenger. The function of these "introns" is not known, nor is the prevalence of their presence in enkaryotic genes. If the system is universal, an mRNA is not a true replica of the gene in the genome. Gene expression is censored at the source, and a stringent censorship may be an unconditional need for the propagation of a meaningful message. We may discover a bureaucracy of censors in the enkaryotic cell not present in bacteria.

But since no system is error proof, one can imagine many slip-ups in the censorship law, with resultant changes in gene expression. With all the cuttings and stitchings needed to process a gene in the chromatin into a shorter, exportable message destined for cytoplasmic translation sites, opportunities for surgical error seem plentiful. But that is the beauty of it, claims Walter Gilbert, chief of Harvard's team of molecular surgeons. Enzymes may be incredibly accurate in cutting out the introns, but mutation can affect the edge of the cut ends so that, when the pieces are stitched together, there will be a slightly altered message: "Single base changes . . ." he said, "if they occur at the boundaries of the regions to be spliced out, can change the splicing pattern. . . .During the course of evolution, relatively rare single mutations can generate novel proteins much more rapidly than would be possible if no splicing occurred" [4].

According to the Gilbertian scheme, then, discontinuous genes offer enhanced chances of changing the genome to meet the challenge of changing circumstances, which appears to have been the grand strategy of evolution. The genome does not just sit around for some mutagen to create a change and wait around some more to see how the change fares in the face of selective pressure. By various recombination maneuvers it can produce new genes and selectively amplify certain genes it already has. It can do this for short-term situations as well to counter a noxious environmental force. Schimke's group at Stanford showed that some mammalian cells can acquire methotrexate resistance by amplifying the gene for dihydrofolate reductase 200-fold [5]. Alternatively, it can step up the rate of transcription of a gene by as much as 1,000-fold in response to some inducing agents.

Polymorphism of the so-called noncoded sequences between and within the genes will certainly be vigorously exploited in the coming years and for a time may overshadow the search for regulatory genes. Clinical geneticists are salivating at the vision of rich gold mines of genetic markers buried in what was only yesterday the no-man's-land of our genome. And no wonder. A year ago Kan reported on the polymorphism of a restriction site adjacent (3' side) to the β-globin structural gene and correlated one variant with the presence of the sickle cell gene [6]. The thalassemias are the next obvious target, and no area of biomedical research is seeing a keener race among the few laboratories equipped to practice the exacting art of restriction mapping. It is a matter of time, probably a short one, before scores of laboratories all over the world will be busily chopping up human genes sent to them from Tom Maniatis's bank of cloned stuff at Caltech.

As far as regulatory genes are concerned, the field will surely be mulling over the hypothesis put forward by Davidson and Britten a few months ago [7]. Their idea is that the repetitive sequences interspersed among the structural genes produce RNA which controls the production of messenger RNA. These repetitive sequences have been found in heterogeneous nuclear RNA, the function of which has never been clear. Certain of these repeat families are found at high levels in the nuclear RNAs of some tissues and at negligible levels in others. This may mean that the call for various repetitive RNAs is determined by the activity or inactivity of the structural genes they subserve according to differentiated states of the tissues or cells.

The Davidson-Britten model will doubtless serve as a paradigm around which the forces of molecular genetics will rally. Repeated sequences, long a DNA in search of a function, have found one—no less than that of the general regulator of gene expression. The conventional notion of regulator genes disappears into the largely uncharted regions of DNA and the motley army of heterogeneous nuclear RNA. Control of genes overshadows their functions by the sheer mass of the apparatus.

The introns are not a part of this model. There are many speculations about their origin and function. One suggestion is that they may be translocatable genes brought into eukaryotic cells by bacteria. The idea has not a germ of evidence at the moment, but translocatable elements in bacteria are worth taking a glance at because they give another peek into the innate pliability of the genetic material. A number of these translocatable elements have been identified in bacteria, including antibiotic resistance factors and some phages. They can move from one place in the genome to another in a rather random fashion, in some cases causing a disruption of gene expression at the sites of their new lodgings. They have been extensively studied because the phenomenon of "jumping genes" is subversive to the long established view of genes' stability as locatable entities. Furthermore, in addition to the fact that some of these translocations are capable of turning on or turning off the genes or a block of genes which are neighbors in their relocated site, the act of translocation itself apparently not infrequently produces minor chromosome rearrangements, particularly deletions [8].

The interest of researchers in jumping genes at this time, however, is in their potential not for anarchy but for order. May these dynamic elements possibly be used by the genome phylogenically and ontogenically—that is, in the evolution of species and differentiation of organisms? Indeed exons may also move around and recombine in the course of processing of messengers, as Crick and others suggest [9]. We are back with Gilbert. If excision and ligation needed to remove introns in the transcription of the gene into mRNA offer a fertile ground for gene rearrangements essential for evolution, so can deletion and other artifacts left behind in the wake of jumping genes. Gilbert has said "Introns are both frozen remnants of history and as the sites of future evolution" [4]. This is more or less what the specialists in jumping genes attribute to their subjects: "Inversions, deletions, and transpositions" wrote Nevers and Saedler, "may serve to create new nucleotide sequences or even new genes at fusion points. These processes may also represent important evolutionary means for reshuffling genetic information, uncoupling genes from old regulatory signals and linking them to new ones" [10].

After all this, one gets the feeling that we no longer need mutation to evolve—at least for quite a while. We already have a huge stock of variant genes which can be recombined endlessly to produce new phenotypes. And we are beginning to learn that our genome contains a built-in system for change and adaptation: we can generate new phenotypes from within and can do so quite rapidly. Some would say we owe to this not only our past but our future; our capacity to transcend humanoid limitations.

Ohno took this idea further by giving us two kinds of evolution: Promethean and Epimethean [11]. Of the two contrasting brothers, Prometheus looked forward, Epimetheus back. Epimethean is evolution by natural selection; Promethean is evolution by—what? By a kind of anticipatory adaptation to the unknown. The prime example is the immune system of the higher vertebrates. The system can deal with a vast range of nonself antigens, old and new, by the combinational possibilities of the heavy and light chains of the variable region and by somatic mutations modifying stem cells during their ontogenic development. The anticipatory repertoire of the system might be expected if the host were to deal successfully with microbes known for their quick adaptability. The second example is the microsomal mixed-function oxigenase system which can merrily metabolize multitudes of new chemicals, including drugs, let loose in the world by the post–World War II technology. Between lymphocytes and P-450 we may be able to disarm a vast variety of pathogens and toxins here now and those yet to come.

The third example is the brain. This may be the touchstone by which we may eventually be able to assess the success or failure of Promethean evolution. There have been many who expressed the fear that the brain is the Achilles heel of man, that the development of the central nervous system of primates is the kind of specialization that will drive the species toward the point of no return. The Promethean hypothesis tends to shed a different light on the problem by emphasizing the unexpected capacity of adaptation to the changing scene. The details of future adaptation are by definition unforeseeable. The brain of the Cro-Magnon man and of Einstein cannot be differentiated anatomically or chemically. What is different is not the brain but the information stored in it and, possibly, the organization of its recall and association. If we had the necessary technology we might be able to see the difference by looking for how neurons are wired, that is, for the number and distribution of synapses. Humans have the same brain as they had 15,000 years ago and are now dealing relatively successfully with an unimaginably altered environment.

We cannot tell about the future partly because we do not know what adaptive tour de force our genome is capable of. We know that we are polymorphic at many gene loci, superpolymorphic at some loci. It is polymorphism that allows most animals to carry a dangerous cargo of lethal and detrimental genes without mishap through their life spans. By reversing the process through intensive inbreeding of experimental animals, geneticists have unmasked innumerable examples of detrimental genes that are a part of our genetic load. Polymorphism must make most individuals biochemically—that is, genetically—unique.

The potential genetic variation in any species may be much greater than we can surmise. Our genome may contain a population of "unused" genes, those not coded for functionally necessary proteins. These could be largely free from selective pressures that weed out functionally incompetent genes. In the Promethean scheme of things, presumably these genes can be mobilized by chance or necessity to play an active role in the ascent up the evolutionary ladder of the species. The fate of organisms with flexible genetic materials may depend on the seeming luxury of maintaining a troop of idle genes. It may become a task of molecular biology in the coming decades to train their powerful instruments in scanning the outer reaches of our genetic cosmos for signs of these "genes of the future."

In the end we are forced to echo the words of Sir Thomas Browne: "We carry with us the wonders we seek without us; there is all Africa and her prodigies in us" [12].

It may be a century before we grope our way into the heart of darkness.

Chromosomes

Somewhere in his writing McKusick compared the dogma of molecular biology with the trinitarian dogma of Christianity. So we have God the Father (DNA) and the only-begotten Son (RNA), but where is the Holy Ghost? Biologists who can be induced to join in the game of this sort would probably say chromosomal proteins. If you ask them to specify, they will probably opt for nonhistones.

Chromosomal proteins are of two kinds: histones enforce silence on DNA; nonhistones allow genes to speak out and send messages. In the present state of molecular biology, chromosomal nonhistone proteins enact the pentecostal role of the Holy Ghost.

The most convincing revelation about the role of nonhistone proteins came from the chromatin reconstitution studies of Gilmour and Paul [13]. Gene expression, it seemed, depended on the presence of tissue-specific nonhistones in the chromosomes. Differentiation obviously involved changes in the population of nonhistones in the chromosomes.

The trouble with nonhistones is that there are so many of them and they are apparently so kaleidoscopic in their comings and goings. They are usually referred to en masse as nonhistones, which is a candid confession of our ignorance. Their alignment in the chromatin is obscure, and their mechanism in permitting gene transcription equally so. Indeed, a few of them may be involved in suppressing gene expression [14].

Doubtless many laboratories are attempting to pull out small groups of nonhistones, characterize them, and raise antisera specific to them. The population of nonhistones on a chromatin may change as rapidly as that of receptors on cell surfaces. Certainly the so-called high mobility group (HMG) chromosomal proteins can readily go in and out of nuclei and seem to achieve a dynamic equilibrium of their presence on and off chromatins. This was what Comings and Harris proposed in 1976 as a probable general characteristic of nonhistones [15].

Histones are another story. We know the five major groups, know their disposition in the chromatin, and now have evidence showing that their genes are probably placed in a neat linked row on chromosome 7 of man [16]. Histones confront nonhistones as the ordered rank of a Roman legion facing the rampant herd of blue-painted Caledonians.

Out of the order brought to histones came the discovery of the nucleosome. One gets the impression that the nucleosome was born almost as a reality rather than as an idea. There seems to have been no usual extended period of gestation when scientists debated the possible reality of the idea preceding Kornberg's article in 1974 [17] which spread the gospel widely among biologists. In no time the belief in the nucleosome as the basic unit of chromatin became universal. These are the beads in the string which we long imagined and then began to see so often in electronmicrograph pictures. The elucidation of the nucleosome is probably the most important conquest made so far in chromosomology.

If the structure of the nucleosome is essentially that of the eight molecules of histones that make up its bulk, it does not follow that the histones determine the architecture of chromatin. Recent reports from Laemmli's group suggest that a network of nonhistones governs the packaging of DNA and histones into chromosomes [18].

Packaging turns one's thoughts to that loner among histones, histone 1 (H1). Worcel's model of the "thick" chromatin fiber has the nucleosome chain bound around by what amounts to a functionally continuous helical chain of H1 molecules, the outer helix to tighten the inner (DNA) helix and keep the whole chromatin condensed [19]. This is of course reversible: H1 can unwind and let the nucleosome chain stretch out. One might think of the play of H1 on nucleosomes as like that of the fingers of the violinist on the strings of his instrument. The counterpoint of H1 and nonhistones draw from DNA the melody we recognize as differential gene activity.

There are some signs on the horizon to indicate H1 may be involved in gene regulation as well as chromosome contraction. The H1 genes are present in all cells, but the relative proportions of different H1 histone subtypes differ in different cells in different stages of the cell cycle. Different H1 histones may impart subtle differences in the structure of interphase chromosomes as they can produce gross changes in that of metaphase chromosomes. This suggests that H1 may be very important in the induction and maintenance of cell differentiation. Hohmann and Cole noted changes in the H1 subtypes when the organ culture of mouse mammary glands underwent differentiation [20]. The same results were seen in vivo in the mammary glands of mice proceeding from pregnancy to lactation.

One possible active area of chromosome research in the coming years is the relation, organizational and temporal, of H1 and DNA in the heterochromatic chromatin, whether it be constitutive as in the centromere regions or facultative as in inactive X.

This leads us into the enigma of X-inactivation in the female of eutherian mammals. It is almost 20 years since Mary Lyon published the hypothesis that bears her name and galvanized the world of genetics [21]. The hypothesis was that early in embryonic life there is a random inactivation of paternally derived X or maternally derived X (X^p and X^m) in the cells of female placental mammals, and that therefore every female is a mosaic inhabited by two populations of cells with different genetic constitutions. Her hypothesis is now as widely accepted as that for evolution, but the mechanism of "Lyonization" remains as obscure as ever.

Lyon's hypothesis has a fascination beyond chromosomology. Consider the temptation of the challenge. One sex says to the other: I have two X's, you have only one. This is inequitable. I will inactivate one of my two X's so that we will be equal. In the face of this overwhelming female gallantry, a number of male investigators have tried to discover the mechanism of the trick. None has succeeded. The eternal feminine has so far eluded them.

To be sure, it is not all the female prerogative. An XXY man inactivates one of his two X's. A boy with XXXXY should have three inactive X's, and so it turned out to be [22]. And the X^p/X^m mosaicism of the female apparently needs a slight modification. It now looks as though in rodents, and possibly in man, almost all cells of the chorion and other extra embryonic membranes contain an active X derived from the mother [23]. At least in these tissues the inactivation is not random. It has been known for some time that the X^p is preferentially inactivated in all cells of the kangaroo and other marsupials [24]. Random inactivation of X^p and X^m was apparently invented after the appearance of placental mammals.

There seems to be something special about X. The Lyon principle was initiated eons ago in some primitive mammal now probably extinct, but has remained the heritage of all subsequent organisms. Because of this, said Ohno, X, unlike other chromosomes, could not have changed, and the genes inhabiting it are the same ones from anteater to man [25]. Remarkably enough, the size of X in relation to the total amount of chromosomes appears to have remained the same in a wide range of species, and their X-linked genes have remained identical. Ohno's law looks as solidly founded as Lyon's. It seems only fair that it was a man who took Mary Lyon's law to a time far antedating the birth of Eve.

With current methods of cytogenetics we are finding more and more people with small deletions, additions, translocations, inversions, and other anomalies. They are likely to be phenotypically normal, but they are to varying degrees at higher risk of having fetal loss or a child with birth defects, in the way the carriers of recessive mutant genes are.

A balanced translocation can be transmitted in a family for generations before surfacing in some unlucky child with partial trisomy and/or other anomalies. Certain clinically neutral chromosome anomalies can persist in families for generations and serve as the family's biological marker, like the well-known Hapsburg lower lip. The males of the Byler family, one of the Amish kindreds intensively studied by McKusick's group, have a distinctively small Y chromosome. Several years ago, McKusick had the opportunity to visit a Swiss canton from whence the founders of the American Bylers migrated in the eighteenth century and looked at the karyotypes of the current generations of the Swiss branch. All the men had the same small Y [26]. This impressive testimony to the marital fidelity of the Byler women—for 200 years, on both sides of the Atlantic!—suggests that some chromosome anomalies can remain in a family like a recessive gene mutation.

In the past decade researchers learned the power and versatility of cell hybridization in biological studies. It was put to many uses, most triumphantly in gene mapping. McKusick aptly pointed out that the human gene map is the anatomy of our genome and we are already well launched in the study of comparative and morbid anatomy of mammalian genomes [27]. No representation of nature is more central to our being than this. We will have a treasury of markers to guide genetic engineers—including the decision not to engineer.

The gene map is the inner picture of the species, its immortal part. The gene map may tell us not just where the genes are but why they are there. It may provide clues to the historical or functional reason for the placement of structural and regulatory genes, of the disposition of single copy genes, repetitive sequences, the vast, intergalactic stretches of spacer DNA that Ohno once called junk, the kitchen midden of evolution but still useful, he thought, as a shield against the mutational overload for functioning genes [28]. Who can say that some Lochinvar of the twenty-first century may not find in these middens the secret of Ohno's Promethean evolution? Is not the iterated sequence specific to the human Y chromosome, isolated and characterized by Boyer's group [29], the first good example of dinosaurs from which we may learn how genes attenuate in our genome?

It is clear that many things are still unclear about chromosomes, including how they are put together. As Ohno pointed out, "From a string of beads, visible under the electron microscope, to a chromosome as seen under the light microscope, needs a great leap of the imagination" [30]. We are seeing these fabulous organelles, vessels of life, by a flickering torch of our inadequate knowledge as vague and awesome apparitions, the Tyger burning bright in the forest of the night, and wonder whose hand and eye may eventually fathom their fearful symmetry.

Cells

Cell culture developed out of tissue culture in the early fifties when a method was found to free cells from tissues. The rediscovery of trypsin to get a flask full of cells was a momentous event, for then you could do with mammalian cells what you had been doing with bacteria. It became possible to count them, grow them, starve them, clone them, mate them, transform them, enucleate and then renucleate them, mutagenize them, lysogenize them, transplant them, grow vaccine viruses on them, vivisect them into minicells, cybrids, and other freaks. After all that and more, if worst comes to worst, said Bob Kruth, you could eat them [31]. Soon biochemists noted that a cell in culture is in many ways a man in miniature, which can be metabolically dissected to reveal molecular characteristics of the donor from whom it came. The belief grew that what *E. coli* did for molecular biology, the fibroblast will do for human genetics.

So it is perhaps not surprising that in the past decade one after another of the departments of anatomy of our medical schools rechristened themselves departments of cell biology. We have shifted our focus from the body politic to the individual citizens that make up the body. The gain in knowledge has been vast and epochal. So much so that the time may soon be ripe for countercurrent research aimed at integrating what we have learned from cell studies into understanding of the whole animal, and in a decade or two we may see the coming of a latter-day Claude Bernard or Walter Cannon to synthesize existing knowledge into new concepts of organismic biology and witness the application to holistic medicine.

The most spectacular coup in cell biology was made by the classical approach of fibroblast analysis. I am referring to the unfolding drama of the receptor behavior and misbehavior in regard to low-density lipoprotein (LDL), the American Ring cycle staged by impresario Joe Goldstein and associates. The leitmotif is already clear: the receptors, like the gods, are a mixed lot, some are weak, corrupt, or otiose; the ultimate prize is more precious than the Rhinegold—the saving of human lives and not only from early coronary heart disease.

Consider the characters and setting as revealed up to now [32, 33]. So far we have been introduced to three types of receptor defect in familial hypercholesterolemia: the first will not bind LDL, the second binds very weakly, the third can bind LDL but cannot get it into the cells. Why can it not get into the cell? That brings us to the arena where the first act of the drama takes place, the all-too-graphically named coated pits. These occupy 2 percent or less of the cell surface but contain 50–80 percent of

LDL receptors. That third kind of receptor either cannot migrate to the pit or is kept out. Why? Possibly because the receptor has a faulty or missing second active site which in our state of ignorance, can only be called the internalization site. So we are given the first intimation of future revelations: there may be other diseases due to receptors' inability to internalize their prey and possibly others due to defective coated pits manifested physiologically as receptor deficiencies. And coated pits are not a specialty of LDL binding. Five others are known to use this device, including transferrin, epidermal growth factor (EGF), and α-2-macroglobulin. Others in the suspect list include nerve growth factor, trans-cobalamin II, and insulin. Cell surfaces may be studded with these elaborate traps to catch the circulating proteins.

The next act of the drama has further surprises. Coated pits loaded with receptor-bound LDL invaginate and form coated vesicles which carry the cargo to lysosomes where digestive enzymes chop up the protein. The surprising part is the structure of the coated vesicle: it is encased in a lattice work of a high molecular weight protein, organized into 12 pentagons and a variable number of hexagons. The geometric design is as elegant as Waterford crystal or the head of a bacteriophage. One wonders why so rococo a vehicle to transport baggage so mundane as LDL to the lysosomal garbage dump? Is this just another of nature's excesses like the peacock's tail or the human brain? It seems unlikely. Barbara Pearse, who isolated coated vesicles and worked out the design of their outer lattice, also did peptide maps of the principal protein, clathrin, from several species and found the protein to be strongly conserved [34]. If structural conservation is a sign of functional indispensability, clathrin may be as important as hemoglobin.

Another facet of the LDL-receptor endocytosis is equally intriguing. Kinetic studies of the process in vitro by Goldstein's group suggest that LDL receptors can be recycled, at least under some conditions [32, 33]. Presumably this can occur independently of de novo synthesis of receptor since, when the latter is shut off by cycloheximide, fibroblasts would still ingest LDL at a uniform rate for more than 6 hours and do so in the absence of detectable pools of ready-made LDL receptors. If we accept reappearance of ingested receptors as recycling, we are still faced with the puzzle: by what Houdini-like agility do receptors escape from the coated vesicle tightly encased in clathrin before being ground to pieces in the lysosomal mill?

The whole intricate endocytosis pathway is still clouded by the shadow of known unknowns. Why, asked Goldstein et al., should protein hormones which can exert their effects at the cell surface undergo the elaborate charade of internalization? Unless, one might suggest, the process is tied in some fashion to homeostatic regulation of receptors to their ligands. It will be helpful indeed to find mutants of one or other elements of this drama, including defective clathrin and deficient re-cycling. These will push the knowledge gained in cell culture to the plane where we can see its biological significance in terms of the whole organism.

The next big push in cell biology will probably center around endothelial cell cultures, mostly because of the strategic locations of these cells between blood and tissues that give a scent of pay dirt for clinically significant findings in the areas of urgent medical and public health problems: hemostasis, thrombogenesis, and atherogenesis. Endothelial cells from human umbilical vein and bovine aortic arch can now be readily grown with the aid of growth factors such as fibroblast growth factor (FGF) and EGF; they can be cloned; plasma rather than serum can be used in the culture medium, thus mimicking the in vivo state; and the culture can be subjected to various physiological or pathological influences for study of their effects on this simulated vascular lining [35]. Endothelial cells are not just stones in the Great Wall of China that line our blood vessels; they make factor VIII antigen, Wichel-Palade bodies, angiotensin-converting factor, prostacyclin, and massive amounts of fibronectin, the sticky glycoprotein that firmly anchors the cells to the vascular wall in the face of a rushing stream of blood that bathes the outer surfaces.

Whether endothelial cell culture will contribute much to the understanding of chronic, progressive, multifactorial diseases such as atherosclerosis is difficult to say. Yet it will be interesting to see how endothelial cells handle LDL (and high-density lipoprotein, HDL, as well) in vitro. Organ culture of blood vessels may be needed, and there has been a recent report on how to make an artificial blood vessel using rat smooth muscle cells overlaid with bovine endothelial cells [36]. What looks like a basal lamina containing a protein-like fibronectin formed between the two. Presumably you could do this with the tissues from the same species. This half a vessel was said to be stable in culture for several months.

The current interest centers around the problem of thrombus formation and the view that there is a lifelong war between circulating platelets and stationary endothelium. This is the field where the coagulation cascade interacts with the prostaglandin cascade. Arachidonic acid is converted by cyclo-oxygenase, apparently present in all cell membranes, into endoperoxides from which platelets can generate thromboxane, a powerful platelet aggregator and vasoconstrictor, while the endothelium can produce prostacyclin with the precisely opposite effect [37]. Every minute break on the vascular wall becomes the battleground of the two forces, each armed with its own prostaglandin; the axe of thromboxane clashing with the sword of prostacyclin.

Needleman's recent review [38] suggests that this may be a simplification, but the fact remains that here are two tissues, created for contrary purposes, fashioned from the same metabolite, weapons designed to further their conflicting aims.

We are viewing from one angle the new world of prostaglandins. Genetics has not entered this arena. Here is an army of agents whose actions are evanescent, seemingly local, and often at war with each other—an army in which we cannot discern a controlling hierarchy or obvious liaison with other systems such as the endocrine and neural. But who can doubt its importance? After all, all mammals are presumably born in the baptismal shower of prostaglandins. At the moment, then, here is fatty acids' challenge to medicine and a reminder that fat is not just for burning.

Development

Development is a wondrous thing, though the end product may be dismaying, partly because we so often do not know the cause of maldevelopment. The developmental history of an idiopathic idiot is as complex as that of the doctor treating him.

No subject in embryology has received more attention in the past 10 years than that of the T-locus of the mouse. The reasons for this are many, but perhaps most important is the fact that mutations in the T-locus genes are often lethal in early embryonic stages when homozygous, and the effect may be due to altered membrane antigens causing failure in cell-cell recognition. That is to say, the locus houses developmental genes, the first such to be recognized in mammals [39]. There are other genes in the T-region unrelated to embryo lethality, and the region is a near neighbor of that most intensively studied gene complex, H-2, on chromosome 17. Man has HLA to match H-2 but where, one asks, is his T-locus equivalent?

The question is natural. After all, HLA is a gross plagiarism of H-2, and H-2 equivalents have been found in all mammals so far studied. Why not another gene complex coding for early embryonic membrane antigens before H-2 comes on the scene? The timing of switch-over from one to the other in the mouse is almost perfect—perhaps too perfect? Still, one is teleologically titillated by the argument of Artzt and Bennett that T-like genes preceded H-2–like ones in evolution as they precede H-2 in the ontogeny of the present-day mouse [40].

No sign of the T-locus equivalent in man has surfaced up to now. In the meantime, the short arm of human chromosome 6, where HLA is located, is gradually being populated by genes, none of which gives a hint of being involved developmentally. Of course, such genes that kill early embryos in homozygotes but are harmless in heterozygotes, except possibly male sterility, would not be easy to unearth in man. When Snell

408 | Samuel J. Ajl and Joseph Mori

suggested in 1968 that the T and H-2 make up a supergene vital to the welfare of the individual mouse and of the species [41] the challenge was heard around the world. We are still waiting for our grapevine to go taut with a whisper that someone, somewhere, has found HLA linkage in, say, familial habitual abortion; it almost did earlier this year when Yunis' group reported an association of human testicular teratocarcinoma and HLA-DW7 [42].

There is one idiosyncracy about the T-locus that may make the mouse situation unique. Perhaps I should remind you that T-locus mutations are not artifacts of laboratory inbreeding; they exist in wild mice and are perpetuated there in spite of their high lethality by their preferential transmission to the offspring by male heterozygotes carrying them. Instead of the expected 1:1 ratio of transmission, the ratio runs as high as 9:1 or more. We have here the only mammalian example I know of what Jim Crow called cheating genes [43]. Many examples of these "segregation distorters" are known in *Drosophila*. The mechanism of distortion is obscure both in the fly and the mouse. But there is in the middle of the T region a group of genes called the A-factor (for abnormal ratio factor) which fixes the odds in the genetic roulette [44]. Furthermore, there is something operating in the T-region which reduces or suppresses crossing-over at meiosis, allowing strings of genes to be transmitted as haplotypes.

In fact there are signs that other examples of mammalian cheating genes may soon share the lonely eminence now occupied by the T-locus. Bennett has cited several suggestive examples [45] and there was a recent report of preferential transmission of a human α_1-antitrypsin variant by male carriers of the mutant [46].

Thus, the T-region is not only an unreal landscape where lethal mutants take refuge but is also inhabited by genes that can bewitch chromosomes into bizarre misbehaviors. It is a daunting subject. Even as Mary Lyon led her troops, in a recent paper, over the labyrinthine ways of the T-region in search of palpable mechanisms behind the mysteries, at the very end she fell back and invoked the most popular figure in biological demonology: Can it be, she asked, the region is haunted by a viral genome [44]?

The teratoma story parallels the T-locus story. Here is a cancer that differentiates and yet remains a cancer. Its pathology is well known—a bouillabaisse of mixed tissues. The cancer's name is well deserved: the sight is a monstrous wonder. It was from an established teratoma line that Dorothea Bennett and collaborators got hold of F9 antigen—present only in sperm, undifferentiated teratoma, and morula-stage mouse embryo—a product of the normal allele of a T-locus mutant, t^{12} [47]. Before long, Elaine Vitetta and her associates showed what looked like remarkable structural similarities between F9 and H-2 antigens [48].

The discovery of F9 stands at the first clear-cut crossroads where the paths of T-locus and teratoma studies intersected.

But the teratoma work was going off on another tangent—the attempt to create chimeric mice by fusion of normal blastocysts with teratocarcinoma cells. This line of work reached the first culmination in papers by Mintz and Illmensee which presented convincing evidence that teratocarcinoma cells in the environment of an early embryo can differentiate into all normal tissue [49, 50]. Tissue distribution of genetic markers of the teratoma line used, including the isozyme of glucose phosphate isomerase, was impressively wide and included sperm in two animals which went on to sire healthy second-generation chimeras. It was concluded that some teratocarcinoma cells are totipotent, like early embryonic cells and, given the right environment, can differeniate into functioning tissues according to the ontogenic schedule of the normal mouse. What forces in the blastocyst can lead these aberrant cells into an integrated pattern of differentiation is not known, nor is the nature of the factors that in the ancestral mouse drove these cells, capable of a different destiny, on to the oncogenic pathway.

The next step Mintz took was a diversion doubtless too tempting to resist. It must have occurred to many people to ask: Why not induce mutation in teratocarcinoma cells, use selective media to pick out the mutant of choice, and then, by the blastocyst route, to create chimeric mice carrying the gene for human mutation? Mintz attempted this in collaboration with Dave Martin and others of San Francisco. Predictably, they chose hypoxanthine phosphoribosyl transferase (HPRT) deficiency. The goal, of course, was to generate the Lesch-Nyhan disease in the mouse.

They have not reached their goal yet. They got the chimeras, and the tissue distribution of the mutant cells was broad: one animal had mutant cells in all 18 tissues examined, except for blood cells. Several had HPRT cells in the gonads but not, apparently, in the germ cells [51].

In a cautionary note on this work, Heath made two interesting points [52]. One was that, when cells from the normal inner cell mass are injected into the blastocyst, one gets a fine-grained mosaicism of the host and transplanted cells, in contrast with the teratomatous chimeras in which the teratoma-derived cells tend to occur in patches. The differences, he hinted, may be important in considering the similarity of the model, if attained, to the human disease. The second point was that the teratoma-derived cells very seldom became germ cells, possibly due to selection against aneuploid gametes.

The fact remains that Mintz and co-workers have succeeded in two different experiments in infiltrating the gonads with cells of teratoma parentage, on one occasion into mature, fertile sperm. The mutant cells they used for the Lesch-Nyhan work were in fact aneuploid—XO with trisomy 6—and one remembers that in the human XO is one of the most common chromosome abnormalities seen in spontaneous abortuses. Mutagens also may produce subtle chromosomal changes that are still inimical to normal gametogenesis.

Conceding all this, the task does not strike us as insurmountable. With a little bit of luck and a little bit of technical innovation, somebody may fairly soon produce carrier females, and homozygous males may then be only a mouse generation away. No royal princeling would receive the kind of care and attention that would be lavished on the world's first Lesch-Nyhan mice. There will be champagne in the laboratory when someone first spots one of the mice compulsively chewing its paws.

But will they? Suppose they do not—what kind of model can one call animals lacking a cardinal clinical feature, even if they possess all the biochemical and cellular markers of the human syndrome? Or if they exhibit a different kind of anomalous behavior—for example, if they starve themselves or mutilate their cage mates? On the other hand, it is possible that such animals may give better leads for unraveling the pathogenesis of the disease. The difference itself can be the key. In fact, Mintz and associates claimed in their paper that the patchy distribution of HPRT⁻ cells in their mosaic mice may make them more advantageous for investigation than "models whose genetic lesion is in all their cells" [53]. Their homozygous males would have the "lesion" in every cell, and yet their disease may differ moderately or radically from the human Lesch-Nyhan syndrome.

In any case, investigators are now as ready to rush into the newly opened field of man-made chimeric mouse models of genetic metabolic diseases as the homesteaders were into the Oklahoma territory. Goldstein's group, in collaboration with Mintz, is now trying to produce LDL-receptor–deficient teratocarcinoma cells by mutagenesis and selection and create a familial hypercholesterolemia model [54]; and the San Francisco group will try a similar feat, aiming for purine nucleoside phosphorylase deficient type of immunodeficiency [55]. Goldstein et al. have reported that teratocarcinoma cells, the ontogenic primitives, are already equipped with high-affinity LDL receptors and can bind, internalize, and metabolize the cholesterol carrier as proficiently as can fibroblasts from the adult human [54].

What animal models could tell us is how an organism survives, succumbs, or succeeds in finding an alternative means to control the biochemical lesion. The predictive potential for these things is limited for cell culture studies. Who would have thought a particular purine dysmetabolism would lead to the brain dysfunction of Lesch-Nyhan? The research community is waiting with great anticipation for investigators to negotiate the next crucial step and present the world with the first man-made animal model of a human disease. But the accolade of history, we think, will be loudest for the investigator who can show what epigenetic or microenvironmental influences determine the fate of early embryonic cells between an animal and a wondrous tumor.

Communicators

By communicators we mean chemical messengers such as hormones, neurotransmitters, growth factors, lymphokines, etc. They are lumped together here because we have not come across a system that can clearly classify the target-cell-active messenger molecules. The coming of the hypothalamic polypeptides was the French Revolution of endocrinology, and we are still living in a post-Bastille confusion. The pituitary was dethroned but has not yet been guillotined. One thing is sure: the brain is as much an endocrine organ as the kidney; both do other things but they also produce hormones.

Neurobiologists are rearranging the anatomy of the brain by grouping nerve cells by their biochemical markers, that is, the neurohumors they produce. The brain is crisscrossed with an intricate network of neurohormone circuits. Many years ago George Bishop, a Washington University neurobiologist, asked, "Where does one come out if he looks at the neuron as a secretory organ . . . to wit, [that] the prime function of a neuron is to produce and apply to other tissues a chemical activator?" [56]. He should be here now.

Thus, we now find that some brain hormones are also made in the gut and some gut hormones in the brain. Somatostatin and cholecystokinin (CCK), for example, are present in both organs. Somatostatin's work in the pancreas is as important as its work in the brain [57] and while we do not know what CCK does in the brain, one suggestion is that it may have something to do with appetite control [58]. If true, its job in the brain is as important as its job in the gut. We are confronted by what that master syncretist, A. G. E. Pierce, calls the "diffuse neuroendocrine system" which, by his count, produces at least 35 physiologically active peptides, including 17 "common" to both central and peripheral locations [59]. If Bishop could examine this system, he would learn that the brain is the largest endocrine organ in the body.

No sooner than we got used to the hypothalamic release factors and a release inhibitor, somatostatin—where, by the way is the growth hormone releaser?—we were faced by another mob of brain messengers which I shall call by the name of its Robespierre, endodorphin. We were in the second stage of the revolution: neuropsychiatry will never be the same again.

It should be recalled that the endorphins were found after the discovery of their receptor cells, and the receptor cells were discovered because they bind morphine and other opiate drugs. When laboratories all over the world raced to look for something endogenous to the brain that might behave like opium and bind to the opiate receptors, Hughes et al. came with enkephalin and Guillemin's group with endorphin [60–63]. The brain, it seems, is endowed with its own anesthesia department. But the brain opiates do other things. When Bloom and co-workers injected rats with endorphins, the animals went into catatonia. They came out of it quickly when injected with a morphine antagonist, naloxone [64]. With that catatonia, psychiatrists began to move in, for that is one of the classic symptoms of some patients with schizophrenia.

What followed has been clouded in confusion, but some things seem more certain than others. Increased amounts of endorphins have been found in the spinal fluid of schizophrenics and manics [65, pp. 19–56]. There followed a rush for controlled trials of schizophrenics with naloxone by dialysis (the drug has a short half-life). The results were mixed [66, pp. 86–95]. No doubt we will get a better bearing when the smog lifts, but some of the later trials, using larger doses, got what was called "encouraging results," especially in the relief of auditory hallucinations [67]. Endorphins apparently could do several things, good and bad: relieve pain and produce catatonia, induce pleasure and provoke seizures. How endorphins are related to one or more kinds of psychosis is not known, but this is an unknown that will doubtless be explored in depth in the coming years.

There are signs indicating that the brain's pharmacy is stocked with other nostrums. Thus, Spector isolated from human spinal fluid and other sources something he calls a morphine-like compound, which is not a peptide like endorphins, but a very small dose of which could stiffen rats into catatonia [68]. But in this case naloxone did not unstiffen the animals in seconds. It took 30 minutes to two hours for the rats to lose their rigidity.

Besides the nonpeptide opiate, we now have what may be the brain's endogenous tranquilizers. Here history repeated itself. First the investigators found in the rat brain what appeared to be specific receptors for that large class of antianxiety drugs known as benzodiazepine, of which valium and librium are the prototypes [69, 70]. Soon others found a protein from rat brain that binds to benzodiazepine receptors and which may also modulate the receptors of the known inhibitory neurotransmitter γ amino butyric acid (GABA) [71, 72]. Morphine and valium—soothers of a ruffled mind! The latest turn of events, though, is that certain purines bind to benzodiazepine receptors and that a valium-like drug and the purines have agonistic and antagonistic effects on the membrane of cultured neurons [73]. What purines? Inosine and hypoxanthine. In the face of this revelation, not a few investigators must have been seized with a wild surmise that we may have here the key to unlock the neurobehavioral secrets of Lesch-Nyhan patients.

Taking a leaf from Bishop, we might ask at this point: Where do we come out if we postulate that subtle derangements in the relationship of endorphins, their precursor molecules, and other neurohormones cause some psychic disorders—to wit, low addiction thresholds, neurotic behavior patterns, suicide propensity, learning disability, temporary sensory loss, depersonalization, anorexia nervosa? We may answer that, if the derangements are caused or abetted by genetic makeup or developmental accidents, we will be forced to deal with a whole new category of birth defects.

It was interesting to learn that endorphin is present in large amounts in human amniotic fluid and that a high concentration of it is present in the pituitary of newborn monkeys [74, 75]. We suppose that a number of investigator groups have studied endorphin levels in human cord blood, and we are beginning to receive reports of such studies [76]. One might expect a plentiful supply of endorphin in late pregnancy, since if anyone needs analgesia it is the perinate on his way past the Pillars of Hercules that is the mammalian birth canal. Every natural birth is probably medicated by the fetus with its own endogenous analgesics from the brain and pituitary.

We have long been interested in that enigma-ridden fetal protein, that is, α-fetoprotein (AFP). It has been noted that AFP has an immunosuppressive effect, and some have proposed that it, with other factors, may play a role in allograft tolerance of the fetus by the mother [77]. Perhaps a better candidate for that role is one of the newly found placental proteins called PSB_1G [78]. It is produced in huge amounts, almost all of its goes to the mother's blood, and production starts very early since it has been detected in blood within 14 days postovulation. Antisera to it caused abortion in monkeys. The PSB_1G concentration rises through pregnancy; that of human AFP falls toward the end. Perhaps the two work synergistically, AFP in the fetus and the other in the mother. One wonders that the fall of AFP levels may signal the mother's endocrine system to get ready for parturition. These proteins, like progesterone, may be subtly integrated into the physiological scheme geared to maintain and terminate pregnancy; birth may be a graft rejection carried out by prostaglandins instead of lymphocytes.

Another thing AFP is supposed to do is to bind estrogen. In rodents AFP binds estradiol [79]; in the human it apparently does not [80]. But if AFP does not do this job, some other fetal protein may well do so [81]. Maternal estrogens do get into the fetus, and a binding protein, so the argument goes, is a tool to keep a homeostatic balance of circulating sex hormones in utero including protection of hypothalamic cells, which in rodents determine during the neonatal period the gender roles of adult sexual behavior [82]. But estrogen-binding hormones may have other functions. For example, out of the 2 million or so girls exposed prenatally to DES, we now have about 400 registered patients with vaginal carcinoma [83]. Why this 400? One might guess they had a genetic defect or developmental delay in the production of a fetal protein that can bind a synthetic estrogen like DES.

All hormones have cell receptors, and the two are inseparable functionally. This goes for growth factors as well. Birth defects due to receptor anomalies are increasingly being recognized. The case can be made that the primary fault in adult-onset diabetes, with or without obesity, lies in insulin receptors [84]. With the steroids the defect may not be in the receptors but in the sites on the chromosome where the hormone-receptor complex attaches to induce gene action [85]. In still others, the defect may be due to the hormone's inability to modulate the number or affinity of the receptors rather than the cell's ability to make them [86].

At a recent endocrine meeting a scientist was overheard saying, "If you are not in the receptor business, you might as well fold up your tent and go home." The next important push in developmental biology will likely surge around the gates of receptorology.

Immunity

Around 1971, Niels Jerne formulated the network theory of immunity. It was all done with his head—a synthesis of existing data and ideas shaped by a daring imagination into a new paradigm of the immune system. The system is a concourse of continuously conversing lymphocytes. An antibody-forming cell generates an immunoglobulin containing the operational segment, variable region, which harbors a marker specific for that cell and its clones—an idiotype. There will be other cells in the system which can recognize and react to that idiotype. And there is a third group, and a fourth and so on. Antigen-antibody reaction may have a specific effect, but the paradigm deprives it of its finality—it is a snatch of conversation in continuous talk. The resemblance to the central nervous system was not overlooked. "Both systems display dichotomies," Jerne wrote, "their cells can both receive and transmit signals, and the signals can be either excitatory or inhibitory . . . lymphocytes can recognize as well as be recognized, and in so doing they too form a network" [87, pp. 52–61].

A few years later in his Harvey Lecture, Jerne used a title more apt than "network" to describe his immune model: he called it a web of V-domains [88, pp. 93–110]. This gives a touch of matter to what seemed like a field of energy; it ties the paradigm closer to physical reality. The V-domains are a triumph of matter, as perfect as Botticelli's Venus and as provocative. We have learned to study the V-domain and to view how its parts are mortised to frame a dazzling shape, as we have already done with the variable regions of immunoglobulin light chains. And the "web" is the right image of Jerne's model; a cobweb of ever-intersecting threads woven by a master spider.

In 1971, the time was ripe for a new synthesis. It was more than 5 years since Good's group formulated the two-component theory [89]. Good has mentioned that at some meeting in the early sixties a young Australian investigator, Noel Warner, kept saying "It takes two to tango" [90]. Well, so there are *two* components. But most people looked at them as parallel, more or less independent forces, and there were diseases like DiGeorge syndrome and Bruton's agammaglobulinemia which fitted the idea that one can be lacking in one without being affected in the other. It gradually dawned on people that the two components are functionally one. The dance of immunity is choreographed for a team. Warner was right; it takes two to *tango*.

The discovery of helper T-cells and supressor T-cells came quickly on the heels of Jerne, and these were powerful arguments for the existence of a hemeostatic drive in the immune system implicit in the network idea of interacting elements.

The T-cell has a function of its own, independent of B, and that is the destruction of nonhistocompatible cells, presumably on the basis of a recognized difference in cell-surface antigens. In a series of elegant studies, Zinckernagel and co-workers showed that the killing of virus-infected cells requires the presence of H-2 antigens on those cells identical to the ones on the killer T-cells [91]. This so-called H-2 restriction phenomenon has raised many questions, including the one about the functional justification of what strikes the outsiders as an unnecessary and paradoxical requirement. A group of lymphocytes known as natural killer (NK) cells, which may or may not be of T origin, can kill without the ceremony of first verifying the H-2 credentials of the victim [92]. But the phenomenon is real, not only in mice with their H-2, but in other species with their respective major histocompatibility complex (MHC). Furthermore, it now seems that T-cells recognize their target antigens always in association with MHC antigens, and this need for double vision applies not only to their killer functions but to helper and suppressor duties as well [93]. This still leaves nonspecialists bewildered: How do we dispose of allograft the first time? Do we have to fall back on NK cells?

We are uncertain about the place of MHC-restriction in the biological warfare of individuals and species, but we are all too aware of the pervasive influence of MHC in human health and sickness as we strive to keep on top of the Niagara of data on HLA-linked diseases. Perhaps the most interesting MHC association uncovered so far is the voracious, indiscriminate appetite for MHC-coded antigens exhibited by Semliki Forest virus. It binds to both H-2 products on mouse cell surfaces and to HLA-products on human cells, as well as to the same products inserted into liposomes, like cloves in a ham. It is said that the stuff can be taken off the membrane and used on a sephadex column to purify MHC antigens [94]. The thought must have occurred to many people—what a bait to fish out MHC!

Polly Matzinger, whose article I am using, goes a little beyond this. The host range of the virus seems almost as wide as toxoplasma's; it can infect mammals, birds, and insects, presumably via specific receptor, MHC antigen or its equivalent. This means either the virus is a virtual virtuoso of receptor hopping or monomaniacally binds to a single receptor so precious as to be preserved intact over a vast stretch of vertebrate phylogeny. Matzinger wants to try the virus on protochordate, colonial ascidians. These animals have cell surface receptors, coded at a single locus, which can serve as recognition signals for free-swimming tadpoles to come together to form a colonial organism, tunicate. Will the virus bind to the scidian antigen? If it does, it will be quite an atavistic feat, considering that the virus uses the mosquito as the vector and the ascidians live in a shallow sea. Of course, mosquitoes can bite them at low tide and some from Jersey probably can bite them under water. In any case, we may know soon since there is apparently a team at Stanford studying just this combination. If the virus pulls out any protein, Lee Hood at Caltech is ready to sequence the stuff from head to toe. We may owe to an arbor virus from Africa the systematics of MHC.

According to Matzinger, that locus controlling cell-cell recognition of ascidians also controls their mating preference. How? Readers of *The Lives of a Cell* will have no doubt: they will answer "by pheromones"—those odoriferous molecules immanent in air, land, and sea, coextensive with life, and its subliminal language. Did not Thomas write in one of his essays, "Chemical signals might serve the function of global hormones, keeping balance and symmetry in the operation of various interrelated working parts . . . designed for the homeostasis of the earth" [95]?

It was Thomas who proposed at the Second International Congress of Immunology in Brighton, England, a possible interrelationship of pheromones with MHC. Not long after, Ted Boyse's laboratory reported a preferential mating in mice by H-2 types [96]. In various combinations of mating strains, sometimes the preference was with the mate of an identical H-2 type, sometimes with an H-2 incompatible mate, but in either case the choice was not random. How do they do it? The best answer at the moment is by smell. Tracking dogs can distinguish each of a hundred men by distinctive odor, but they cannot distinguish identical twins.

At the first of the annual cell-cell interaction conferences the March of Dimes held with the Scripps Foundation, Boyse gave a talk entitled "Immunogenetic Aspects of Biologic Communication" [97]. The next day a headline in the *San Diego Union* ran: "Dogs can smell your genes." Not perhaps the antigens themelves but possibly the microbes which prefer to live on the hosts with certain antigens. Or, as Thomas suggests, we may all have pheromones, like insects. Presumably antigens came after smells. Some mammals have both. Rabbits have 100 million olfactory receptors, and these increase by training. The recognition by smell, that is, by olfactory receptors, may be as sensitive as recognition by lymphocyte antigen receptors.

If HLA and smell are two facets of the same sensing system, we have the first bridge between the immune system and the central nervous system. Jerne is doubtless overjoyed with the prospect that Sloan-Kettering may soon bring his two networks under one tent.

Metabolism

Garrod planted the seedlings, named them inborn errors of metabolism, and predicted there would be an orchard there one day. It took 50 years for the trees to grow, but when the orchard bloomed it dazzled the world. No group of human diseases received such a thorough and rapid explication of its etiologies as inborn errors of metabolism. No wonder young researchers flocked to biochemical genetics; the methods were there and constantly improving, the concepts were there and constantly proliferating, the patients were around if you had luck and a nose for the strange. Were not maple syrup urine disease and isovaleric acidemia gifts of the pheromones? It was a great party there, under the cherries covered with snow. That was in the 1950s and early 1960s.

It is summer now; the flowers are gone, but the fruits are ripening; some have been picked. It is a time for consolidating following breakthroughs. A major way of doing that in genetics is to chop the diseases into subdiseases. Genetic heterogeneity seems to be the rule rather than the exception for most inborn errors of metabolism. Differences in enzyme concentrations, activities, or tissue distribution have been found in many. When we cannot do that or do not know the underlying defect in the first place, we like to classify them by age of onset. There must be something basically different between the infantile and adult forms of Gaucher's disease. Sometimes what had been regarded as two different diseases on clinical grounds were shown to be deficient in the same enzyme, as in Hurler and Scheie diseases, except that one had more enzyme than the other. A few years ago McKusick resurrected the notion of a genetic compound, that is, an individual with two different mutant alleles at one locus [98]. No sooner did he propose this than he and others found the prototype example: the Hurler-Scheie compound. These patients had symptoms not as severe as Hurler's and not as mild as Scheie's, the same enzyme deficiency [99]. Presumably, here the two genes modified the expression of each other like gin and vermouth.

If one introduces the factor of environmentally induced modifications, the problem rises to a higher order of complexity. One can carry this line of thinking to a point of no return as Barton Childs did in the keynote address to our Fifth International Congress on Birth Defects at Montreal. He said: "The number of . . . genotype-environment combinations (which in birth defects must include both mother and fetus) could be very large so that the precise description of cause of the same malformation could differ from case to case and may be, in fact, unique in each" [100].

Some may think this is a license to investigative nihilism. But, when biochemical genetics turns to multifactorial diseases, the complexity of the problems it faces may approach the potential diversity sketched by Childs. One must agree that how the genetic and environmental tangents cross and recross in our lives to produce defects and diseases, whether these interactions occur prenatally, as in cleft palate or heart defects, or postnatally, as in diabetes or hypertension, is the most formidable problem facing birth defects research.

The starting point of this campaign is surely human polymorphism. The work of Harry Harris and others in elucidating the extent and diversity of human polymorphism may be as important in guiding the direction of research in the coming decades as Garrod's Leonardo-like vision has been up to now. Actually, in his later Oxford years, Garrod was apparently preoccupied with the question of predisposition to common ailments in later years, as Blass and Gibson pointed out in their paper on the Wernicke-Korsakoff syndrome [101]. Garrod thought such a predisposition involved in inborn errors of specific proteins and that requires the help of a specific environment to achieve phenotypic expression: let us say, severe alcoholism. One's understanding of multifactorial diseases has not gone much beyond this.

Blass and Gibson's patients had abnormal transketolase, a thiamine-requiring enzyme. That is the trait that marked them off from other drunks who would drink as much and eat as poorly as Blass' patients but remained free from the distressing symptoms of Wernicke-Korsakoff syndrome. Here then is a high-risk population vis à vis a special environmental agent. "The time is ripe," Blass and Gibson wrote on another occasion, "for the application of modern methods of biochemical genetics to the nature-nurture problem in alcoholism" [102].

Childs touched on this question in a wider context in his presidential address to the American Society of Human Genetics. He lamented the absence or scarcity of intellectual and organizational interchange between geneticists, epidemiologists, and public health officials. He was skeptical of a ready rapprochement between these sectors of medicine in the near future largely because of the indifference of prevention-oriented physicians to the genetic point of view, which regards every human being as unique, shaped by his genes and a lifetime of special experiences. The traditional view of epidemiology, on the other hand, tends to see its subjects as a mass of ciphers, essentially identical. "In the emphasis on extrinsic causes of disease," said Childs, "and the prescription of abstinence or moderation of living habits equally for all, the genes are ignored, and the inference is that nurture is more important than nature" [103].

The armory of human genetics is now stocked with methods and concepts of great potential value to public health. A few examples must suffice: vitamin-dependency syndromes are prime examples from which we can learn the causes of human variability in the need for essential nutrients; pharmacogenetics is still the best—perhaps the only—discipline for elucidating in detail gene-environment interaction; HLA is a treasury of genetic markers for common diseases, and we may find other loci coding for arrays of cell surface markers to which disease predispositions can be linked; cell culture can uncover receptor defects from diabetes to hypercholesterolemia; chromosome studies have disclosed a small group of people with high cancer risk, not only in chromosome fragility syndromes but with specific chromosome anomalies as with deletions in Wilm's tumor and retinoblastoma [104]. We may be on the eve of an explosion of findings relating chromosome characteristics and cell behavior with cancer or precancerous states. This can be good, bad, or both.

For example, there is the Philadelphia chromosome (Ph¹), long known to be the hallmark of chronic myelogenous leukemia (CML). Harnden thinks it is a marker for a preexisting abnormal clone of bone marrow cells [105]. Many years ago Hirschhorn came upon a leukemia-ridden family from the Bronx in which the grandfather had a Ph¹-positive CML [106]. His healthy daughter in her early thirties also had Ph¹ and so did her young boys in the first decade of life. Can we inherit such clones? Can we fit this into Knudson's two-hit theory of carcinogenesis, the Ph¹ clones being equivalent of the cells that had received the first hit and poised on the brink of transformation [107]? Possibly, if the first hit had struck a germ cell of one of the ancestors. Some bearers of this genetic gift will not develop cancer in their lifetimes, having escaped the second hit. Others, however, will not and the question is: Is it worthwhile to search for such high-risk people, and what can we do when we find them?

At present the answer to the first part of the question is no, because we do not have an effective treatment. But even if we had effective therapy, we would be at a loss to prevent leukemia because we do not know where that second arrow of Knudson's is going to come from. So we would have a test developed by the geneticists to pick out the cancer prone, the public health doctors ready to use it to screen people, the oncologists ready to treat any fish caught in the screening net—and a hundred thousand or so Ph[1] positive men and women trying to forget their appointments in Samarra in drink, drugs, debauchery, or dementia. One can hear the doctor saying to his healthy, Ph[1]-positive patient: "As you know, we have found a cytological marker indicating a high risk for leukemia. I know this has given you much anxiety. The four times a year checkup we are urging you to have, I am afraid, may add to your anxiety load. But . . . the intimation of mortality is better than mortality itself, don't you think, Mr. Knudson?"

If preventive medicine ever develops a battery of truly predictive tests and gets ready to practice them on the population at large, it would be well also to have ready an effective and widely available form of preventive psychiatry.

Conclusion

So the journey's end.

Some may say it was more like a junket: a quick tour of hot spots, spectaculars, and a few out-of-the-way esotericas, and where was an overview to tie all these together, seen in the light of the eternity of birth defects and research on them?

Well, I *like* hot spots and am not so fond of overviews with the smell of eternity. I am like that old college classmate of Samuel Johnson's who, years later, met Dr. Johnson in London and confessed that he, too, wanted to be a philosopher but added, "I don't know . . . cheerfulness always kept breaking in."

What I tried was to give you a selected group of studies in the main areas of research we support, areas of interest to me—hot spots—that also give the scent—pheromone—of unmistakable future developments, including unexpected developments. I, too, would have liked to weave these together into a coherent picture—an overview. But I doubt I could have done justice in such a short presentation.

What stopped me short was the remembrance of David McCord's couplet called "Epitaph for a Waiter," which goes:

> By and by
> God caught his eye.

I just had a sensation of catching your eyes and the message was as clear as God's.

But I would like to leave this thought with you. You will be gone from here in a couple of days and undoubtedly our paths will cross, diverge, and cross again in the future, if not in eternity. But wherever you are, whatever your research activity, and whatever its outcome, you will always remain close to our hearts, and to come across your name in a paper will elevate our awareness for the moment and perhaps fill us with a surge of nostalgia for our days on this island.

REFERENCES

1. MARTIN, G. Personal communication, 1979.
2. THOMAS, L. Hubris in science? *Science* 200:1459–1462, 1978.
3. EBERT, J. D. Birth defects: the present and the future. In *Birth Defects: Proceedings of the Fourth International Conference,* edited by A. G. MOTULSKY and W. LENZ. Amsterdam: Excerpta Medica, 1974.
4. GILBERT, W. Why genes in pieces? *Nature* 271:501, 1978.
5. ALT, F. W.; KELLEMS, R. E.; BERTINO, J. R.; et al. Selective multiplication of dihydrofolate reductase genes in methotraxate-resistant variants of cultured murine cells. *J. Biol. Chem.* 23:1357–1370, 1978.
6. KAN, Y. W., and DOZEE, A. M. Antenatal diagnosis of sickle cell anemia by DNA analysis of amniotic fluid cells. *Lancet* 2:910–912, 1978.
7. DAVIDSON, E. H., and BRITTEN, R. J. Regulation of gene expression: possible role of repetitive sequences. *Science* 204:1052–1059, 1979.
8. KLECKNER, N. Translocatable elements in procaryotes. *Cell* 11:11–23, 1977.
9. CRICK, F. H. C. Split genes and RNA splicing. *Science* 204:264–271, 1979.
10. NEVERS, P., and SAEDLER. H. Transposable genetic elements as agents of gene instability and chromosomal rearrangements. *Nature* 268:109–115, 1977.
11. OHNO, S. Promethean evolution as the biological basis of human freedom and equality. *Perspect. Biol. Med.* 19:527–532, 1976.
12. BROWNE, SIR THOMAS. *Religio Medici.* London, 1643.
13. PAUL, J., and GILMOUR, R. S. Organ-specific restriction of transcription in mammalian chromatin. *J. Mol. Biol.* 34:305–316, 1968.
14. KOSTRABA, N. C., and WANG, T. Y. Inhibition of transcription in vitro by a nonhistone protein isolated from Ehrlich ascites tumor chromatin. *J. Biol. Chem.* 250:8938–8942, 1975.
15. COMINGS, D. E., and HARRIS, D. C. Nuclear proteins. II. Similarity of nonhistone proteins in nuclear sap and chromatin, and essential absence of contractile proteins from mouse liver nuclei. *J. Cell. Biol.* 70:440–442, 1975.
16. CHANDLER, M. E.; KEDES, L. H.; COHN, R. H.; et al. Genes coding for histone proteins in man are located on the distal end of the long arm of chromosome 7. *Science* 205:908–910, 1979.
17. KORNBERG, R. D., and THOMAS, J. O. Chromatin structure. *Science* 184:865–871, 1974.
18. PAULSON, J. R., and LAEMMLI, U. K. The structure of histone-depleted metaphase chromosome. *Cell* 12:817–828, 1977.
19. WORCEL, A., and BENYAJATI, C. Higher order coiling of DNA in chromatin. *Cell* 12:83–100, 1977.
20. HOHMANN, P., and COLE, R. D. Hormonal effects on amino acid incorporation into lysine-rich histones. *Nature* 223:1064–1066, 1969.
21. LYON, M. F. Gene action in the X-chromosome of the mouse (*Mus musculus* L) *Nature* 190:372–373, 1961.
22. MILLER, O. J. The sex chromosome anomalies. *Am. J. Obst. Gynecol.* 90:1078–1139, 1964.
23. TAKAGI, N., and SASAKI, M. Preferential inactivation of the paternally-derived X chromosome in the extraembryonic membranes of the mouse. *Nature* 256:640–642, 1975.

24. Cooper, D. W.; Van de Berg, J. L.; Sharman, G. B.; et al. Phosphoglycerate kinase polymorphism in kangaroos provide further evidence of paternal X inactivation. *Nature New Biol.* 230:155–157, 1971.

25. Ohno, S. The mammalian genome in evolution and conservation of the original X-linkage groups. In *Comparative Mammalian Cytogenetics.* 1969.

26. McKusick, V. A. Personal communication, 1976.

27. McKusick, V. A. Anatomy of human genome. Wilhelmine E. Key lecture. 30th annual meeting of the American Society of Human Genetics, Minneapolis, October 4, 1979. In press.

28. Ohno, S. Personal communication, 1974.

29. Kunkel, L. M.; Smith, K. D.; Boyer, S. H.; et al. Analysis of human Y chromosome-specific reiterated DNA in chromosome variants. *Proc. Natl. Acad. Sci. USA* 74:1245–1249, 1977.

30. Ohno, S. Molecular human cytogenetics. *Nature* 266:589–591, 1977.

31. Krooth, R. S. The future of mammalian cell genetics. In *New Directions in Human Genetics,* edited by D. Bergsma and J. German. New York: National Foundation–March of Dimes, 1965.

32. Goldstein, J. L.; Anderson, R. G. W.; and Brown, M. S. Coated pits, coated vesicles, and receptor-mediated endocytosis. *Nature* 279:679–685, 1979.

33. Brown, M. S., and Goldstein, J. L. Receptor-mediated endocytosis; insights from the lipoprotein receptor system. *Proc. Natl. Acad. Sci. USA.* 76:3330–3337, 1979.

34. Pearse, B. M. F. Clathrin: a unique protein associated with intracellular transfer of membrane by coated vesicles. *Proc. Natl. Acad. Sci. USA.* 73:1255–1259, 1976.

35. Gospodarowicz, D.; Moran, J.; and Baum, D. Control of proliferation of bovine vascular endothelial cells. *J. Cell Physiol.* 91:377–386, 1976.

36. Jones, P. A. Construction of an artificial blood vessel wall from cultured endothelial and smooth muscle cells. *Proc. Natl. Acad. Sci. USA* 76:1882–1886, 1979.

37. Moncada, S., and Vane, J. R. Arachidonic acid metabolite and the interactions between platelets and blood-vessel wall. *N. Engl. J. Med.* 300:1142–1147, 1979.

38. Needleman, P. Prostacyclin in blood vessel–platelet interactions: perspectives and questions. *Nature* 279:14–15, 1979.

39. Pious, D. Cell surfaces, genetics, and congenital malformations. *J. Pediatr.* 86:162–164; 1975.

40. Artzt, K., and Bennett, D. Analogies between the T/t complex and the major histocompatibility complex. *Nature* 256:545–547, 1975.

41. Snell, G. D. The H-2 locus of the mouse: observations and speculations concerning its comparative genetics and its polymorphism. *Folia Biol.* 14:335–358, 1969.

42. De Wolfe, W. C.; Einarson, M. E.; and Yunis, E. J. HLA and testicular cancer. *Nature* 277:216–217, 1979.

43. Crow, J. F. Genes that violate Mendel's rules. *Sci. Am.* 240:134–146, 1979.

44. Lyon, M. F.; Evans, E. P.; Jarvis, S. E.; et al. T-haplo-types of the mouse may involve a change in intercalary DNA. *Nature* 279:38–42, 1979.

45. Bennett, D. Developmental antigens and differentiation: do T/t locus mutations occur in species other than the mouse? In *Birth Defects: Proceedings of the Fifth International Conference,* edited by J. W. Littlefield and J. de Grouchy. Amsterdam: Excerpta Medica, 1978.

46. CHAPUIS-CELLIER, C., and ARNAUD, P. Preferential transmission of the Z-deficient allele of alpha₁-antitrypsin. *Science* 205:407–408, 1979.
47. ARTZT, K.; BENNETT, D.; CONDAMINE, H.; et al. Surface antigens common to mouse cleavage embryos and primitive teratocarcinoma cells in culture. *Proc. Natl. Acad. Sci. USA* 70:2988–2992, 1973.
48. VITETTA, E. S.; ARTZT, K.; BENNETT, D.; et al. Structural similarities between a product of T/t-locus isolated from sperm and teratoma cells and H-2 antigens isolated from splenocytes. *Proc. Natl. Acad. Sci. USA.* 72:3215–3219, 1975.
49. MINTZ, B., and ILLMENSEE, K. Normal genetically mosaic mice produced from malignant teratocarcinoma cells. *Proc. Natl. Acad. Sci. USA.* 72:3585–3589, 1975.
50. ILLMENSEE, K., and MINTZ, B. Totipotency and normal differentiation of single teratocarcinoma cells cloned by injection into blastocysts. *Proc. Natl. Acad. Sci. USA.* 73:349–353, 1976.
51. DEWEY, M. J.; MARTIN, D. W., JR.; MARTIN, G. R.; et al. Mosaic mice with teratocarcinoma-derived mutant cells deficient in hypoxanthine phosphoribosyltransferase. *Proc. Natl. Acad. Sci. USA.* 74:5564–5568, 1977.
52. HEATH, J. K. Man-made mouse. *Nature* 271:610–611, 1978.
53. WATANABE, T.; DEWEY, M. J.; and MINTZ, B. Teratocarcinoma cells as vehicles for introducing specific mutant mitochondrial genes into mice. *Proc. Natl. Acad. Sci. USA.* 75:5113–5117, 1978.
54. GOLDSTEIN, J. L.; BROWN, M. S.; KRIEGER, M.; et al. Demonstration of low density lipoprotein receptors in mouse teratocarcinoma stem cells and description of a method for producing receptor-deficient mutant mice. *Proc. Natl. Acad. Sci. USA.* 76:2843–2847, 1979.
55. MARTIN, JR., D. W. Personal communication, October 1949.
56. HAMBURGER, V. Changing concepts in neurobiology. *Perspect. Biol. Med.* 18:162–178, 1975.
57. UNGER, R. H.; IPP, E.; SCHUSDZIARRA, V.; et al. Hypothesis: physiologic role of pancreatic somastatin and the contribution of D-cell disorders to diabetes mellitus. *Life Sci.* 20:2081–2086, 1977.
58. STURDEVANT, A. L., and GOETZ, H. Cholecystokinin both stimulates and inhibits human food intake. *Nature* 261:713–715, 1976.
59. PIERCE, A. G. E., and TAKOR, T. Embryology of the diffuse neuroendocrine system and its relationship to the common peptides. *Fed. Proc.* 38:2288–2294, 1979.
60. GOLDSTEIN, A.; LOWNEY, L. I.; and PAL, B. K. Stereospecific and non-specific interactions of the morphine congener levorphanol in subcellular fractions of mouse brain. *Proc. Natl. Acad. Sci. USA.* 68:1742–1747, 1971.
61. PERT, C. B., and SNYDER, S. H. Opiate receptors: demonstration in nervous tissue. *Science* 179:1011–1014, 1973.
62. HUGHES, J.; SMITH, T. W.; KOSTERLITZ, H. W.; et al. Identification of two related pentapeptides from the brain with potent opiate agonist activity. *Nature* 258:577–579, 1975.
63. GUILLEMIN, R.; LING, N.; and BURGUS, R. Endorphins, peptides, d'origine hypothalamique et neurohypophysaire a activite morphincomimetique: isolement et structure moleculaire de l'α-endorphin. *C. R. Acad. Sci.* 282:783–785, 1976.
64. BLOOM, F.; SEGAL, D.; LING, N.; et al. Endorphins: profound behavioral effects in rats suggest new etiological factors in mental illness. *Science* 194:630–632, 1976.

65. WAHLSTROM, A.; JOHANSSON, L.; and TERENIUS, L. Characterization of endorphines (endogenous morphine-like factors) in human CSF and brain extracts. In *Opiates and Endogenous Opioid Peptides,* edited by H. W. K. KOSTERLITZ. Amsterdam: North-Holland, 1976.

66. LEFF, D. N. Endorphins. *Med. World News* (January 9, 1978).

67. Editorial. Enkephalins: the search for a functional role. *Lancet* 2:819–820, 1978.

68. BLUME, A. J.; SHORR, J.; FINBERG, J. P.; et al. Binding of the endogenous nonpeptide-like compound to opiate receptors. *Proc. Natl. Acad. Sci. USA.* 74:4927–4931, 1977.

69. SQUIRES, R. F., and BRAESTRUP, C. Benzodiazepine receptors in rat brain. *Nature* 266:732–734, 1977.

70. BRAESTRUP, C., and SQUIRES, R. F. Specific benzodiazepine receptors in rat brain characterized by high-affinity (H3) diazepan binding. *Proc. Natl. Acad. Sci. USA.* 74:3805–3809, 1977.

71. MARANGOS, R. J.; PAUL, S. M.; GREENLAW, P.; et al. Demonstration of an endogenous competitive inhibitor (s) of [3H] diazepam binding in bovine brain. *Life Sci.* 22:1893–1900, 1978.

72. IVERSEN, L. L. GABA and benzodiazepine receptors. *Nature* 275:477, 1978.

73. SKOLNICK, P.; MARANGOS, P. J.; GOODWIN, F. K.; et al. Identification of inosine and hypoxanthine as endogenous inhibitors of (3H) diazepam binding in the central nervous system. *Life Sci.* 23:1473–1480, 1978.

74. GANTRAY, J. P.; JOLIVET, A.; VIELH, J. P.; et al. Presence of immunoassayable β-endorphin in human amniotic fluid: elevation in cases of fetal distress. *Am. J. Obstet. Gynecol.* 129:211–212, 1977.

75. SILMAN, R. E.; HOLLAND, D.; CHARD, R.; et al. The ACTH "family tree" of the rhesus monkey changes with development. *Nature* 276:526–528, 1978.

76. STARK, R. I. Personal communication, October 1979.

77. YACHNIN, S. Demonstration of the inhibitory effect of human α-fetoprotein on in vitro transformation of human lymphocytes. *Proc. Natl. Acad. Sci. USA.* 73:2857–2861, 1976.

78. Editorial. New placental proteins. *Br. Med. J.* 2:1044, 1977.

79. URIEL, J.; NECHAND, B. DE; and DUPIERS, M. Estrogen-binding properties of rat, mouse and man fetospecific serum proteins. *Biochem. Biophys. Res. Commun.* 46:1175–1180, 1972.

80. SWARTZ, S. K., and SOLOFF, M. S. The lack of estrogen binding by human alpha-fetoprotein. *J. Clin. Endocrinol. Metab.* 39:589–591, 1974.

81. ATTARDI, B., and RUOSLAHTI, E. Fetoneonatal estradiol-binding protein in mouse brain cytosol is alpha-fetoprotein. *Nature* 263:685–687, 1976.

82. PLAPINGER, L.; McEWEN, B. S.; and CLEMENS, L. E. Ontogeny of estradiol-binding sites in rat brain. II. Characteristics of a neonatal binding macromolecule. *Endocrinology* 93:1129–1139, 1973.

83. HERBST, A. L.; COLE, P.; NORUSIS, M. J.; et al. Epidemiologic aspects and factors related to survival in 384 registry cases of clear cell adenocarcinoma of the vagina and cervix. *Am. J. Obstet. Gynecol.* 135:876–886, 1979.

84. ROTH, J.; NEVILLE, D. M., JR.; KAHN, C. R.; et al. Hormone resistance and hormone sensitivity. *N. Engl. J. Med.* 296:277–278, 1977.

85. CHAN, L., and O'MALLEY, B. W. Steroid hormone action: recent advances. *Ann. Intern Med.* 89:694–701, 1978.

86. TATA, J. R. Modulation of hormone receptors. *Nature* 269:757–758, 1977.

87. JERNE, N. K. The immune system. *Sci. Am.,* July 1973.

88. JERNE, N. K. The immune system: a web of v-domains. *Harvey Lecture,* 70 Ser., pp. 93–110, 1974–1975.

89. Cooper, M. D.; Peterson, R. D.; and Good, R. A. Delineation of the thymic and lymphoid systems in the chicken. *Nature* 205:143–146, 1965.

90. Good, R. A., and Gabrielsen, A. E. (eds.). *The Thymus in Immunobiology.* New York: Hoeber, 1964.

91. Zinckernagel, R. M. Major transplantation antigens in host responses to infection. *Hosp. Pract.* 13:83–90, 1978.

92. Beverley, P., and Knight, D. Killing comes naturally. *Nature* 278:119–120, 1979.

93. Robertson, M. Recognition, restriction and immunity. *Nature* 277:14–15, 1979.

94. Matzinger, P. Semliki forest virus: key to the MHC? *Nature* 277:14–15, 1979.

95. Thomas, L. Vibes. In *The Lives of a Cell.* New York: Viking, 1974.

96. Yamazaki, K.; Boyse, E. A.; Mike, V.; et al. Control of mating preferences in mice by genes in the major histocompatibility complex. *J. Exp. Med.* 144:1324–1335. 1976.

97. Boyse, E. A., and Cantor, H. Immunologic aspects of biologic communication: a hypothesis of evolution by program duplication. In *The Molecular Basis of Cell-Cell Interaction,* edited by R. A. Lerner and D. Bergsma. New York: Liss, 1978.

98. McKusick, V. A.; Howell, R. R.; Hussels, I. E.; et al. Allelism, nonallelism and genetic compounds among the mucopolysaccharidoses. *Lancet* 1:993–996, 1972.

99. Kajii, T.; Matsuda, I.; Ohsaw, A. T.; et al. Hurler-Scheie genetic compound (mucopolysaccharidosis 1H-1S) in Japanese brothers. *Clin. Genet.* 6:394–400, 1974.

100. Childs, B. Priorities for research on birth defects. In *Birth Defects: Proceedings of the Fifth International Conference,* edited by J. W. Littlefield and J. de Grouchy. Amsterdam: Excerpta Medica, 1978.

101. Blass, J. P., and Gibson, G. E. Abnormality of a thiamine-requiring enzyme in patients with Wernicke-Korsakoff syndrome. *N. Engl. J. Med.* 297:1367–1370, 1977.

102. Blass, J. P., and Gibson, G. E. Porphyria: cutanea tarda or variegata? (letter). *N. Engl. J. Med.* 298:1317, 1978.

103. Childs, B. Persistent echoes of the nature-nurture argument. *Am. J. Hum. Genet.* 29:1–13, 1977.

104. Riccardi, V. M.; Sujansky, E.; Smith, A. C.; et al. Chromosomal imbalance in the aniridia-Wihns tumor association: 11p interstitial deletion. *Pediatrics* 61:604–610, 1978.

105. Harnden, D. G. Cytogenetics of Human Neoplasia. In *Genetics of Human Cancer,* edited by J. J. Mulvihill, R. W. Miller, and J. F. Fraumeni. New York: Raven Press, 1977.

106. Hirschhorn, K. Cytogenetic alterations in leukemia. In *Perspectives in Leukemia,* edited by W. Dameshek and R. M. Dutcher. New York: Grune & Stratton, 1968.

107. Knudson, A. G. Heredity and cancer in man. *Prog. Med. Genet.* 9:113–158, 1973.

SOME GENETIC ASPECTS OF THERAPEUTIC ABORTION*

JAMES V. NEEL, M.D., Ph.D.†

My role in this program is to discuss the genetic indications for interruption of pregnancy. To those of you expecting simple, clear-cut formulas, I must apologize. From my vantage point, with respect to the issue of pregnancy interruption, your specialty has one piece of an increasingly complex problem, a problem which raises some of the more difficult biomedical issues toward whose solution our society is groping. We most assuredly will not settle your piece of the problem in the course of this panel discussion, but perhaps together we can put it in a light which will suggest certain courses of action.

For some twenty-six years now the University of Michigan has maintained a Heredity Clinic. The limitations on staff, time, and space being what they are, we have not attempted to see all the potential problems in genetic counseling referred to the University Hospital but rather, both for service and training purposes, to consult each year with a limited number of people with representative questions requiring such counseling. One of the issues arising from time to time has been whether a given probability of congenital abnormality in a child is an indication for interrupting a pregnancy. Each time this question has fallen to me, I have been troubled by the near total absence of guidelines.

For present purposes, we will define a possible "genetic indication" for terminating a pregnancy as an appreciable probability that the unborn child will, either at birth or subsequently, be found to have a major abnormality in whose etiology genetic factors are of established importance. The most common situation involves serious congenital defect. In passing, let us recognize that the etiology of congenital defect is extremely com-

* Paper delivered April, 1967, in a panel discussion, "The Place of Therapeutic Abortion in Modern Obstetric Practice," at the annual meeting of the American College of Obstetrics and Gynecology in Washington, D.C. Published with the permission of *Obstetrics and Gynecology*.

† Department of Human Genetics, University of Michigan Medical School, Ann Arbor.

plex, with both environmental and genetic factors of various types involved, but we will not go further into this subject now. Begging for the moment the legality of abortion on genetic grounds—a subject we come to shortly—let me at this juncture point out the wide range of recurrence risks with which one may be confronted in genetic counseling. Five examples, any one of which might prompt the question of the propriety of terminating a pregnancy, will suffice:

1. First child with anencephaly; no other instance of congenital defect known in family; mother again pregnant. Here the recurrence risk for this or other severe defect of the central nervous system is about 4 per cent [1].

2. Two children with severe central nervous system defect; immediate family history otherwise negative for congenital defect; mother again pregnant. Now the recurrence risk is 10 per cent [2].

3. Parents second cousins; first child normal; second child developed infantile amaurotic idiocy at six months. On the well-justified assumption of simple recessive inheritance, the recurrence risk is 25 per cent.

4. For the fourth example, we turn to a problem seen fourteen years ago. The first child of normal-appearing parents had died of a complex cardiac malformation; the second had been found to have cystic fibrosis of the pancreas. Moreover, a rising anti-Rh titer was encountered during that second pregnancy, although there was no clinical erythroblastosis; the mother was Rh-negative and the father was found very probably to be R_1R_2, so that all subsequent children would be Rh-incompatible. Our calculation of the probability of a normal outcome, should the mother again become pregnant, was 36 per cent, computed as the product of three probabilities: normality for congenital heart disease (95 per cent), normality for cystic fibrosis (75 per cent), and non-occurrence of Rh disease (50 per cent). This illustration was chosen not only to bring out the complexity sometimes encountered in counseling but also the rapid evolution of knowledge. Successive developments in the anticipation and treatment of erythroblastosis would now lessen greatly the likelihood of severe disease on this score, while the prophylactic use of an anti-D serum during the pregnancies of predisposed women has the possibility of preventing the disease altogether [3, 4].

5. The final example concerns pregnancy in a woman with the dominantly inherited disorder termed Huntington's chorea. The child will have a 50 per cent chance of inheriting the disease, and there is the additional consideration of the inability of the mother to care for the child. While Huntington's chorea and the other dominantly inherited degenerative diseases of the nervous system can scarcely be termed "congenital defect," we must recognize the narrow margin which sometimes separates "inherited predisposition" and congenital defect.

At this point we should take note of the emerging possibilities for converting these a priori probabilities to figures which will present a more satisfying prospect for meaningful intervention. In the case of a woman who has a family history of hemophilia and has produced a hemophilic son, one-half of her future sons but none of her daughters will be expected to be affected. The risk regardless of sex is 1 in 4. If sex-chromatin studies on cells obtained by amniocentesis reveal the fetus to be a girl, the woman can be completely reassured (unless her husband is hemophilic), but if it is a boy the risk is now 50 per cent. One can visualize an even greater change in the odds, to certainty, in case one could apply newly described techniques for the culture of cells obtained by amniocentesis [5] to a pregnant woman with a translocation which predisposes her offspring to Down's syndrome. That is, given this woman's abnormal karyotype had become known because of a previous child with mongolism, one could hope to detect the presence of mongolism in future pregnancies at a very early stage. However, amniocentesis during the first trimester is not without its risks, nor is it at all assured the specimen will contain the cells necessary for diagnosis or culture. But assuming the necessary technical advances, it seems clear we can or shortly will be able to define a spectrum of risk to an unborn child ranging from a few per cent to virtual certainty. Incidentally, there is an important philosophical difference between these genetic risks and the rubella risks in that the latter are non-recurrent.

We must also take note of developments which could make the distinction between contraception and abortion somewhat academic. Hardin [6], in a thought-provoking argument for a virtually complete abrogation of existing legislation regarding the termination of early pregnancies, has in particular emphasized how the development of the so-called "morning after" pill, capable either of interfering with implantation or causing sloughing of the early embryo, will bring to a focus the moral issues involved in *any* termination of pregnancy. But a point which he did *not* make was that the Intrauterine Device, which apparently acts through preventing implantation of the blastocyst, and which is as of this writing surely the most effective means of control of fertility, involves the very same issue, the only difference being that the IUD requires (hopefully) one decision, the "morning after" pill many.

The foregoing, then, are examples of the kinds of situations in which the question of inducing an abortion on genetic grounds is apt to arise. They will obviously stem in many instances from failure of family planning; for the possible other genetic consequences of such planning the reader is referred to the recent excellent paper of Matsunaga [7]. Let us turn now to the practicalities of therapeutic abortion. To the best of my knowledge, there are only two states in which the prospect of an abnormal child is recognized as a legal indication for abortion. On the other hand, there are many states in which, stemming largely from the data on the adverse effects on the fetus of maternal rubella during the first trimester, with risks to the fetus no greater than some of the genetic risks just considered, there has been tacit recognition that such prospect was sufficient grounds for an abortion. But at a time when in one state physicians are being threatened with legal action because they did induce an abortion and in another state being sued because they failed to abort, in each instance for virtually the same indications, it is clear that some clarification of the indications for inducing an abortion is urgent.

Under these circumstances, undoubtedly along with many of you, I have been interested in the section on abortion in the Model Penal Code drawn up by the American Law Institute. This section, usually with some modifications, has now been adopted by Colorado, North Carolina, and California, and is being considered by several additional states. The pertinent sentence reads as follows: "A licensed physician is justified in terminating a pregnancy if he believes there is a substantial risk that continuance of the pregnancy would gravely impair the physical or mental health of the mother or that the child would be born with grave physical or mental defect, or that the pregnancy resulted from rape, incest, or other felonious intercourse." I assume, although the section does not make it explicitly clear, that the consent of the individual involved is a prerequisite. With this reservation, I favor general adoption of such a code, which would represent an important clarification of our present confusion. However, there is an ambiguity to that key sentence which generates a certain uneasiness. How great must a risk be to be termed substantial, and where does one draw the dividing line between grave and non-grave? In drawing up this suggestion, the Institute was undoubtedly concerned not to infringe on the judgment of the physician. However, in view of the temper of the times, we may anticipate that enactment of this legislation will be followed by a whole series of precedent-setting legal decisions. Specifically, where a pregnancy terminates in an abnormal child, will the physician now be vulnerable to legal harassment because where there was reason to suspect such an outcome he failed to advise the mother accordingly, or sufficiently strongly?

It seems appropriate at this time to suggest that this distinguished society set up a committee to establish guidelines on these indications. Such a committee could delineate a series of situations which in its collective judgment involved "substantial" risk of "grave" defect. These guidelines would have no legal basis, yet could not fail to influence both medical practice and court opinion. The Royal Medico-Psychological Association of Britain, in its recent policy statement on therapeutic abortion [8], felt that no "spelling out in detail" was necessary. While in general I agree with this viewpoint, from the genetic standpoint the spectrum of risk is so wide, and genetic knowledge even today so poorly disseminated, that I believe the practicing physician would welcome some guidelines. Since clearly a value judgment is involved, it would seem appropriate that any committee which might be established reach well beyond the medical and allied professions, to include experts, theologians, and social scientists. Although the physician certainly bears a special responsibility in these matters, in my opinion he will do himself a disservice, and only delay the clarification we seek, by failing to recognize the need to enlist all the best thinking on this subject.

Among the many difficult value judgments to be reached, one in particular has impressed itself on us over the years. This is the question of weighing immediate versus delayed impact of a genetic defect. The nature of medical practice being what it is, the immediate seems more pressing and has a way of pre-empting our attention. But consider, for example, the woman who has given birth to two anencephalic children. There is a 1 in 10 chance that her next child will be a frightful monster, but the monster will almost surely die shortly after birth and seven normal fetuses will be aborted for every one such birth prevented by abortion. Consider, on the other hand, the apparently normal young woman whose father has Huntington's chorea. There is a 1 in 2 chance she has inherited the gene herself and, if so, another 1 in 2 chance she will transmit it to any child. There is thus a 1 in 4 probability any child to which she gives birth will some thirty or forty years later develop progressive degenerative disease of the central nervous system, with prolonged emotional trauma to the family and a long and expensive hospitalization. Or consider a marriage in which both parents have either the sickle-cell trait or thalassemia minor, with a 1 in 4 probability that any child will have a chronic, debilitating anemia, with death at a relatively early age. Any one of us can document instances in which the prolonged care of such individuals as these just mentioned has entailed exorbitant cost to some agency or institution, with the ultimate result—loved though he or she may be—a very marginal performance in our complex society. While I do not for one moment wish to place a price tag on a human life, I cannot help wondering how that same sum spent on normal children might advance the interests of society. Which of the foregoing situations best justifies an interruption of pregnancy? Do any? Returning as well to our five earlier examples, where should the dividing line be drawn?

The model code also groups pregnancies resulting from rape and incest in a juxtaposition of apparent equivalence. Although both are morally abhorrent, from the biological standpoint incest is in a very different category from rape. The child of rape faces the biological risk of any illegitimate child, but the child of the incest carries the added risk of high inbreeding. It has been very difficult to get an accurate evaluation of these risks. Recently we have published the results of a small series of such children [9]. Among eighteen children of father-daughter or brother-sister relationships in which the mother's pregnancy became known prior to the birth of the child, there were six who on follow-up six months after birth had either died or had major defects, whereas among eighteen children born to a similarly ascertained, matched series of unwed mothers, none had died and only one had a major congenital defect. Carter [10] has recently reported an even smaller series, in which the results were quite similar. Taken at face value, these biological effects are greater than would be predicted from the results of first-cousin marriages [11] and perhaps reflect the non-random selection that enters into this situation; but whatever the reason, the risk would seem to qualify as "substantial."

It is appropriate in a discussion of this type to recognize that except in those instances where a parent has manifest disease, or one parent (or both) has a defined genetic carrier state, we usually cannot identify high-risk families until after the birth of a defective child. Since most defective children are single events within the sibship and born to normal parents, even a rather liberal attitude toward therapeutic abortion on genetic grounds could result in only a fractional decrease in the frequency of congenital defect.

At the outset, we recognized that this general problem was only one aspect of a much larger issue. The term "genetic engineering" is frequently heard these days. Although I have been unable to determine with certainty who first coined the term, it apparently stems from the theoretical possibility of direct intervention into the genetic material that our expanding knowledge of DNA creates. Where such intervention activates or represses the expression of genes in somatic tissue, it is in effect not too different from, for example, hormone therapy. But where it actually alters genes, especially in germinal material, a new principle would be introduced. If directed mutation is involved, then there are implications for the gene pool of the next generation, and in many respects genetic engineering becomes one approach to what has commonly been termed "eugenics." The issues raised by the prospect of eugenic measures are many and controversial. So it seems only appropriate to point out that the decision to induce an abortion on genetic grounds also has implications for the gene pool of the next generation and so falls under the broad category of eugenic measures. Until we understand the genetic structure of human populations far better than is the case at present, and until we have a modest experience with legalized abortion, it is fitting to proceed cautiously and conservatively.

REFERENCES

1. C. O. CARTER. The inheritance of common congenital malformations. *In:* A. G. STEINBERG and A. G. BEARN (eds.). Progress in medical genetics, vol. 4, p. 59. New York: Grune & Stratton, 1965.
2. C. O. CARTER and J. A. FRASER ROBERTS. Lancet, 1:306, 1967.
3. R. B. McCONNELL. Annu. Rev. Med., 17:291, 1966.
4. C. O. CARTER. *In:* J. F. CROW and J. V. NEEL (eds.). 3d Int. Congr. Human Genetics, Proc. Baltimore: Johns Hopkins Press, 1967.
5. M. W. STEEL and W. R. BREG. Lancet, 1:383, 1966.
6. G. HARDIN. Perspect. Biol. Med., 10:1, 1966.
7. E. MATSUNAGA. J. Amer. Med. Ass., 198:533, 1966.
8. Royal Medico-Psychological Association J. Amer. Med. Ass., 199:167, 1967.
9. M. S. ADAMS and J. V. NEEL. Pediatrics, 40:55, 1967.
10. C. O. CARTER. Lancet, 1:436, 1967.
11. W. J. SCHULL and J. V. NEEL. The effects of inbreeding on Japanese children. New York: Harper & Row, 1965.

GENETIC SCREENING IN NEWBORNS:
VOLUNTARY OR COMPULSORY?

LORETTA KOPELMAN*

Testing of infants for some genetic disease or inborn errors of metabolism is required by law in 43 states [1]. The intention of legislators in providing funds for these compulsory screening programs is the early detection, treatment, or prevention of disease. Most states which do not have compulsory programs have voluntary ones with high compliance rates [1]. Still, there is no uniform policy in the United States concerning which disorders should be tested for, the proper disposition of information of positive test results, the ownership of the blood sample taken from the infant, the rights of parents to refuse their infant's participation, and the risks of stigmatization, labeling, discrimination, and invasion of privacy which can result from the discovery of a nontreatable or late-onset disease.

The National Academy of Sciences (NAS) has expressed concern about these programs, writing that in some cases tests other than those mandated by law are undertaken [1, p. 106].

In some cases, almost countless dried blood spots are stored away against the day when additional tests could be done. That such tests might be done would, of course, be unknown to the person from whom the blood was taken. This raises certain legal questions with regard to the ownership of such samples, the consent needed for running tests on them, and the disposition of information. If, for example, it should become possible to do a test for Huntington's Chorea on such a sample, what should be done about the information? Should positive results be relayed to the parents or guardians of the person from whom the blood was taken, or, if the person is old enough, should it be transmitted to him? Little thought has been given to the impact of such tests on the public, and stores of blood, which will be regarded as a rich potential source of research purposes, raise important ethical and legal questions which must be faced.

Compulsory screening programs of infants began in the United States in the mid-1960s when the decision was made in many state legislations

*Associate professor of philosophy and medicine, Department of Pediatrics, East Carolina University School of Medicine, Greenville, North Carolina 27834.

to provide funds for the required testing of infants for the treatable condition of phenylketonuria (PKU). The motive for this legislation was a compassionate one, namely, to prevent retardation through early detection and treatment (i.e., control of diet). It would be more accurate to say that PKU was later found to be treatable through diet, because at the time of enactment of this legislation it was not clearly established that this method of prevention was sound. Currently about 3,000,000 or 90 percent of all infants born in the United States are tested for PKU [1, p. 23] at a cost of $0.26 to $3.00 per test [1, p. 298]. About 200 infants with PKU are identified each year in this way.

Such compulsory health programs are an example of a paternalistic interference on the part of the state over the lives of its citizens. (Other examples are laws requiring inoculations, banning of saccharine in prepared foods, wearing of seat belts, blood tests for venereal disease upon applying for marriage licenses, and the wearing of safety helmets when driving motorcycles.) In enacting paternalistic legislation, the state bears a heavy burden of proof to show that such interference is justified. It must demonstrate the exact nature of the harmful effects to be avoided or the beneficial consequences to be achieved and the probability of their occurrence [2]. Thus, it is appropriate to ask in any mandatory screening program of newborns what benefit is to be achieved, or harm to be avoided, and who is to benefit. Such programs may benefit the infants if, as in the case of PKU, the condition is treatable. Hence, such a burden of proof could be met by demonstrating that the screening program will benefit the child directly by the early detection and treatment or prevention of disease. In some cases, however, the intended benefits are not for the child but for the parents or the family. For example, discovery of sickle-cell disease or trait, Down's syndrome, or hemophilia might influence reproductive choices the parents might later make. Another intended benefit of a screening program might be to neither the infant nor the parents directly but to the community in the form of new information and therapies which might be developed as a result of such testing. There is a need to learn of the distribution of genotypes among identifiable groups, to determine the physical or behavioral correlates of specific genotypes, and to learn the effectiveness of possible therapies in the prevention or treatment of genetic disorders. In addition to the interests in having testing done to benefit the infant, the parents, and society, some might be inclined to point out the researchers' interests in getting grants, publishing papers, gaining recognition, and being promoted. However, none of us is cynical enough to suppose this influences the kinds and numbers of tests done!

If we can agree that testing should not be done simply because it can be done (irrespective of the consequences to the infant), then, of the many competing tests that can be done, what tests should be done?

Lappe and Roblin [3, p. 15] summarize the view of many when they urge that the justification for state or federally supported mass testing programs increases directly with several considerations. First, it increases directly with the frequency of the disorder. More common problems should be given first priority. Second, it increases directly with the severity of the disease. Such ranking would serve the purpose of using funds most effectively for health benefit to the public. Third, it varies directly with the potential consequences if it is undetected and untreated. Fourth, it should vary with the efficacy of current available therapies. If there is mass screening for a disorder, then there should be therapy available. Finally, the total cost of the program should be considered, including testing for all those equally at risk, and funds should be provided for counseling of families and follow-up treatment.

Mass screening programs, especially compulsory ones funded through the state or federal government, should be for treatable disorders. However, in such a rapidly changing field as medicine, judgments about the treatability and hence about the value of testing change. Thus, a means should be provided for keeping such programs current about the relative benefits or risks of testing. In order to do this, Maryland has established a commission to determine which tests should be administered in their newborn screening program. The NAS recommends this as well as voluntary programs [1].

Most states which have legislation requiring testing for PKU and other disorders during the newborn period stipulate that the parents have the right to refuse that their infant be included in the program [1, p. 106]. "Of the 43 states with mandatory statutes five stipulate that testing will not be performed if the parent objects and 31 permit parental objections if made on religious grounds" [1, p. 50]. However, according to the NAS, the reality of the matter is that parents are rarely given the opportunity to refuse and that the blood sample is drawn routinely before the infant is discharged. Sometimes parents receive information of the testing and the right to refuse as the mother leaves the hospital after the test has been done [1, p. 50]. The NAS urges the cooperation and consent of parents in testing, proposing voluntary rather than compulsory programs. It cites the high compliance rate (96–97 percent) in states, such as Delaware and North Carolina, which have voluntary programs to show that noncompulsory programs can be very successful.

Genetic research is, of course, important, and it is convenient to do additional tests on infants being tested for treatable conditions. It is likely that nontherapeutic research and state-supported "service programs" which test for nontreatable conditions serve an important first step in prevention or provision of therapy. But unknown or future benefits do not transform such programs and/or research into therapeutic ones. These programs and the controversies surrounding them are a special instance of a general controversy: Under what circumstances is it justifiable to use minors who cannot give consent for themselves in nontherapeutic research studies and programs? The view which has gained wide acceptance in the Department of Health, Education, and Welfare and in the academic community is that nontherapeutic studies and programs could sometimes be conducted on minors but only (1) with the consent of guardians and (2) if there is no risk, or no more than a minimal risk, for the child who is involved in the study [4]. Consent must be adequately informed and unpressured or freely given. A reliable means, such as a review board, must be provided for estimation of the nature and likelihood of the risk [5]. Let us consider this with regard to testing for sickle-cell disease.

Sickle-cell disease is a condition many states mandate must be tested for during the newborn period. For example, Kentucky, Rhode Island, South Carolina, and New York currently require testing for it [1]. There is no present agreement that there is any therapeutic benefit for the infant to be derived from such testing. Some argue that informing the physician or parents that an infant has sickle-cell disease enables them to cope better with the later medical crises that may occur. But this questioned benefit must be balanced against the damaging effect to the parent-child relationship which may result from informing new parents that their apparently healthy infant has this late-onset fatal condition. This is not a trivial problem since it is well established that the first few months are critical ones in the bonding of the parents to the child.

In the state of New York, discovery of sickle-cell trait is a coproduct of the test used to determine sickle-cell disease. Testing for sickle-cell disease, but not sickle-cell trait, is required by New York State [6]. Results must be recorded on the infant's chart [7]. However, this state routinely collects and records positive tests for both. We know this is so because they also routinely inform hospitals and physicians of the results of these tests. No guidelines are provided about what to do with this information—for example, whether or not the physicians are required to inform parents that their child has either condition—nor do they provide any funds for counseling or follow-up. One can suppose that some physicians inform and counsel parents; some merely inform, and some do neither. We do know that this information becomes part of the permanent medical hospital record of the infant. In the past, insurance companies have demanded, as a precondition to insuring individuals, full access to medical records and, upon discovering conditions such as sickle-cell disease or sickle-cell trait, have denied insurance to individuals. Moreover, discrimination in jobs (especially those involving flight) have resulted when it was learned that individuals have sickle-cell trait. Parents have mistakenly believed that infants with sickle-cell trait had a debilitating illness or that the activities of these children needed to be restricted.

Risk, then, must be thought of as something more than physical risk at the time of testing. It involves the risks of labeling, the invasion of privacy, loss of self-esteem, and discrimination. Moreover, why should testing be required by law unless or until there is a clearly established preventative or therapeutic benefit to the infant that outweighs risks to them? Legally and morally, persons have the right to be consulted about whether they or their wards will be part of the programs or research not designed for their own benefit. Consent from parents might justify nonbeneficial programs and research where there was no risk to the infant. However, where there is risk to the infant, it cannot be justified (e.g., according to federal guidelines) even if the parents are willing to give their consent for the study.

Compassionate intentions aside, if ignorance is no excuse for failure to obey the law, it surely should not be an acceptable excuse for framing laws at variance with federal guidelines and guaranteed rights. The inappropriate character of some state laws is illustrated by the New York State Public Health Law, article 25 (Conklin Bill) [6].

It shall be the duty of the administrative officer or other person in charge of each institution caring for infants twenty-eight days or less of age and the person required . . . to register the birth of a child, to cause to have administered to every such infant or child in its or his care a test for phenylketonuria, homozygous sickle cell disease, branched-chain ketonuria, galactosemia, homocystinuria, and adenosine deaminase deficiency, histidemia and such other diseases and conditions as may from time to time be designated by the commissioner in accordance with rules or regulations prescribed by the commissioner.

The law does not state that the tests are intended to benefit the infant or even the family; it does not state that the benefits must outweigh the risks to the infant; it makes no provision for parental consent and in only one circumstance permits parental objection to such testing. The second 1974 amendment allows members of religious sects whose teachings and tenets are contrary to the testing to refuse administering the test [6]. This law provides no funds for counseling of families and treatment of infants with these disorders; it does not state the nature of the legal obligation of physicians or hospitals to inform guardians of positive test results; it requires tests for two conditions where there is currently no demonstrated therapeutic benefit, namely, sickle-cell disease and maple syrup urine disease. (The Committee on the Fetus and the Newborn of the American Academy of Pediatrics on Screening argues that screening for this disorder cannot be justified on all infants for a therapeutic purpose [1, p. 302].) Moreover, it records and keeps as a permanent record not only this potentially sensitive information but routinely tests, records, and informs physicians and hospitals of positive tests for sickle-cell trait, which is not mandated by the law. Testing for conditions such as sickle-cell disease and sickle-cell trait, on balance, may involve risks to the infant of discrimination, invasion of privacy, prejudicial treatment, and loss of self-esteem. If there is no therapeutic benefit to some of the testing, then, this bill seems incompatible with article 24-A of the New York Public Health Law, enacted in 1975, which requires written consent from the legal guardian if a minor participates in a research study. The Committee on Medicine and Law of the Bar of the City of New York goes so far as to state that "neither a minor nor anyone else on his or her behalf can validly consent to a treatment from which the child cannot benefit" [8].

It is easy and convenient to test newborns for additional conditions when testing for treatable ones. Important research and development of therapies may result from it. However, extension of programs to nontherapeutic testing requires consent and a clear demonstration that no risk to the infant is involved [9]. Some of the state-supported or required "service" programs have not been sufficiently concerned about the risks of stigmatization, invasion of privacy, and labeling of infants; nor have they seemed to consider that their statutes may be inconsistent with federal restrictions about consent and risk involving studies and programs on minors which are not clearly beneficial to them. The researchers or special interest groups who have pressed for legislation to extend programs beyond clearly demonstrated therapeutic ones should realize that important ends, such as the goals of research and therapeutic advance for other members of the community, do not justify any means to achieve those ends. One is not justified in volunteering or sacrificing the welfare of another, even for a really good cause! Whenever possible, nontherapeutic testing should be done on adults, so that they may judge risks and benefits for themselves and thereby consent or refuse to participate accordingly. Some prospective studies may have to give way to retrospective ones, such as the Witkins XYY study from Denmark [10].

To conclude, a clear demonstration of the therapeutic benefits of testing should precede state-supported mass testing of infants. The use of state monies to test for treatable disorders such as PKU seems justifiable; and, while voluntary programs are successful and seem preferable, required testing might be justified if there is a clearly demonstrated therapeutic benefit to the infant being tested, and if funds are provided for counseling and treatment. However, it seems untenable to require testing for nontreatable conditions or to allow testing for conditions which involve risk to the infant. No one has the right to subject others to procedures which on balance involve risk to them. The guardian's role, whether the state or the parent, is always to act in the infants' best interests and to safeguard their well-being.

REFERENCES

1. National Academy of Sciences. Genetic screening: programs, principles and research. Washington, D.C.: National Academy of Sciences, National Research Council, 1975.
2. G. Dworkin. *In:* S. Gorovitz, A. L. Jameton, R. Macklin, J. O'Connor, E. V. Perrin, B. P. St. Clair, and S. Sherwin (eds.). Moral problems in medicine, p. 185. Englewood Cliffs, N.J.: Prentice-Hall, 1976.
3. M. Lappe and R. O. Roblin. *In:* D. Bergsma (ed.), and M. Lappe, R. O. Roblin, J. H. Gustafson, and N. W. Paul (assistant eds.). Ethical, social and legal dimensions of screening for human genetic disease, vol. **10,** no. 6, p. 1. White Plains, N.Y.: National Foundation–March of Dimes, 1974.
4. Department of Health, Education, and Welfare, Commission for the Protection of Human Subjects. Proposed policy. Paper distributed at the Rights of Children as Research Subjects Conference, Urbana-Champaign, October 14–16, 1976.
5. Department of Health, Education, and Welfare. Protection of human subjects, p. 30648. Federal register 39, no. 165. Washington, D.C.: Government Printing Office, 1974.
6. New York State Public Health Law. Section 2500-a. Laws of 1964, chap. 785, as amended; laws of 1974, chap. 943, sec. 1; laws of 1974, chap. 1016, sec. 1.
7. New York State Law. NYCCR, pt 69.3(d), 1974.
8. Committee on Medicine and Law. Rec. Assoc. Bar City New York, **31:**694, 1976.
9. L. Kopelman. Perspect. Biol. Med., **21:**196, 1978.
10. H. A. Witkins, S. A. Mednick, F. Schulsinger, E. Bakkestrom, K. O. Christiansen, D. R. Goodenough, K. Hirschorn, C. Lundsteem, D. R. Owen, J. Philip, D. R. Rubin, and M. Stocking. Science, **193:**547, 1976.

FUTURISM: INTRODUCTION

Horowitz distinguishes between somatic mutations occurring during the life span of an individual from conception to death and the transmission of mutant genes across the generations as they lead to genetic disease. Assuming the greater significance of the latter, he proposes that medical genetics would profit by the detection of heterozygous individuals to advise parents at risk on the probability of a mutant child, by identifying the biochemical lesion caused by mutations and by proposing appropriate therapy such as direct protein replacement for the absent or malfunctioning enzyme or replacing the mutant gene with a normal gene by transformation or viral vector.

Tatum approaches the future of medicine by discussing the potential gain in studying the molecular biology of pathogenic viruses; genetic, metabolic, and developmental defects; and cancer. The common denominator in this approach is the unity of basic processes and concepts of molecular biology and the structure and function of nucleic acids. Tatum could not have anticipated the impact of cloning, recombinant DNA, and nucleotide sequencing on the course of molecular biology and its effects on medical science.

Fraser views the rapid advances in recombinant DNA technology as the shape of things to come in the production of drugs, hormones, and antibiotics. This original article presents a short course (with glossary) in genetic engineering and illustrates the rapid merger of laboratory and production line in a high-technology industry.

In Project K, Crick proposes a radical approach to speed developments in molecular biology: a concerted, heavily funded effort by a central laboratory complex with a large diverse resident and visiting staff to investigate one species: *Escherichia coli* strain K. Such a microbial Manhattan Project is expected to yield greater returns than the dispersal of funds to many investigators. To give the proposal made to the European Conference of Molecular Biology in 1967 wider publicity, Crick submitted the proposal to *Perspectives*. While the idea of such a venture is breathtaking, yeast geneticists might prefer another microbial species for such financial and intellectual attention.

PERSPECTIVES IN MEDICAL GENETICS

N. H. HOROWITZ*

Introduction

This year marks the 100th anniversary of Mendel's famous paper on inheritance in peas. It is particularly fitting, therefore, that the organizers of this meeting have decided to hold a colloquium dealing with genetic problems in medicine. To me has fallen the assignment of discussing perspectives in this field. To obtain a perspective, one can look either forward or backward. In this centennial year, many speakers will look backward over the long road that genetics has traveled since Mendel delivered his report to the Natural History Society of Brünn. The view is an inspiring one. The units that we now call "genes" were, in Mendel's paper, mere mathematical abstractions—entities that he invented in order to account for the numerical results of his breeding experiments. Today, the genes are real chemical substances, located in special organelles of the cell. The chemical structure of genes and their mode of replication are, for the most part, known, and their mode of action is understood. These advances in our knowledge of the genes have revolutionized biology. The creation of molecular genetics, starting from the laws of Mendel, is surely among the greatest accomplishments of twentieth-century science.

The spectacular progress of modern genetics has convinced some thoughtful biologists that we are on the threshold of a new era in which man will be able to control—and, indeed, to *remake*—his own heredity [1]. Control of his heredity would give man the capability of directing his own development and evolution. The prospects are, to say the least,

* Professor of Biology, California Institute of Technology. This paper was presented at the Third Panamerican Symposium on Pharmacology and Therapeutics held in San Luis Potosi, Mexico, August 4–7, 1965. It will be published in Spanish by *Excerpta medica* (Spanish edition). Results from the author's laboratory were obtained under Grant GB-444 from the National Science Foundation.

staggering. Even the remote possibility of such power compels us to pause and consider the situation. In this paper, I should like to examine the medical genetic aspects of this question.

The Scope of Medical Genetics

Medical genetics in its most narrow sense refers to genetic defects that are transmitted from parent to offspring. These defects are inherited in a Mendelian way, and they are usually congenital. Typical examples are galactosemia, phenylketonuria, sickle-cell anemia, and hemophilia. Also in this category of congenital afflictions are those, such as mongolism, which result from chromosomal aberrations of one kind or another.

Congenital defects such as these form only a small part of the genetic diseases of man, however. Much the larger part fall into a category that, for convenience, I shall refer to as *somatogenetic* diseases. These are defects that result from somatic mutations acquired during life. It is admittedly difficult, if not impossible, to prove that any disease is caused by somatic mutation, since, by the nature of the case, we are precluded from applying the breeding tests by which gene mutations are operationally defined. Nevertheless, there are strong reasons for thinking that somatic mutation is an important cause of death in man. It has long been suspected that cancer is caused by somatic mutations. Indeed, this must be true almost by definition, since cancer cells are permanently altered cells that transmit their abnormalities to their daughters through an indefinite number of mitotic generations. The real problem of cancer etiology is to delineate the chain of events that terminates in the appearance of the mutated cell [2]. It is, of course, well known that cancer can be produced by ultraviolet light and X-rays, which are powerful mutagens. The fact that the incidence of leukemia in man increases with the dose of ionizing radiation in an apparently linear fashion [3, 4] suggests that, as in the case of gene mutations, leukemia can be caused by a single hit in the chromosomes. It is known that some cancers have a viral origin. Viral diseases, too, come under the general definition of somatogenetic afflictions, since viruses produce their effects by insinuating their own genomes into the genetic machinery of the cell.

Also of great interest in connection with somatogenetic effects is the evidence that ionizing radiation accelerates the aging process in animals

[5, 6]. Experiments in a number of laboratories have shown that mice exposed to X-rays or neutrons show a shortening of their life span in proportion to the absorbed dose. Treated animals undergo degenerative changes that closely resemble those accompanying normal senescence. At the same time, they show a greatly increased frequency of chromosomal aberrations in their liver. Significantly, the radiomimetic drug, nitrogen mustard, does not shorten the life span, although it is known to produce chromosomal breaks. The difference apparently resides in the fact that whereas radiation affects all cells, nitrogen mustard affects only dividing tissues—e.g., bone marrow. The implication of these findings appears to be that the process of aging results from the accumulation of mutations in non-dividing tissues. Mutations in dividing cells, except when they result in cancer, are apparently lost in the competition with normal cells [6].

This view of the aging process is supported by cytological findings which show that chromosomal aberrations accumulate with age in unirradiated mice and human beings [7]. The rate of accumulation is surprisingly high. In an experiment reported by Curtis [6], 22 per cent of the liver cells of normal one-year-old mice were found to carry visible chromosomal abnormalities. This result suggests an answer to one of the strongest objections to the mutation theory of aging; namely, that mutation rates, as measured in germ cells, are too low to explain senescence. The cited evidence suggests that dominant lethal mutations—i.e., gross chromosomal aberrations—accumulate in somatic tissues at a much higher rate than would be predicted from measurements carried out on germ cells.

In summary, the scope of medical genetics includes not only inherited genetic defects, but also somatogenetic changes of the kind that apparently result in cancer and senility. The greatest challenge to medical genetics of the future clearly lies in the latter area. Let us now consider the prospects for new advances in the prevention and treatment of both categories of genetic disease.

Congenital Genetic Diseases: Galactosemia

To illustrate present and future possibilities for controlling congenital genetic diseases, let us take the specific example of galactosemia. This is a simple Mendelian recessive condition characterized by a block in the metabolism of galactose, as a result of which galactose-1-phosphate accumulates in the blood cells. The normal pathway of galactose metabolism in newborn children is as follows:

REACTION	ENZYME
1. Galactose+ATP→galactose-1-phosphate+ADP....................	Kinase
2. Galactose-1-phosphate + uridine diphosphate glucose ⇆ uridine diphosphate galactose + glucose-1-phosphate.......................	Transferase
3. Uridine diphosphate galactose ⇆ uridine diphosphate glucose.........	Epimerase

Exogenous galactose serves as a metabolic fuel by conversion to glucose through the above series of reactions. In galactosemia, the transferase is missing [8].

The effects of this disease, which are serious, can be completely avoided by removing milk, the chief dietary source of galactose, from the diet. Sufficient galactose is synthesized in the body by the epimerase reaction to satisfy the requirement for brain galactolipids. This treatment is perfectly satisfactory, provided that the disease is recognized early enough. It is of interest, however, to consider other possibilities that could become useful in this or other genetic diseases in the future.

Detection of Heterozygotes

The red cells of individuals who are heterozygous for the galactosemia gene contain only one-half the normal amount of transferase, although these individuals are phenotypically normal [9]. The overlap between the heterozygous and normal ranges of transferase activity is slight, and heterozygotes can be detected with at least 95 per cent certainty. Thus, if it were desirable to do so, such individuals could be recognized before marriage and advised not to marry other heterozygotes. In this way, the incidence of galactosemia could, in principle, be reduced to the vanishing point. By eugenic measures—i.e., preventing heterozygotes from having children—the mutant gene could be practically eliminated from the population. Such measures are, of course, not necessary for so rare and easily controlled a disease as galactosemia, but they might be advisable in other cases.

One genetic defect for which it would be desirable to be able to detect the heterozygotes is cystic fibrosis of the pancreas. This lethal disease is considered to be one of the most serious clinical problems among children in the United States, where it occurs in approximately one out of every two thousand births [2]. There is no specific treatment. It would be advantageous in this case if heterozygotes—who, by the Hardy–Weinberg distribution, comprise 4 per cent of the U.S. population—could be warned against marrying other heterozygotes. This disease is also interesting because it illustrates one of the dangers inherent in attempts to eliminate a mutant gene by eugenic measures. The high frequency of a lethal defect such as cystic fibrosis of the pancreas suggests that the mutant gene is being maintained in the population by virtue of a selective advantage of the heterozygotes over normal individuals. (The classic example of such selection is in sickle–cell anemia, heterozygotes for which are resistant to malaria.) Elimination of such a gene by overzealous eugenic measures might thus mean the loss of an important genetic resource.

In any case, the ability to detect heterozygotes would represent an important advance in the control of many genetic diseases, and we can anticipate further progress in this direction as our knowledge of the biochemistry of these diseases improves.

Prospects for Molecular Control of Genetic Diseases: Protein Replacement

Let us now consider the prospects for controlling genetic disease at the molecular level. According to present evidence, most, if not all, gene mutations are expressed in the loss or alteration of specific proteins; the symptoms of disease are secondary to these protein losses or alterations. If the specific protein could be restored, the disease would be cured. Insulin therapy is the preeminent example of the treatment of a genetic disease by this means. The brilliant success of this treatment does not mean, however, that other genetic diseases can be conquered so easily. Insulin is a special case. As a hormone, it is normally secreted into the blood-stream, from which it is readily taken up by its target cells. In addition, the near-identity of structure that characterizes the insulins of different species has minimized the immunological problem that would otherwise be encountered.

The equivalent treatment in the case of galactosemia, for example, would be the administration of the transferase enzyme. To be effective, the enzyme would have to penetrate the cells where it is required, and it would have to be excluded from cells where its presence might be harmful. At the present time, there is no way of satisfying these requirements. With further advances in our knowledge of enzyme structure, it may become possible to synthesize modified enzymes capable of being taken up by cells and lacking antigenicity. Such modified structures might consist of the active center of the natural enzyme with just sufficient superstructure to stabilize it and give it the desired size and charge. The problem of insuring selective uptake would still remain, however.

Enzyme Induction and Repression

Another conceivable means for controlling genetic disease is through induction and repression of enzyme synthesis. Taking galactosemia again as a model, it is known that older patients develop a tolerance for galactose. There is evidence [10] that this tolerance is due to the activity of a pyrophosphorylase which removes galactose-1-phosphate by the following reaction:

$$\text{Galactose-1-phosphate} + \text{uridine triphosphate} \rightleftarrows$$
$$\text{uridine diphosphate galactose} + \text{pyrophosphate}$$

This enzyme is not present in newborn children, but it appears later in life. If its synthesis could be switched on earlier, the disease might be prevented [2]. Alternatively, if the production of galactokinase, the enzyme that makes galactose-1-phosphate, could be switched off, the same result would follow, since it is probable that the sugar phosphate, rather than galactose itself, is the toxic substance in this disease. It is interesting to note that newborns have a higher level of kinase activity than adults; the adult activity is attained only at three to six years of age [11]. This means that, as with the pyrophosphorylase, there is a natural mechanism for changing the enzyme level. The question is how to activate these mechanisms at will.

In bacteria, the rates of enzyme synthesis are governed by specific repressors whose activity is, in turn, determined by the concentration of low-molecular-weight substrates and end-products in the environment. The repressors themselves are almost certainly macromolecules, either proteins or RNA—it is not certain which.

In higher organisms, the situation is more complex. Except for the liver, which, by virtue of its portal circulation, is exposed more or less directly to the variable contents of the intestine, the cells of vertebrates enjoy a nearly constant chemical environment. Enzyme synthesis is regulated by internal mechanisms, of which the hormones are an important part [12]. Much remains to be learned about these mechanisms. In a study of biochemical differentiation in Neurospora, which may be a model system for genetic regulation in other Eucaryotes, we have obtained evidence suggesting that the synthesis of tyrosinase and L-amino acid oxidase is repressed in growing cultures by a rapidly turning over protein (or proteins). Partial inhibition of general protein synthesis by amino acid analogs, actidione, or starvation leads to the formation of these enzymes. In the normal life cycle of Neurospora, both enzymes are produced during sexual differentiation [13].

The regulation of gene activity in higher organisms is currently under study in many laboratories. We can anticipate that the solution of this problem will open new possibilities for controlling genetic defects in man.

Transformation, Transduction, and Directed Mutation

Finally, we consider the most radical of all possible therapies for genetic disease; namely, altering the genotype. Means for accomplishing this are, conceivably, transformation, transduction, and directed mutation. Transformation and transduction are well-known mechanisms of genetic exchange in bacteria. In transformation, the cells to be altered are exposed to DNA extracted from a genetically different strain of the same species. A small fraction—at most a few per cent—of the treated cells take up some of the foreign DNA and incorporate it into their own chromosomes. In transduction, the result is similar, but the transfer in this case is mediated by virus particles which have the special property of being able to transport bits of host-cell DNA from one cell to another.

Despite numerous attempts—many of them unpublished—no corroborated example of transformation or transduction has been found in any non-bacterial species. A report of transformation of human cells in tissue culture which appeared several years ago remains to be confirmed [14]. It may well be that the organization of the genetic material into chromosomes in higher organisms, as opposed to the relatively simple DNA structure found in bacteria, precludes the possibility of their transformation. Despite the unpromising outlook, however, attempts to induce transformation and transduction in higher species will undoubtedly continue. Success in this project would still leave some major problems to be solved before these methods could be used in therapy, but its importance for the genetic analysis of somatogenetic defects would be very great indeed.

A more promising, but still controversial and poorly understood form of transformation uses RNA instead of DNA as the transforming agent. It has been reported recently from two laboratories that RNA isolated from mature uteri of mice and rats induces morphological and enzymatic changes when introduced into immature uteri of these species [15, 16]. The effect is presumably due to a phenotypic transformation, such as might be expected if messenger-RNA were being transferred. Alternatively, the introduced RNA could be activating the DNA of the recipient cells, allowing them to form their own messenger. Further studies on this system can be anticipated with interest.

The idea that genetic diseases may be cured by means of directed mutations is a recurrent one in medical-genetic discussions. Its roots perhaps go back to the Lamarckian belief that beneficial changes in heredity can be induced at will. Ancient though it may be, it is still the farthest from actual accomplishment of any of the procedures that I have discussed so far. The induction of mutations in a directed manner has not been accomplished even in bacteria, not to speak of higher species. At least we can see today more clearly than ever before, the difficulties of the problem. As Muller and others have pointed out [1], a mutagen capable of inducing a specific mutation in a specific gene would have to be of the same order of complexity as the gene itself, since otherwise it could not recognize the latter among thousands of similar genes and react with it to the exclusion of all others. In order to produce such a mutagen, we would therefore have to know the structure of the target gene, and we would also have to know precisely where in this structure we would have to induce a chemical change in order to attain the desired result. All this is beyond our present capability, but the fact that we are able to state the problem is a significant accomplishment and is, perhaps, the first step in its eventual solution.

Conclusions

In conclusion, the discoveries of modern genetics have opened many new prospects for controlling or curing genetic defects in man. Of the two categories of genetic disease, congenital defects are the more amenable to control. In particular, it should be possible with increasing success in the future to detect heterozygous carriers of serious genetic defects. The greatest challenge to medical genetics, however, is in the prevention or cure of somatogenetic diseases, since these affect all of us. Unfortunately, these afflictions are not yet well defined genetically or biochemically; they are probably complex, involving many genes in most cases. The immediate objective is therefore to acquire a better understanding of these defects, and it is here, rather than in the area of therapy, that molecular genetics offers the most hope at present. By applying the concepts and methods of microbial genetics to the study of somatogenetic defects in cell cultures, as a number of investigators are now doing, it may be possible to obtain the fundamental knowledge that must precede rational therapy.

REFERENCES

1. T. M. Sonneborn (ed.). The control of human heredity and evolution. New York: Macmillan Co., 1965.
2. A. G. Knudson, Jr. Genetics and disease. New York: McGraw-Hill, 1965.
3. E. B. Lewis. Science, 125:965, 1957.
4. ———. Ibid., 142: 1492, 1963.
5. A. C. Upton. In: B. L. Strehler (ed.). The biology of aging, p. 318. Washington, D.C.: American Institute of Biological Sciences, 1960.
6. H. J. Curtis. Science, 141:686, 1963.
7. P. A. Jacobs, W. M. Court Brown, and R. Doll. Nature, 191:1178, 1961.
8. H. M. Kalckar, E. P. Anderson, and K. J. Isselbacher. Proc. Nat. Acad. Sci. U.S.A., 42:49, 1956.
9. R. K. Bretthauer, R. G. Hansen, G. Donnell, and W. A. Bergren. Proc. Nat. Acad. Sci. U.S.A., 45:328, 1959.
10. K. J. Isselbacher. J. Biol. Chem., 232:429, 1958.
11. G. N. Donnell, W. G. Ng, W. R. Bergren, J. Melnyk, and R. Koch. Lancet, 1: 553, 1965.
12. N. H. Horowitz and R. M. Metzenberg. In: J. M. Luck (ed.). Annual review of biochemistry, vol. 34, p. 527. Palo Alto, Calif.: Annual Reviews, Inc., 1965.
13. Unpublished experiments from the author's laboratory.
14. E. H. Szybalska and W. Szybalski. Proc. Nat. Acad. Sci. U.S.A., 48:2026, 1962.
15. S. J. Segal, O. W. Davidson, and K. Wada. Abstracts, 6th International Congress of Biochemistry, p. 733, 1964.
16. A. M. Mansour and M. C. Niu. Proc. Nat. Acad. Sci. U.S.A., 53: 764, 1965.

MOLECULAR BIOLOGY, NUCLEIC ACIDS, AND
THE FUTURE OF MEDICINE*

E. L. TATUM†

It seems particularly fitting, in this symposium, to try to assess the probable and possible impact of molecular biology and its important component, the structure and function of nucleic acids, on the future of medicine. Although I attempt specific predictions with considerable trepidation, not being qualified in either fortunetelling or medicine, I do so in the firm belief that the findings and concepts of molecular biology will play a leading role in the future of medicine. Medicine, after all, primarily involves the application of biological concepts and understanding to the health and welfare of man.

After considering various alternative presentations, I have decided first to point out some of the principal problem areas facing medical science in which molecular biology is most immediately concerned. I will then do my best to predict some developments in these areas over the next ten to twenty years.

Before doing this, however, I feel a real obligation to mention a problem that is peripheral in one sense, but that in another is related to and even supersedes most others in basic importance to man. This is the world problem of population and the means of arriving at and maintaining an effective balance between the population explosion and natural resources. Although this problem is outside the main thread of this symposium, it is actually of primary importance. Unless it can be solved, and a new "dark

* This paper was presented at the dedication symposium of the new research laboratories of Merck Sharp and Dohme in New York, May, 1966. Columbia University College of Physicians and Surgeons was a joint sponsor. All the papers presented will be published in book form by Merck & Co. Cost of publication of this article has been covered by the National Foundation.

† Rockefeller University, New York, New York.

age" avoided, scientific progress can hardly be maintained, and the application to man of new findings will be defeated by sheer numbers.

However, let us assume that this problem will be solved, and proceed to problem areas in medical science. Some of the most important of these are being discussed by the other symposium speakers, auto-immune diseases by Sir Macfarlane Burnet, brain function by Dr. Schmitt, and degenerative diseases by Sir George Pickering. Accordingly, I will concentrate primarily on the following areas in which progress seems to me to be particularly directly related to the concepts of molecular biology and dependent on nucleic acid research. These major areas are: (1) viruses and virus diseases; (2) hereditary metabolic defects, enzymatic, and regulatory; (3) developmental, congenital, and structural defects; and (4) cancer. What developments can be predicted in these general areas during the next twenty years or so?

In the field of viruses and viral diseases, it can be anticipated fairly confidently that the study of viruses—bacterial, plant, and animal—will continue to hold as important a place in molecular biology and genetics as has been true during the past decade or so. Most, if not all, viral diseases will be conquered either through immunological means or by the design and synthesis of specific antiviral chemicals. With this, and with definitive understanding of the roles of viruses in human problems involving development and with regulation of cell growth, will come effective prevention and hence control of human problems attributable to viral disease. Finally, it can be anticipated that viruses will be effectively used for man's benefit, in theoretical studies in somatic-cell genetics and possibly in genetic therapy.

In the area of metabolic disorders, the recognition of the genetic basis of many more disorders can be anticipated. The specific enzymatic defects will be identified in many more instances, as already has been done for PKU (phenylketonuria), galactosemia, certain amino acidurias, and abnormalities in hemoglobin and other serum proteins. Rapid, simple, and sensitive methods for the detection of carriers and for the early diagnosis of affected individuals will be developed, thus facilitating both more effective eugenic measures and more effective therapy by dietary and other means. We can even be somewhat optimistic on the long-range possibility of therapy by the isolation or design, synthesis, and introduction of new genes into defective cells of particular organs.

In the field of developmental biology, as the consequence of a better understanding of the molecular and spatial-temporal sequences involved in differentiation and development, we can foresee the effective prevention and alleviation of developmental errors, such as congenital malformations, whether these be due to genetic defects or to faulty gene expression or regulation or are indirectly related to gene activity via hormone production or target-organ receptivity.

Perhaps in no area is the foreseeable rapid accumulation of basic information more pertinent to molecular biological research and concepts, or more promising for the future, than in the area of neoplasia—cancer. I feel that we can reasonably anticipate that the basic causes of many, if not all, forms of cancer will be established within the next few decades. All suspected causes, viral, mutational, or regulatory failures, center on cell genetics, and on nucleic acid structure and function. Hence, we can be reasonably optimistic of the development, first, of effective preventive measures and, later, of curative therapy. These will come by epidemiological, immunological, and chemotherapeutic means, by modification and regulation of gene activities, or by means of gene repair or replacement.

Let us now explore the basis for my optimism, which rests, first, on the general validity of the concepts of molecular biology and the significance of nucleic acid structure and function therein and, second, on the "state of the art" in nucleic acid research and on the exponential rate at which knowledge in this and other areas of biology is increasing.

It is now generally accepted that the basic unit of heredity in all forms of life is DNA, deoxyribonucleic acid, consisting of the now familiar complementary stranded double helix of Watson and Crick. This structure uniquely possesses all the qualifications essential for genetic function: specificity through purine and pyrimidine base sequence; mutation through alterations in base sequence; replication by the enzymatic assembly of new strands on the two parental strands as templates; and translation into cell function by transcription of one strand into a complementary strand of messenger RNA (ribonucleic acid). The messenger RNA is transferred from the nucleus to the cytoplasm and there directs the assembly of amino acids into enzymes and other proteins in the ribosomes, with the sequence of base triplets in the gene thus specifying a particular sequence of amino acids in the protein product.

The past several years have seen the fleshing out of these bare conceptual bones in considerable detail and with some remarkable observations and phenomena.

Genetic material is frequently, or even perhaps usually, present in the form of circular molecules, such as in bacterial "chromosomes" in bacteriophages, in polyoma virus, and perhaps even in chromosomes of higher forms. It would appear that this circularity has a control function in the replication process, which, at least in bacteria, appears to start at a particular location and proceed in one direction at a single growing point which moves along the chromosome until all the genes have been copied and the process is completed [1]. The circularity may also serve in part to protect DNA from enzymatic attack on free ends of the molecule.

The evidence suggests that a double helical form of nucleic acid is essential for its replication. For example, replication of a single-stranded phage DNA (ØX174), of single-stranded RNA of the phage f2, and of several plant and animal viruses has been shown to involve the enzymatic formation of a double-stranded replicative form. Only one of these (−strand), the complement to the parental strand, would then serve as template for the synthesis of new viral RNA (+strand). An apparent exception to the requirement for replication of a double-stranded structure, the in vitro replication of a biologically active bacterial virus RNA, was recently reported by Spiegelman and co-workers [2]. However, later experiments by Weissman and Feix [3] strongly suggest the presence of a double-stranded replicating form of RNA in this system as well.

Strandedness would appear also to be important in repair of damaged DNA, as with ultraviolet radiation. It should be recalled that ultraviolet-produced thymine dimers are split in photoreactivation but are excised and the strand repaired in a dark reaction. Bacterial mutants deficient in the excision reaction are known, and the ultraviolet-damaged double-stranded replicative form of ØX174 is capable of dark repair, whereas the single-stranded form of the bacteriophage DNA is not. Incidentally, it might also be pointed out here that exo-nuclease III may well be involved in the excision reaction and therefore might be absent in the "excision negative" bacterial mutants just mentioned.

Strandedness and circularity seem also to be important in determining the template specificity in RNA synthesis on either DNA or RNA templates. In vitro, cellular RNA polymerase can use either DNA or RNA templates, either single or double stranded. With double-stranded DNA, both strands usually serve as templates. However, in in vivo m-RNA synthesis and in vivo synthesis of viral RNA, only one of the double strands of the nucleic acid appears to serve as template. This also seems to be true in vitro in several other instances involving transcription to RNA of double-stranded nucleic acids related to bacterial viruses. That circularity may be involved in determining single-strand transcription is suggested by the experiments of Hayashi, Hayashi, and Spiegelman [4] with ØX174 replicating form DNA. In vitro, only one strand of the circular form was transcribed, but both strands of the open linear form were so used.

In connection with the structure of DNA and its transcription, an important field of investigation was initiated a few years ago with the discovery by Reich and co-workers that actinomycin D primarily acts by binding specifically to the amino group of guanine in the minor groove of double-stranded DNA. In so doing, it prevents the DNA from functioning as a template for cellular RNA synthesis but does not affect RNA-dependent synthesis of viral RNA [5]. Other studies [6] have shown that chromomycin and related antibiotics react similarly and that ethidium bromide and daunomycin bind to DNA and inhibit its function non-specifically, perhaps, like proflavin, by intercalation between adjacent base pairs. Bhuyan and Smith [7], in similar studies, have shown that nogalomycin binds to DNA, probably to either adenine or thymine residues. In contrast, another antibiotic, tubercidin, an analogue of adenosine, apparently is incorporated into both RNA and DNA [8], as are some other base analogues. Such studies in general not only provide an understanding of the molecular basis of activity of these antibiotics but lead to the recognition and use of valuable tools in investigating the structure and functioning of nucleic acids.

The second step of gene transcription involves not only m-RNA but two other classes of RNA, both of which are gene determined. Transfer or soluble RNA (s-RNA) attaches activated amino acids at its CCA terminal end and transfers them to the growing polypeptide chain in the ribosome, which contains at least two ribosomal structures containing RNA (18S and 23S). Each amino acid is carried by at least one specific s-RNA. Some evidence suggests that the DNA loci responsible for ribosomal and transfer RNA are bunched and may be transcribed more or less as a group, whereas the loci for m-RNA are more widely and randomly distributed [9].

Each s-RNA has two separate recognition sites, one specific for its amino acid and one specific for the corresponding base triplet or codon on the m-RNA. The first complete base-sequence analysis was reported just this last year by Holley and collaborators for yeast alanine s-RNA [10]. When sequences of other s-RNA molecules are established, as can be anticipated fairly soon, it should be possible to define structurally both recognition sites, the anticodon, and the amino acid recognition site.

The final step in protein synthesis, the assembly of polypeptide chains, takes place in the ribosomes. These consist of two different-sized subunits and function most effectively as aggregates or polysomes, with each ribosome involved in transcription of a different section of the m-RNA molecule that holds the ribosomes together in the aggregate. In some instances, at least, a single polycistronic m-RNA molecule may code for several polypeptide chains.

Protein synthesis on the ribosome is subject to regulation and control in a number of ways. One of the important recent questions has been the punctuation in the genetic code. Are there start and stop signals? It seems that there are! The start signal for most *Escherichia coli* proteins appears to be a particular nucleotide sequence which specifies at least N-formyl-methionine, and perhaps even N-formyl-methionyl-alanyl-serine. After completion and release, one or more of the N-terminal groups is enzymatically removed. The evidence for this process comes from the elegant work of Webster, Engelhart, and Zinder [11] on the in vitro synthesis of f2 virus coat protein and from related findings of Adams and Capecchi [12].

That there are also stop signals seems equally probable. These may function in a manner analogous to the termination of peptide synthesis by puromycin, which is added to the growing carboxyl end of the chain, stopping further additions and causing release of the incomplete chain from the ribosome. At least two possible nucleotide triplets (UAG and UAA), which seem not to code for any amino acid, have been suggested by Sarabhai, Stretton, Brenner, and Bolle [13] as giving a signal for peptide chain termination in some strains of E. coli.

Ribosomal protein synthesis in vitro can also be modified by conditions or substances that bind to ribosomes, disassociate their subunits, or otherwise affect their structure or binding of s-RNA. Examples are basic compounds, such as spermidine, and the antibiotics lincomycin and, probably, chloramphenicol. Interestingly, organic solvents such as alcohol and certain salts may change the specificity of recognition between m-RNA codon and the s-RNA anticodon. Related to this is the finding that suppressor mutations may involve alterations in the recognition site of a particular s-RNA and thus restore normal transcription. Another very important related finding is the discovery by Davies, Gilbert, and Gorini [14] that streptomycin alters the specificity of m-RNA codon recognition by s-RNA, and hence disarranges normal transcription, but can function as a "suppressor substance" in correcting faulty transcription due to gene mutation.

In view of the widespread interest in and knowledge of the intricacies of the triplet genetic code, it should here suffice to point out that, as the result of the brilliant pioneering work of Nierenberg, Ochoa, and Khorana, and their collaborators, we now have an almost complete key to the triplet codons for all amino acids. This accomplishment has required particularly the techniques of nucleic acid chemistry and biochemistry, in producing the necessary oligonucleotides of defined sequence and length. I would only remind you, in addition, that the universality of the code has now been fairly convincingly established by work with viruses, bacteria, plants, and animals. This is particularly pertinent to our later consideration of genetic engineering.

Another area of nucleic acid research that should be mentioned here is that of mutation, which in its simplest form represents the substitution of one base for another in a DNA triplet. We already know that certain substitutions are more frequent than others and can selectively be made still more frequent by the incorporation into DNA of base analogues which alter base pairing specificities or, under more natural conditions, by the presence of a "mutator" gene. These observations, together with knowledge of the code, make the prospects of directed mutation somewhat more hopeful for the future.

We perhaps should here also remind ourselves that genes and gene functions in living organisms are not isolated entities and phenomena in a test tube but are subject to regulation or control mechanisms which turn them on, or turn them off, either separately or in operon groups, in feedback response to repressor or activator molecules. Developmental geneticists believe that gene activation and repression are of primary significance in processes of differentiation and development. The more we learn about these repressors and activators and how they work, the more optimistic we can be about the prospects of controlling and correcting faulty developmental processes.

Finally, in this general discussion, both for aesthetic completeness and because of its possible role in infective processes and in development, I want to mention extrachromosomal inheritance. The last few years have seen a considerable clarification of the physical basis of extrachromosomal genetic phenomena. Cell organelles such as mitochondria and plastids, and entities such as infective bacterial episomes, replicate and divide independently of the nucleus, can mutate, and control certain typical characteristics of the cells in which they exist. These characteristics are inherited in a non-Mendelian pattern, often completely maternally. It is intellectually satisfying that all such entities investigated have been found to contain double-stranded, helical, high-molecular-weight DNA. It is particularly gratifying to me that it is now known that *Neurospora* mitochondria grow and divide [15], that they contain DNA [16], and that a cytoplasmic character in *Neurospora* is transmitted from cell to cell by pure isolated mitochondria, as shown in our laboratories [17]. Drs. Luck and Reich have most recently [18] produced evidence consistent with the semiconservative replication of *Neurospora* mitochondrial DNA, as is true for bacterial, plant, animal, and some viral DNA. They have also shown, as followed by a species-specific density difference, that this mitochondrial DNA is inherited maternally. A role of mitochondrial DNA in maternally inherited respiratory-deficient mutants of *Neurospora* has very recently been suggested by Woodward and Munkres [19] to involve the control of mitochondrial structural protein. It is of considerable interest that chemotherapy is of potential value in controlling the replication of extrachromosomal DNA, as suggested by the effects of agents such as proflavin, streptomycin, and nitrosoguanidine in micro-organisms.

In the light of the foregoing survey of some of the high spots of nucleic acid research pertinent to molecular biology, let us refocus our attention on my earlier predictions as to the future of medicine. I hope that I have succeeded in making clear, first, the central role played by nucleic acids in biology and genetics, so that the basic phenomena in the various problem areas, virus infection, metabolic diseases, developmental defects, and neoplasia, are actually interrelated, and, second, the fact that these basic phenomena involve changes from normality because of interference with, or changes in, the normal sequences of molecular biology that link gene to enzyme.

Virus infection introduces new genetic material and interferes with the functioning of host genes. Mutation changes genes, generally for the worse, as in inborn metabolic diseases characterized by single enzyme defects.

Developmental defects may be due to mutant genes, as for clubfoot, harelip, etc.; to virus infection, as with *Rubella* or probably adenovirus; to drugs such as actinomycin or thalidomide; or to prenatal environmental factors such as a riboflavin deficiency. Some of these environmental factors probably act by changing the spatial-temporal sequence of gene activation and expression necessary for normal development. Certain developmental defects may be due to an extra chromosome as in mongolism or Turner's and Klinefelter's syndromes. These chromosomal effects appear to be due to defective gene expression as the consequence of chromosome imbalance. It is of considerable interest that some recent data on mongolism in Australia support the possibility that some unknown infective agent may affect oogenesis, causing the non-disjunction that leads to the extra chromosome 21. If this proves to be true, the significance for preventive therapy is obvious.

Certain heritable developmental abnormalities are accompanied by and even attributable to hormone deficiencies, as for thyroxin in cretinism or growth hormone in a type of dwarfism. The basis of hormone activity itself is still unsettled. However, evidence has been accumulating recently that many, particularly the sex hormones and the insect hormone ecdysone, act by regulating gene activity in the target cells, leading to the production of specific m-RNA, as shown for animals by the recent experiments of Kidson and Kirby [20]. It should be pointed out, however, that other modes of action have not been ruled out, for instance, that cell-membrane permeability is involved.

The last area I have emphasized, that of neoplasia or cancer, is at the same time one of the greatest problems facing medicine and one of the most complex. In an intellectual sense this complexity, centering on the variety of phenomena implicated in the causation and manifestations of cancer, is reason for optimism, since it provides the opportunity of understanding the process of carcinogenesis, "spontaneous," viral, or chemically or physically induced. All of these would appear basically to involve genetic material. Similarly, there is cause for optimism in understanding and treating the manifestations, since they would appear to involve various steps in gene expression.

This general thesis is supported by the increasing evidence that many forms of cancer in animals are indeed due to particular viruses in a phenomenon basically analogous to lysogenesis in bacteria. Most simply, it would appear that viral genes so introduced are integrated into the host-cell genome, transforming them into tumor cells, with characteristic new properties such as loss of contact inhibition, change of morphology and growth requirements, and new tumor or viral antigens. We are coming closer to the clarification of the role of viruses in human cancer through detailed studies of the "transformation" process in cultured cells, induced by human and animal viruses; through studies showing the need for a "helper" virus acting with a tumor virus in viral carcinogenesis in animals; and through epidemiological and other studies in human cancer, such as malignant lymphoma and leukemia.

The possibilities of prevention and therapy would seem to be many and diverse, ranging from prevention of infection, replication, expression, and integration of viral genetic material, perhaps as with antiviral agents such as HBB, guanidine, and IUDR, to specific chemotherapeutic suppression of the transformed tumor cells, or even their reconversion to normality. Effective attacks on these fronts will lean heavily on our knowledge of the molecular events involved, that is, on knowledge in molecular and nucleic acid biology.

Many, including myself, have talked and written about the application of the newer knowledge of molecular genetics and biology to the improvement of man's life, heritage, and health in terms of engineering. In these terms, eugenic engineering operates at the level of existing genes and involves purposeful, conscious effort to decrease the prevalence and expression of undesirable genes. These efforts will be effective in proportion to the numbers of detrimental genes that can be identified; to the development of effective methods for their detection in the hidden, carrier state; and, most important, to the general acceptance by individuals of their social responsibility not to perpetuate these genes.

Eugenic engineering operates at the level of gene expression on the phenotype of the individual. It is already widely used in medicine, if not so recognized, as in the administration of vitamins, hormones such as insulin, and thyroxine. Better-recognized examples include the prevention of the harmful accumulation of toxic materials associated with genetic defects by dietary restrictions such as of phenylalanine in PKU, of galactose in galactosemia, or of certain branched amino acids in maple-sugar urine disease or the recently described isovaleric acidemia. Such diseases are of special interest and significance in that they characteristically involve brain function and mental retardation. This, and the known actions of drugs such as LSD, forecast the eventual understanding of the organic basis of other mental illnesses, such as schizophrenia, and their successful treatment.

Another much less developed type of eugenic engineering would make use of the concepts of gene regulation and control, by way of feedback regulation by the administration of compounds yet to be discovered. For example, the activities of harmful dominant genes in theory could be repressed as desired, or inactive genes could be turned back on or derepressed as needed, even in utero at critical periods of development. As already pointed out, hormore therapy may actually represent this type of gene regulation.

I would define genetic engineering as the alteration of existing genes in an individual. This could be accomplished by directed mutation or by the replacement of existing genes by others. In principle, and in respect to possible ways of accomplishing this replacement, there are only minor technical differences between genetic engineering as applied to genes in germinal and in somatic cells, although the net result would be considerably different. Precedents for the introduction or transfer of genes from one cell to another exist in microbial systems and are now being tried with mammalian cells in culture. Isolated DNA as such is physically taken up and integrated into the recipient bacterial genome in the transformation process, or in transduction is transferred from one bacterium to another by a virus, followed by integration. If this can be done successfully with animal cells, it will facilitate the development of a mammalian somatic-cell genetics. It will also bring us considerably closer to successful genetic engineering. It is pertinent to point out that, for the phenotypic correction of most genetic errors, only one of the two inactive genes in a diploid cell need be replaced by an active gene and that this may need to be done only in a certain critical number of cells in the particular organ in which the gene function is needed. Hence, it can be suggested that the first successful genetic engineering will be done with the patient's own cells, for example, liver cells, grown in culture. The desired new gene will be introduced, by directed mutation, from normal cells of another donor by transduction or by direct DNA transfer. The rare cell with the desired change will then be selected, grown into a mass culture, and reimplanted in the patient's liver. The efficiency of this process and its potentialities may be considerably improved by the synthesis of the desired gene according to the specifications of the genetic code and of the enzyme it determines, by in vitro enzymatic replication of this DNA, and by increasing the effectiveness of DNA uptake and integration by the recipient cells, as we learn more about the factors and conditions affecting these processes.

An even more speculative biological possibility for genetic engineering may be suggested, stemming from the finding by Harris [21] that, in culture, mammalian cells, even from different species, can be caused to fuse by exposure to an as yet unidentified component of certain animal viruses. These "hybrid" cells can survive and grow, retaining all or part of the chromosome complements of the two "parental" cells. The possible applicability of this approach to the introduction of new genetic material for purposes of genetic engineering seems obvious.

Let me now try to summarize. I have attempted to point out and discuss some of the high spots in molecular biology and nucleic acid research that are particularly promising for the future of medicine. It is apparent that multidisciplinary concepts, approaches, and techniques are essential to the fulfilment of the potentialities which can now be only speculated on. This is particularly apparent if we try to list some of the areas and techniques involved and indicate some needs for further information and development.

One major area involves the design and synthesis of compounds for chemotherapeutic use. In relation to viruses, these will include agents affecting their adsorption, penetration, and replication. In relation to gene function, transcription to m-RNA, and protein, agents already exist and more surely can be designed that will even more specifically affect DNA and RNA structure and particular functions. Other "suppressor substances" can be imagined, which, like streptomycin, will change nucleic acid transcription at the level of protein synthesis and thereby correct genetic errors. Still others, patterned after repressor or regulator substances yet to be isolated and identified, will be able to modulate the activities of specific genes.

With the structures of key enzymes established, the DNA code completely established, and with suitable synthetic methods available, genes can be synthesized to order. The time may soon come when a few molecules of such synthetic genes, or of genes isolated in pure state from nature, will be replicated enzymatically in vitro. With more complete knowledge of the biological processes and techniques involved in DNA uptake and integration, these DNA's can be incorporated into chromosomes. Genetic engineering will then be just around the corner.

Such speculations as these may be considered by some as too idle daydreaming for a serious symposium. Yet the phenomena of molecular biology which are now almost taken for granted were not even dreamed of a very few years ago! So, to paraphrase,

> The time has come, it may be said,
> To dream of many things;
> Of genes—and life—and human cells—
> Of Medicine—and kings—

REFERENCES

1. K. G. LARK. Bacteriol. Rev., 30:3, 1966.
2. S. SPIEGELMAN, I. HARUNA, I. B. HOLLAND, G. BEAUDREAU, and D. MILLS. Nat. Acad. Sci. (U.S.), Proc., 54:919, 1965.
3. C. WEISSMAN and G. FEIX. Nat. Acad. Sci. (U.S.), Proc., 55:1264, 1966.
4. M. HAYASHI, M. N. HAYASHI, and S. SPIEGELMAN. Nat. Acad. Sci. (U.S.), Proc., 51: 351, 1964.
5. E. REICH. Symp. Soc. Gen. Microbiol., 16:266, 1966.
6. D. C. WARD, E. REICH, and I. H. GOLDBERG. Science, 149:1259, 1965.
7. B. K. BHUYAN and C. G. SMITH. Nat. Acad. Sci. (U.S.), Proc., 54:566, 1965.
8. G. ACS, E. REICH, and M. MORI. Nat. Acad. Sci. (U.S.), Proc., 52:493, 1964.
9. B. J. MCCARTHY and E. T. BOLTON. J. Mol. Biol., 8:184, 1964.
10. R. W. HOLLEY et al. Science, 147:1462, 1965.
11. R. E. WEBSTER, D. L. ENGELHARDT, and N. D. ZINDER. Nat. Acad. Sci. (U.S.), Proc., 55:155, 1966.
12. J. ADAMS and M. R. CAPECCHI. Nat. Acad. Sci. (U.S.), Proc., 55:147, 1966.
13. A. S. SARABHAI, A. O. W. STRETTON, S. BRENNER, and A. BOLLE. Nature, 201:13, 1964.

14. J. Davies, W. Gilbert, and L. Gorini. Nat. Acad. Sci. (U.S.), Proc., **51**:883, 1964.

15. D. J. L. Luck. Nat. Acad. Sci. (U.S.), Proc., **49**:233, 1963.

16. D. J. L. Luck and E. Reich. Nat. Acad. Sci. (U.S.), Proc., **52**:931, 1964.

17. E. G. Diacumakos, L. Garnjobst, and E. L. Tatum. J. Cell Biol., **26**:427, 1965.

18. E. Reich and D. J. L. Luck. Nat. Acad. Sci. (U.S.), Proc., **155**:1600, 1966.

19. D. O. Woodward and K. D. Munkres. Nat. Acad. Sci. (U.S.), Proc., **55**:872, 1966.

20. C. Kidson and K. S. Kirby. Nature, **203**:599, 1964.

21. H. Harris and J. F. Watkins. Nature, **205**:640, 1965.

THE FUTURE OF RECOMBINANT DNA
TECHNOLOGY IN MEDICINE

THOMAS H. FRASER*

We are now on the verge of a medical revolution owing to the development of recombinant DNA technology. In the last 2 years this field has made some giant strides toward the realization of its potential. The idea of producing large amounts of animal, viral, and human proteins in microorganisms has progressed from plausible theory to proved reality. Although there is still a great deal of creative work to be done in order to make a significant impact on medicine, the field at this time looks more promising than ever. Even some of the most optimistic predictors of a few years ago have had to shorten the time frame in which they foresaw practical applications of recombinant DNA technology.

As most readers know, recombinant DNA technology is the art of cutting and splicing pieces of DNA. The three major components involved in DNA splicing are a host cell or organism, a vector molecule containing a selectable genetic marker and capable of replicating within the host, and the DNA to be cloned. The almost limitless possible combinations of hosts, vectors, and genes give this technology incredible power both to solve important scientific problems and to directly add new weapons to the physician's armamentarium.

Many of the new products to be produced will result from the development of microbial factories for the synthesis of specific polypeptides and proteins. Other products will include antibiotics for combating infectious diseases.

Within the next several years, virtually unlimited amounts of human hormones, human interferon, human blood proteins, specific human

*Research scientist, Infectious Diseases Research, Upjohn Company, Kalamazoo, Michigan 49001.

antibodies, and viral antigens (for use as vaccines) will be made available as a result of the application of recombinant DNA technology. As important as these products will be, they represent only the first generation of medicinals that will be produced by microbial factories. The acquisition of new knowledge in many of the basic medical sciences is proceeding at an explosive pace. As more clinical researchers become sensitized to the opportunities that recombinant DNA technology offers for synthesis of large amounts of specific proteins and polypeptides, it is virtually certain that rational therapies using these proteins and polypeptides will be tested and refined. At this time it is feasible to attempt the cloning of any animal or viral gene into a microorganism if a probe for that gene is available.

Gene Cloning and Screening

The first step in the synthesis of a foreign protein in a microorganism is to obtain and clone the gene coding for a protein of interest; several different approaches have been successfully carried out. The general strategy of most approaches is to first obtain DNA preparations enriched for a specific gene and then biochemically insert the DNA into a cloning vector and propagate it in *E. coli*. A certain proportion, depending upon the enrichment of the original DNA preparation, of the transformed *E. coli* clones will then carry the gene of interest. The successful isolation of a gene, therefore, ultimately resides in the unambiguous identification of that gene cloned in either *E. coli* or another host.

There are only two possible sources of a gene: (*a*) an organism or virus in which it is naturally found and (*b*) *de novo* chemical synthesis. Chemical synthesis has been successfully employed in making human somatostatin [1] and human insulin [2] genes that direct protein synthesis in *E. coli*. These are extremely significant accomplishments, and they demonstrate that oligonucleotide synthesis will play an important role in the development of medical applications of recombinant DNA technology. There are, however, limitations to gene synthesis by chemical means. Although the techniques in this field have improved rapidly, there will always be size limits beyond which it is not practical to synthesize a gene chemically. In addition, it is necessary to know the entire amino acid sequence of a protein prior to initiating chemical synthesis of the gene coding for that protein. In an attempt to circumvent these limitations, a technique has been developed in which a chemically synthesized oligonucleotide representing only a small portion of a given gene can be used to isolate the entire gene from an organism in which it is naturally found [3]. Either chromosomal genes or messenger RNAs can be identified by hybridization with this short oligonucleotide (12–16 bases) whose sequence is deduced from a partial amino acid sequence of the desired protein. Among vertebrates, this approach has been successfully applied to the detection of hog gastrin mRNA [4]. The synthesis of such a probe depends upon the prior determination of a unique sequence of five or six amino acids within the desired protein. The number of different nucleotide sequences which could theoretically code for these amino acids (due to degeneracy of the genetic code) should be small so that all possible sequences can be synthesized, if necessary. It is already clear that this technique will be very useful for the isolation of specific genes and for detection of these genes in bacterial clones.

Another very powerful group of gene screening methods is based upon expression of the protein coded for by a cloned gene. Several genes from the lower eukaryotes, *Saccharomyces cerevisiae* and *Neurospora crassa,* have been cloned in *E. coli* and identified either by functional complementation of *E. coli* mutations or by immunoassay [5–8]. In these procedures the chromosomal DNA from the lower eukaryotic organism is cut up and directly inserted into a DNA cloning vector which is then used to transform *E. coli.* The resulting transformant clones are then screened for expression. While this procedure works very well for the cloning of at least some genes from lower eukaryotic organisms, the expression of genes from viruses and higher eukaryotic organisms is complicated by the presence of intervening DNA sequences within viral and animal structural genes (see [9] for a review). Since prokaryotes apparently do not have the enzymatic machinery necessary to remove intervening sequences [10], these sequences must be removed from a gene prior to cloning in a bacterium such as *E. coli* if expression of the entire gene is required, although it may be possible to express immunologically reactive fragments of the gene product without removal of the intervening sequences. One way to remove the intervening sequences is to isolate messenger RNA, from which the sequences have already been cleaved. Then RNA-dependent DNA polymerase (reverse transcriptase) may be used to make a DNA copy (cDNA) of the RNA. After synthesis of a complementary second strand, this DNA may be cloned in a microorganism and screened for in vivo expression. As with the expression assays for genes isolated from lower eukaryotes, the clones containing genes from higher eukaryotes may be identified either by function of the cloned gene product or by immunoassay. An example of assay by function involves cloning of the mouse dihydrofolate reductase gene into *E. coli;* expression of this gene in transformants increases resistance to trimethoprim [11]. Some recently described radioimmunoassays are capable of detecting less than 1 pg of protein [12, 13]. Thus, if one could induce a bacterium to make (and not degrade) an average of less than one higher eukaryotic protein molecule per cell, the presence of a particular gene in a bacterial clone could be established.

In order for cloned genes to be expressed in *E. coli,* even at low levels, the foreign gene must be recognized by the bacterial cellular machinery, so that it is first transcribed into messenger RNA and then translated into protein. This requires recognition by bacterial RNA polymerase of a promotor on the DNA and recognition by bacterial ribosomes of a Shine-Dalgarno sequence on the 5' side of an initiation codon in the mRNA [14, 15], although the Shine-Dalgarno sequence may not be absolutely necessary for low-level expression. These considerations indicate that, in many cases, expression of a foreign gene in *E. coli* will be dependent upon the DNA cloning vector into which the gene has been inserted. Thus, some vectors have been designed so that foreign DNA inserted into a specific restriction site will be transcribed under the control of a promoter present in the vector. In addition, the cloning vector may initiate protein synthesis near the restriction site, which would allow bacterial ribosomes to read into the RNA transcribed from the cloned DNA. The transcription and translation of several different higher eukaryotic genes and an animal viral gene—including rat proinsulin [16], rat growth hormone [17], chicken ovalbumin [18, 19], mouse dihydrofolate reductase [11], and hepatitis B viral antigen [20]—have amply demonstrated that there are no inherent barriers to this expression.

Except for those rare instances when a functional assay can be devised for a cloned animal gene product, the success of gene detection by expression is dependent upon the availability of a specific antibody against the protein of interest. In some cases pure antigens are not available, and antibodies must be raised against a complex mixture of cellular macromolecules, thereby compromising the validity of an immunoassay. The development of hybridoma technology (see [21] for a review), however, offers a means to synthesize large amounts of a monoclonal antibody specific for virtually any protein. Thus, the availability of a specific antibody will probably not be a serious problem in the use of methods involving screening cloned genes based upon their in vivo expression. Although some of the more sensitive immunoassays will not work with a single monoclonal antibody, there is at least one that will [8].

By use of the screening methods dependent either upon synthesis of a specific oligonucleotide hybridization probe or upon in vivo gene expression, an investigator may conveniently screen thousands of microbial clones. Thus, there is no need to extensively enrich the DNA preparation that is originally cloned for a particular gene. In addition, clones may be screened by hybridization with radioactively labeled preparations of purified messenger RNA or its reverse transcript. However, if a specific hybridization probe, antibody, or functional assay is not available, there are several other, more laborious, methods of screening for the presence of a desired gene. One method is hybrid-arrested cell-free translation [22], based upon the observation that, in a cell-free protein synthesis system, RNA that is hybridized with complementary DNA is unable to direct the synthesis of proteins. Thus, a preparation of messenger RNA is copied to make cDNA and cloned into a microorganism. The foreign DNA is removed from individual clones and hybridized to the original mRNA, which is then translated in a cell-free system. The disappearance in the translation cocktail of either a specific protein band on a gel or immunoreactivity with a specific antibody identifies the clone containing DNA that codes for that protein or antigen. It is also possible to hybridize specific cloned fragments with cellular messenger RNA, followed by elution of the mRNA, translation in a cell-free system, and identification (by gel electrophoresis or immunoassay) of the protein that is synthesized [23].

Another method of identifying cloned genes is to sequence the inserted DNA. The DNA sequencing techniques of Maxam and Gilbert [24] and Sanger, Nicklen, and Coulson [25] have made this a reasonable method for identification of genes, and it has been used successfully with the rat growth hormone gene [26] and human chorionic sommatomommatropin gene [27]. This method usually requires gene enrichment prior to cloning, since sequencing DNA from a large number of different transformants is very time-consuming.

Thus, there are a number of different techniques that may be brought to bear upon a specific gene cloning problem, either individually or in various combinations. Although the success of each project cannot be guaranteed, the confidence of molecular biologists in their ability to clone virtually any gene appears justified both by their accomplishments so far and by this variety of powerful techniques.

Microbial Factories

While the cloning of a gene coding for a protein of medical interest is a necessary first step in the development of a micobial factory for production of that protein, the practical future of this technology will be dependent upon the ability to synthesize large amounts of biologically active proteins that can be economically produced and purified. Several factors that must be considered include the efficient transcription and translation of a cloned gene, secretion of the product into the growth medium, posttranslational modification of the polypeptide chain coded for by the gene, and choice of a suitable host organism. Although we certainly do not have all the answers that are needed in these areas, many of the results that have been reported are very encouraging.

The fact that a number of different higher eukaryotic and animal viral proteins have been expressed in E. coli has clearly demonstrated that the bacterial transcription and translation apparatuses can recognize the coding sequences for these proteins. In all cases, however, a bacterial promoter has been required to initiate transcription, and in most cases the protein products have been fused to segments of *E. coli* proteins. There are several reasons why investigators have found it advantageous to engineer these fusions. An important advantage that can be gained by fusion of coding sequences is an increase in the efficiency of translation initiation directed by the fused mRNA. This increase in efficiency is expected because the protein synthesis initiation signals in higher eukaryotes and animal viruses are different from those in *E. coli,* and fusion to a cloned gene of a gene fragment coding for the amino terminal end of an *E. coli* protein should result in protein synthesis initiation on the bacterial gene. Once initiated, the bacterial ribosome will translate the entire fused mRNA. The key requirement in constructing a fused gene of this type is to maintain the reading frame from the bacterial translation initiation site into the cloned gene, so as not to generate a frameshift mutation. This strategy has been successfully employed to synthesize between 20,000 and 30,000 molecules per cell of both rat growth hormone [17] and chicken ovalbumin [18, 19] in *E. coli.*

The presence of a segment of bacterial protein may not be acceptable when proteins produced in microorganisms are used in clinical situations. The host organism's protein sequences may inhibit the biological activity of the protein, or they may induce antibody formation in a patient. It may not be necessary, however, to include any bacterial gene segments that code for amino acids in order to achieve efficient translation. Ptashne's laboratory [28, 29] has shown that when a fragment of *lac* operon DNA containing the Shine-Dalgarno sequence from the *E. coli* *lac Z* gene is fused near the initiation codon for a lambda phage gene, the resulting hybrid ribosome binding site efficiently initiates translation of the phage gene. As they have pointed out, this should, in principle, work for expression of any foreign gene in *E. coli.*

Although the preceding strategy may solve the problem of inefficient translation of some genes in *E. coli,* fusion to *E. coli* proteins has also been shown to be useful in the prevention of degradation of some human proteins in *E. coli* after they have been translated. Some genes, such as those that have been chemically synthesized to code for the human insulin A and B chains [2] and human somatostatin [1], were fused to a large segment of the *lac Z* gene, which codes for β-galactosidase. It was shown by immunoassay that these hybrid genes directed synthesis of the desired protein. In the case of somatostatin, when the synthetic gene was fused to a smaller segment of the β-galactosidase gene, no protein was detected. It is thought that the large β-galactosidase fragment, consisting of approximately 1,000 amino acids, physically surrounds and protects the human polypeptides from attack by *E. coli* proteases. After breaking open the cells and inactivating the proteases, the human proteins can be neatly cleaved from the β-galactosidase by cyanogen bromide treatment, which cuts after a methionine residue whose codon is inserted between the *E. coli* and the human protein. Although the hybrid protein product in these cases can be easily cleaved, the overall synthesis is extremely inefficient. While about 20 percent of the total cellular protein consists of β-galactosidase-insulin A or B chain hybrid, at most about 0.5 percent of the cellular protein can be either of the desired insulin chains. Thus, while it is clear that removal of the gene fragment coding for a large *E. coli* protein sequence could lead to potentially higher levels of product synthesis, this must be coupled with some other means of preventing degradation of the animal protein by bacterial proteases. It is possible that this could be accomplished by cloning in bacterial mutants defective in proteolytic functions.

Fusion to a bacterial gene has also been used to stimulate secretion of a foreign gene product that would normally be degraded in *E. coli*. This strategy was first employed by Villa-Komaroff et al. to express the gene coding for rat proinsulin [16], which is normally rapidly degraded by *E. coli* proteases. They inserted the rat gene into a gene carried on the plasmid pBR322 [30] which codes for penicillinase. Since the penicillinase is normally secreted into the *E. coli* periplasmic space, they reasoned that the fusion of this secreted protein to the rat proinsulin would result in secretion of the hybrid protein and possibly protect the rat proinsulin from degradation by intracellular proteases. Although they were able to immunologically detect insulin, the yield was only about 100 molecules per cell. It is not clear that this procedure will be generally useful, since insertion of the rat growth hormone gene into the same site in the penicillinase gene did not lead to secretion of the resulting hybrid protein into the periplasmic space [17]. While it is probable that fusion of genes to the penicillinase gene in pBR322 may be useful for screening of cloned genes, it is unlikely to be useful in production of most gene products due to difficulties in specifically removing the penicillinase fragment from the desired protein.

As is apparent from the preceding discussion, an important area of concern in the microbial production of clinically important proteins deals with yield. How much foreign protein can be made in *E. coli,* and what are the factors that limit production? Fraser and Bruce [18] have designed a plasmid (pUC1001) carrying the chicken ovalbumin gene fused to *lac* control elements which, when inserted into *E. coli*, directs the synthesis of large amounts of chicken ovalbumin fused to several amino acids of β-galactosidase. Although 1.5 percent of the cellular protein is ovalbumin, this level is considerably lower than the level theoretically attainable in this system. It has been determined that there are between

25 and 30 copies of the pUC1001 plasmid per chromosome equivalent in *E. coli*. It is known that β-galactosidase, with one gene copy per chromosome equivalent, is expressed as 2 percent of the total *E. coli* protein in fully induced cultures [31, p. 398]. Therefore, considering the gene copy numbers as well as the differences in molecular weight between ovalbumin and β-galactosidase, one would expect approximately 19 percent of the total protein in the pUC1001-containing bacteria to be ovalbumin. In fact, although a significant amount of ovalbumin is synthesized, it is less than 10 percent of the theoretically predicted amount.

There are several possible explanations for lower-than-theoretical levels of ovalbumin in cells carrying the pUC1001 plasmid, including inefficient transcription, inefficient translation, and proteolytic degradation of ovalbumin molecules.

It has been reported that a *lac* promoter identical with the promoter in pUC1001 functions with full efficiency when cloned into a multicopy plasmid [28]. Thus, it is also likely that transcription initiation is fully efficient in the case of pUC1001. Inefficient translation of the ovalbumin mRNA may result from either inefficient initiation or inefficient elongation. When the *E. coli* β-galactosidase gene was inserted into a plasmid and high copy number mutants of that plasmid were selected, it was found that more than 50 percent of cellular protein synthesis was engaged in production of β-galactosidase [32]. Therefore, it is likely that protein synthesis initiation is not limiting in this system. The rate of elongation of mRNA translation will be dependent upon many factors, one of which is the availability of aminoacyl-tRNAs with anticodons complementary to the codons on the mRNA. The sequence data from several prokaryotic and eukaryotic genes, including ovalbumin, have revealed nonrandom utilization of synonymous codons [33–35]. Although not enough data have been collected to permit any generalizations, it is probable that some of the tRNA isoaccepting species required in relatively large amounts for the translation of ovalbumin are present in *E. coli* in low concentrations, thus limiting synthesis.

The third explanation for lower-than-theoretical ovalbumin production is proteolytic degradation. This may occur either with nascent polypeptide chains on polysomes or with mature ovalbumin molecules, or both. Self-digestion data with *E. coli* cell extracts suggest that the mature ovalbumin molecules are stable to proteolytic digestion in vitro, but the data do not extend to the stability of nascent polypeptide chains.

Whatever the currently limiting factors in the microbial synthesis of higher eukaryotic proteins are, it is virtually certain that they will be identified and ultimately circumvented. I would speculate that the limit to production of these proteins in microorganisms may be in the range of 50 percent of the total cellular protein. If, however, a protein were secreted into the medium, it may be possible to continuously engage about 50 percent of the protein synthetic machinery in its production.

The ovalbumin model system has also been instructive in the area of secretion. It has been found [18] that the ovalbumin molecules synthesized in *E. coli* carrying the pUC1001 plasmid are secreted into the periplasmic space. These findings suggest that common features exist in the secretion recognition mechanisms of chicken oviducts and *E. coli*. Work reported by Lingappa et al. [36] has shown that part of the chicken ovalbumin molecule functions as a signal sequence for secretion and that the mechanism of secretion recognition is similar to that for other secreted animal proteins. This result and the fact that chicken ovalbumin synthesized in *E. coli* is secreted indicate that perhaps all eukaryotic secretory proteins may be actively secreted by any microorganism, as long as the signal sequence is present and the microorganism has an apparatus to secrete proteins.

Many proteins isolated from higher eukaryotes and animal viruses are posttranslationally modified. Common posttranslational modifications include glycosylation, phosphorylation, proteolytic cleavage, and acetylation. With the exception of proteolytic cleavage, it is not known in many cases what role these modifications play in the observed biological activities of these proteins. It seems unlikely that the foreign proteins synthesized in microorganisms will be postranslationally modified by microbial enzymes in the same way that they would normally be modified by the enzymes of the organism in which they were naturally found. The PAS staining [37] of ovalbumin made in *E. coli* has indicated that these molecules are not glycosylated, as is the case with ovalbumin synthesized in chicken oviducts. If particular posttranslational modifications are required for biological activity, this could be accomplished either in vitro by enzymatic reaction or in vivo by cloning into the microbial host organism those genes that code for enzymes which catalyze posttranslational modifications.

Consideration of the various elements involved in the development of microbial factories indicates that the key to this development resides with the host organism that is chosen to harbor the cloned genes. Ultimate yields, and perhaps the ability to make a product at all, will be a function of the host's ability to transcribe and translate a foreign gene and its inability to degrade the protein product. Growth requirements of the host must be favorable to production and purification of the product. Specific enzymes required for posttranslational modification, if naturally present in a host, could be a distinct advantage. So far, however, the only organism which has been intensively studied as a potential host for microbial factories is *E. coli*. This has been due both to scientists' familiarity with this particular microorganism and to restrictions in guidelines for recombinant DNA research. Risk-assessment studies have now gathered enough information so that the guidelines of many countries currently allow considerably more latitude in the choice of host organisms than they have in the past.

Although we know a great deal about the molecular biology of *E. coli*, it does have some disadvantages as a host for production of recombinant DNA products. Perhaps the most serious problem, which applies to other gram-negative organisms as well, is that the lipopolysaccharide component of the outer cell membrane, also known as endotoxin, has a large number of deleterious biological activities and is active in nanogram amounts. The major problem in the production of a product for human use, therefore, is removal of trace amounts of contaminating endotoxin from the preparations. This removal is likely to be prohibitively complex and expensive for the purification of many of the proteins that could be produced by microbial factories. Although it may be possible to genetically delete the ability of an organism to synthesize endotoxin, this will be a difficult problem. Much of the biological activity of lipopolysaccharide has been traced to the lipid A component, and, while mutants defective in the other components of lipopolysaccharide have been easily isolated, only conditional lethals have been found in the lipid A component [38].

A great deal of effort is now being devoted to the development of cloning systems in lower eukaryotes. In addition to easing the endotoxin problem, there are several probable advantages to producing an animal protein in a simple eukaryotic organism such as yeast. For example, it is likely that the mRNAs of many vertebrate genes will be translated in yeast without having to rely on fusion to yeast gene segments, since the translation initiation mechanisms may be similar. Furthermore, if the composition of tRNA populations turns out to be different between eukaryotes and prokaryotes, the translation of foreign animal genes may be more efficient in yeast than in a prokaryote. A transformation system has now been worked out in yeast [39], and several cloning vectors have been constructed [40–42]. Although a yeast system looks promising at this time, a lot of work will have to be done before we can determine whether it will be a better host for production of recombinant DNA products than *E. coli*. Other potential host organisms that are under investigation include *Bacillus subtilis* and various *Streptomyces* species.

While there are clear challenges ahead in the development of microbial factories, particularly in searching for host-vector systems that will allow high-level protein production and that will simplify purification, several proteins, such as human interferon and human growth hormone, are currently so difficult to obtain that even relatively low-level production in *E. coli* will be a marked improvement over present production methods.

Antibiotics

Another area of application of recombinant DNA technology is in the development of new antibiotics. This area is not as far advanced as the use of microbial factories, owing largely to the difficulties of developing host-vector cloning systems in antibiotic-producing microorganisms. So far several plasmids that could serve as starting material for construction of cloning vectors have been isolated from *Streptomyces* [43, 44], and an efficient transformation system using one of these plasmids has been reported [45]. An obvious contribution of recombinant DNA technology would be in the titer improvement of known antibiotics. While this may not appear particularly significant from the medical point of view, it is a fact that many new and potentially useful antibiotics are not developed as products because they are synthesized at low levels by the organisms that produce them. Thus, the effect of titer improvement will be both to make production more economical and to make more new antibiotics available.

Existing antibiotics may be made more potent or more effective against resistant organisms by transplanting genes coding for specific enzymes into antibiotic-producing microorganisms, thereby making *in vivo* derivatives of the original antibiotic. It is also possible that entire banks of genes coding for enzymes that make antibiotic subunits could be transferred from one organism to another, yielding new classes of antibiotics, although this may ultimately be done more efficiently by protoplast fusion.

It is difficult to predict exactly when the impact of recombinant DNA technology will be felt in the development of new antibiotics, due to the lack of an extensive molecular biological data base in antibiotic-producing microorganisms. However, acquisition of important knowledge, such as structure-function analysis of specific genes and control regions, especially those involved in antibiotic synthesis, will be made much easier by the application of recombinant DNA technology to the problems, independent of its application to development of new antibiotics.

Epilogue

Predictions regarding the ultimate impact of recombinant DNA technology on medicine would not be wise to make at this time. While a great deal of attention has been focused on the microbial factories and antibiotic production, it must not be overlooked that the results from basic research made possible by recombinant DNA technology will almost certainly have important applications in the areas of development, aging, cancer, neurobiology, endocrinology, and immunology.

Many of these applications may result in the identification of clinically important human proteins that could then be mass-produced by microbial factories. Any consideration of the future of recombinant DNA technology in medicine must, therefore, take into account the powerful new methods which basic researchers now have at their disposal due to this technology.

Appendix

Glossary for the Nonspecialist

clone: A group of cells which have been propagated asexually from a single cell. DNA is cloned by inserting it into a cell which is capable of such propagation.

codon: A set of three adjacent nucleotides in mRNA that codes for either an amino acid or termination of protein synthesis. The four different nucleotides can form 64 different codons.

frameshift: If a nucleotide or nucleotides are added to or deleted from a DNA sequence, the messenger RNA transcribed from this DNA will give rise to an altered protein. If the number of added or deleted nucleotides is not evenly divisible by three (the number of nucleotides in a codon), the translational reading frame will be shifted, giving rise to a frameshift mutation.

hybridization: In nucleic acid biochemistry, this term refers to the noncovalent binding of parts or all of polynucleotide chains (i.e., DNA or RNA) to each other or themselves due to regions of paired bases.

lac operon: A group of three physically linked *E. coli* genes and their transcription and translation control elements that are involved in lactose metabolism.

plasmid: A circular DNA molecule in either eukaryotes or prokarotes that is replicated and maintained extrachromosomally.

polysome, polyribosome: A group of two or more ribosomes actively translating the same messenger RNA molecule and thus linked to each other.

promoter: A region of DNA where the enzyme RNA polymerase can bind and initiate transcription.

protoplast fusion: A method for generating genetic recombinants between cells by first stripping them of their cell walls. It is particularly useful in forming recombinants between cells that do not normally exchange genetic information.

reading frame: During translation, messenger RNA is decoded sequentially, three bases (one codon) at a time. Thus, a given mRNA sequence may be read in three different frames, depending upon where within the sequence decoding is initiated.

restriction endonuclease: An enzyme which will cut a DNA molecule at specific internal sites. The cutting sites are determined by a sequence of base pairs (usually four or six) which is different for different restriction enzymes.

Shine-Dalgarno sequence: A short stretch (approximately three to six nucleotides) on a messenger RNA molecule that is complementary to the 3' end of the intermediate-sized ribosomal RNA in bacteria and apparently plays an important role in initiation of translation.

transcription: The synthesis of RNA from a DNA template catalyzed by RNA polymerase.

transformation: The acquisition of new, heritable characteristics in a cell following uptake of DNA. The term can be ambiguous when applied to animal cells, since cell biologists refer to the conversion of normal to cancerous cells in culture as transformation.

translation: The assembly of a protein molecule from its constituent amino acids directed by a messenger RNA molecule and taking place on the ribosome.

vector: In general terms, a piece of DNA into which a second piece of DNA can be bound in order to propagate the second piece of DNA in a host organism or cell. In practice, the vector is usually a plasmid or viral DNA molecule which is capable of replicating itself in the host and contains a selectable genetic marker.

REFERENCES

1. K. Itakura, T. Hirose, R. Crea, A. D. Riggs, H. L. Heyneker, F. Bolivar, and H. W. Boyer. Science, **198:**1056, 1977.
2. D. V. Goeddel, D. G. Kleid, F. Bolivar, H. L. Heyneker, D. G. Yansura, R. Crea, T. Hirose, A. Kraszewski, K. Itakura, and A. D. Riggs. Proc. Natl. Acad. Sci. USA, **76:**106, 1979.
3. J. W. Szostak, J. I. Stiles, C. P. Bahl, and R. Wu. Nature, **265:**61, 1977.
4. B. E. Noyes, M. Mevarech, R. Stein, and K. L. Agarwal. Proc. Natl. Acad. Sci. USA, **76:**1770, 1979.
5. K. Struhl, J. R. Cameron, and R. W. Davis. Proc. Natl. Acad. Sci. USA, **73:**1471, 1976.
6. B. Ratzkin and J. Carbon. Proc. Natl. Acad. Sci. USA, **74:**487, 1977.
7. D. Vapnek, J. A. Hautala, J. W. Jacobson, N. H. Giles, and S. R. Kushner. Proc. Natl. Acad. Sci. USA, **74:**3508, 1977.
8. R. A. Hitzeman, A. C. Chinault, A J. Kingsman, and J. Carbon. *In:* R. Axel and T. Maniatis (eds.). Eukaryotic gene regulation, p. 57. New York: Academic Press, 1979.
9. F. H. C. Crick. Science, **204:**264, 1979.
10. O. Mercereau-Puijalon and P. Kourilsky. Nature, **279:**647, 1979.
11. A. C. Y. Chang, J. H. Nunberg, R. J. Kaufman, H. A. Erlich, R. T. Schimke, and S. N. Cohen. Nature, **275:**617, 1978.
12. S. Broome and W. Gilbert. Proc. Natl. Acad. Sci. USA, **75:**2746, 1978.
13. H. A. Erlich, S. N. Cohen, and H. O. McDevitt. Cell, **13:**681, 1978.
14. J. Shine and L. Dalgarno. Proc. Natl. Acad. Sci. USA, **71:**1342, 1974.
15. J. A. Steitz and K. Jakes. Proc. Natl. Acad. Sci. USA, **72:**4734, 1975.
16. L. Villa-Komaroff, A. Efstratiadis, S. Broome, P. Lomedico, R. Tizard, S. P. Naber, W. L. Chick, and W. Gilbert. Proc. Natl. Acad. Sci. USA, **75:**3727, 1978.
17. P. H. Seeburg, J. Shine, J. A. Martial, R. D. Ivarie, J. A. Morris, A. Ullrich, J. D. Baxter, and H. M. Goodman. Nature, **276:**795, 1978.
18. T. H. Fraser and B. J. Bruce. Proc. Natl. Acad. Sci. USA, **75:**5936, 1978.
19. O. Mercereau-Puijalon, A. Royal, B. Cami, A. Garapin, A. Krust, F. Gannon, and P. Kourilsky. Nature, **275:**505, 1978.
20. C. J. Burrell, P. Mackay, P. J. Greenaway, P. H. Hofschneider, and K. Murray. Nature, **279:**43, 1979.
21. F. Melchers, M. Potter, and N. L. Warner. Current topics in microbiology and immunology: lymphocyte hybridomas, vol. **81.** Berlin: Springer-Verlag, 1978.
22. B. M. Paterson, B. E. Roberts, and E. W. Kuff. Proc. Natl. Acad. Sci. USA, **74:**4370, 1977.

23. M. M. Harpold, P. R. Dobner, R. M. Evans, and F. C. Bancroft. Nucleic Acids Res., **5:**2039, 1978.
24. A. Maxam and W. Gilbert. Proc. Natl. Acad. Sci. USA, **74:**560, 1977.
25. F. Sanger, S. Nicklen, and A. R. Coulson. Proc. Natl. Acad. Sci. USA, **74:**5463, 1977.
26. P. H. Seeburg, J. Shine, J. A. Martial, J. D. Baxter, and H. M. Goodman. Nature, **270:**486, 1977.
27. J. Shine, P. H. Seeburg, J. A. Martial, J. D. Baxter, and H. M. Goodman. Nature, **270:**494, 1977.
28. K. Backman and M. Ptashne. Cell, **13:**65, 1978.
29. T. M. Roberts, R. Kacich, and M. Ptashne. Proc. Natl. Acad. Sci. USA, **76:**760, 1979.
30. F. Bolivar, R. L. Rodriguez, P. J. Greene, M. C. Betlach, H. L. Heyneker, H. W. Boyer, J. H. Crosa, and S. Falkow. Gene, **2:**95, 1977.
31. J. H. Miller. Experiments in molecular genetics. Cold Spring Harbor, N.Y.: Cold Spring Harbor Laboratory, 1972.
32. D. H. Gelfand, H. M. Shepard, P. H. O'Farrell, and B. Polisky. Proc. Natl. Acad. Sci. USA, **75:**5869, 1978.
33. L. McReynolds, B. W. O'Malley, A. D. Nisbet, J. E. Fothergill, D. Givol, S. Fields, M. Robertson, and G. G. Brownlee. Nature, **273:**723, 1978.
34. F. Sanger, G. M. Air, B. G. Barrell, N. L. Brown, A. R. Coulson, J. C. Fiddes, C. A. Hutchinson III, P. M. Slocombe, and M. Smith. Nature, **265:**687, 1977.
35. A. Efstratiadis, F. C. Kafatos, and T. Maniatis. Cell, **10:**571, 1977.
36. V. R. Lingappa, D. Shields, S. L. C. Woo, and G. Blobel. J. Cell Biol., **79:**567, 1978.
37. R. M. Zacharius, T. E. Zell, J. H. Morrison, and J. J. Woodlock. Anal. Biochem., **30:**148, 1969.
38. V. Lehmann, E. Rupprecht, and M. J. Osborn. Eur. J. Biochem., **76:**41, 1977.
39. A. Hinnen, J. B. Hicks, and G. R. Fink. Proc. Natl. Acad. Sci. USA, **75:**1929, 1978.
40. J. D. Beggs. Nature, **275:**104, 1978.
41. K. Struhl, D. T. Stinchcomb, S. Scherer, and R. D. Davis. Proc. Natl. Acad. Sci. USA, **76:**1036, 1979.
42. C. Gerbaud, P. Fournier, H. Blanc, M. Aigle, H. Heslot, and M. Guerineau. Gene, **5:**233, 1979.
43. H. Schrempf, H. Bujard, D. A. Hopwood, and W. Goebel. J. Bacteriol., **121:**416, 1975.
44. V. S. Malik. J. Antibiot., **40:**897, 1977.
45. M. J. Bibb, J. M. Ward, and D. A. Hopwood. Nature, **274:**398, 1978.

PROJECT K: "THE COMPLETE SOLUTION OF E. COLI"*

F. H. C. CRICK†

It is convenient to consider future developments in molecular biology (in the widest sense) under three headings: (*a*) studies on cell components, (*b*) studies on unicellular organisms, and (*c*) studies on multicellular organisms. The latter, although of great importance, will not be dealt with here. The division between cell components (which may come from any sort of cell) and organisms is admittedly arbitrary and is only introduced here to make the discussion easier. In practice, most work on complete organisms is supplemented by studies on the components of that organism.

It is first necessary briefly to take stock of the present position. As far as classical biochemistry is concerned, many enzyme reactions are known, and for a minority of these the action of the pure enzyme is understood in outline. For no case have the details of the enzymatic action been firmly established in chemical terms. Within the field of molecular biology (in the narrow sense) we now understand in outline the synthesis of the nucleic acids and of proteins, their interrelation in the genetic code, and a little about their control mechanisms.

It seems likely that future progress will take place in several broad areas:

1. The more detailed test-tube study of the structure and chemical action of biological molecules (especially proteins). Typical of such studies will be the detailed action of enzymes (already getting very close with the solution by X-ray crystallography of the structure of several enzymes), the way proteins fold themselves up (a backward field), the radiation damage to molecules, especially to DNA, and many other topics. It is characteristic of these studies that they involve the application of complicated and advanced methods of physical chemistry to biological molecules, and often

---

* The idea arose in conversation with Dr. Sydney Brenner, who invented the title "Project K" and whom I have to thank for useful discussions on the topic. This short paper was originally circulated in a European Molecular Biology Organisation (EMBO) document [1] toward the end of 1967. It still seems to me to be an attractive scheme for people of the right temperament, and since EMBO is now unlikely to take it up I thought that it might be useful to give the idea wider publicity.

† MRC Laboratory of Molecular Biology, Hills Road, Cambridge, CB2 2QH, England.

require rather large amounts of (pure) material. We may also expect the chemical synthesis of model compounds to play an important part.

2. The filling-in of the broad outlines already established, for example, the biochemical mechanism of protein synthesis, the unwinding of DNA, the mechanism of genetic recombination (which is probably related to DNA repair mechanisms), and, on the classical biochemical side, the exploration of more metabolic pathways and especially their interrelationships.

3. Work on subjects of fundamental importance which are little studied at the moment, for example, the structure and function of cell membranes,[1] the mechanism of cell division, and the biochemistry of spore formation.

4. The study of control mechanisms at all levels, in particular the interrelation of the known mechanisms, leading to an appreciation of the economy and "design" of the cell.

5. The behaviour of natural cell populations and their population genetics, leading to the consideration of the evolution of the cell.

The above discussion is necessarily sketchy, but it clearly brings out three important points: (a) an enormous amount of work remains to be done without ever going to multicellular organisms; (b) important problems exist at all levels of complexity; and (c) there is likely to be an increasing demand for large amounts of pure cell components present in the cell in rather small amounts. For these reasons, it seems certain that in spite of the obvious opportunities awaiting the study of organisms having many cells, a major effort will almost certainly continue to be applied to single-cell organisms, in particular to bacteria.

The point of this paper is to argue that such work should be concentrated on one organism (probably *Escherichia coli*, K12) and that a case exists for centralizing many aspects of such work in a "central laboratory."

The major reasons for wanting to have the "complete solution" of a bacterial cell have been listed above. In addition, there is the intellectual satisfaction of having a single living cell "completely" explained. Of course, it is unlikely that the work will ever be pushed to the point that every possible detail about the cell is known. It does not seem very probable, for example, that all the various proteins of the cell (which may number several thousand) will all have their amino acid sequences and stereochemical structure determined. By "complete" one means complete in the intellectual sense, implying that nothing appears to remain which further experiment could not easily explain using well-established facts and ideas.

It is clear that if the cell is going to be considered as a well-integrated chemical factory, information from many different laboratories will have

[1] The understanding of cell membranes has advanced greatly since this was written.

to be pooled. This might argue for a central laboratory to act as a focus for such work, but there are additional reasons of a technical nature which make the case even stronger.

In the first place, the technique of studying the action of a gene by picking up specific mutants of it is likely to continue to be widely used. Now for a limited class of genes it is possible to devise special selective techniques, but this cannot be done for the majority. For many genes, however, it is possible to produce "conditional lethal" mutants. These are usually of two classes: (1) temperature-sensitive mutants, which grow at one temperature but not at another, and (2) suppressible mutants. Unfortunately, when producing these mutants one cannot usually obtain mutants of just the gene being considered, but mutants in many different genes. These must then be screened to obtain the class of mutants in which one is interested. The rest are usually discarded, which is clearly a wasteful process. It would be a great advantage if such mutants were characterized as far as possible so that they could be made available to other workers who might wish to study them in more detail. A central laboratory for producing mutants, and for receiving mutants from other laboratories, which would then be classified, stored, and made available to others would be an obvious help to everybody in the field.

Another reason for a central laboratory would be the production of cells on a large scale. Again, at the present time the tendency is for each laboratory to grow a large culture for one particular chemical component in the cell and to discard the rest. This is wasteful and will become more so as larger batches are needed in order to obtain sufficient supplies of the rarer molecules in the cell. Whereas some growing could be done commercially, this will not be enough in the long run, as large batches of special mutants will eventually be required for certain pieces of work.

All this suggests that there is a case for a central laboratory to coordinate and assist experimental work going on in many different places. It remains to discuss the choice of a suitable organism. There is, of course, no reason why eventually a central laboratory might not deal with several organisms, or alternatively that several such laboratories be started, each with its own special organism. However, in the first place it would seem sensible to start with one only.

The obvious requirements are: (1) the organism should be reasonably small to reduce the complexity of the problem; (2) it should be easy to handle and able to grow on a relatively simple, defined medium; and (3) the same strain should be used by most workers. Yeast is probably too big, and has the complication that many different yeasts are in use (e.g., "bakers' yeast" and "brewers' yeast"). It has the advantage, however, of easily forming stable diploids. The pleuropneumonia-like organisms (PPLO), though small, are difficult to handle and need a complicated growth medium. The obvious choice is *Escherichia coli* (probably the K12 strain for genetic reasons), but *Bacillus subtilis* and possibly *Salmonella* would also have to be considered. The main point is that several reasonably satisfactory organisms are already well known. The final choice between them could be left open at this stage.

A central laboratory might well contain the following groups: (1) a genetic group for developing new and rapid methods of genetic mapping and screening; (2) a group to develop instruments for the automation of experiments; (3) a biochemical genetics group to produce, receive, classify, and supply mutants of all possible genes; (4) a fractionation group for developing more and better methods of fractionation; (5) a production group for supplying very large batches of partly fractionated material; and (6) a group to study control mechanisms and the general economy and design of the cell. To this could usefully be added various associate groups and visiting workers studying areas of growing interest (such as membranes or cell division), who would find the facilities provided by the rest of the laboratory an attraction and who in turn would point out the material most needed at any particular time.

Postscript: Since the above was written, some of the last-enumerated suggestions have been taken up, in particular (3) and (5). However, as far as I know, no one is at the moment trying to set up a central laboratory, and the idea of making our knowledge of *E. coli* (say) as complete as possible has not been announced by any group as their acknowledged target.

REFERENCE

1. Proposal for a European laboratory of molecular biology. Document CEBM 68/31E, European Conference of Molecular Biology, November 1967.

INDEX

Twins—*Continued*
  and IQ, 251–252
  nature-nuture analysis in, 244–245

Udenfriend, S., 268
United Nations Scientific Committee on
  the Effects of Atomic Radiation, 312
Universal genetic code, 61
Urea, in treatment of sickle cell anemia,
  278–279
Urey, H. C., 165

V-domains, web of, 415–416
Villa-Komaroff, A., 475
Viruses and viral diseases, 456–457, 463–464
Voelker, R. A., 316
Vogel, F., 306, 313

Wahl, R., 174
Warner, N., 416
Watkins, W. M., 260
Watson, J. D., 29, 32, 61, 75–76, 86–87,
  89, 116, 158–161, 210, 235, 457
Watson-Crick model of DNA, 29, 32, 61–
  66, 86–87, 158–161, 182, 210, 235,
  457
Weaver, W., 73–74
Webster, R. E., 460
Weidel, W., 192
Weismann, A., 7, 9–10, 13–16, 19, 87,
  100, 153
Weissman, C., 458
Wells, I. C., 260
Wernicke-Korsakoff syndrome, 419
Wheeler, J. A., 326

Williams, R. J., 265
Wilson, A., 223
Wilson, E. B., 19, 21, 32, 86
Witkins XYZ study, 440
Woods, P. S., 304
Woodward, D. O., 462
Worcel, A., 402
Wright, S., 35–36, 38, 41, 45, 47–50, 55,
  59, 93, 206
  anecdotes about, 133–135
  awards and honors, 122
  family and personal life of, 121–124,
    130
  works of
    animal breeding, 124–125, 127–130,
      132–133
    evolution, 130–133
    mammalian genetics, 127–129
    philosophical, 133
    scientific, 124–133
Wyatt, G. R., 184

Xeroderma pigmentosum, 315

Yamaguchi, O., 316
Yarbrough, K. M., 371
Y chromosomes, 361, 404

Z form of DNA, 63, 211
Zimmer, K. G., 75, 310, 320
Zinc therapy, in sickle cell anemia, 284–
  288, 290
Zinder, N. D., 460
Zoarces, 111
Zooganus mirus, 99–101